W9-CSE-912

Table of Values

Value and Units	Item	Symbol or Abbreviation
General		
10^{-8} cm	$\equiv$ 1 angstrom	Å
2.998×10^2 volts	$\cong$ 1 statvolt	
2.998×10^{10} cm sec^{-1}	Speed of light in vacuum	c
6.673×10^{-8} dyn cm^2 gm^{-2}	Gravitational constant	G
Thermal		
1.3805×10^{-16} erg deg^{-1}	Boltzmann constant	k_B
8.314×10^7 ergs mole^{-1} deg^{-1}	Gas constant	R
1.013×10^6 dynes cm^{-2}	Atmospheric pressure	
22.4×10^3 cm^3 mole^{-1}	Molar volume at STP	
2.69×10^{19} cm^{-3}	Loschmidt number	n_0
6.022×10^{23} mole^{-1}	Avogadro number	N_0
4.184×10^7 ergs	$\equiv$ 1 calorie	cal
1×10^7 ergs	$\equiv$ 1 joule	J
Atomic		
6.626×10^{-27} erg-sec	Planck's constant	h
1.054×10^{-27} erg-sec	Planck's constant/2π	$\hbar$
$\left.\begin{array}{l} 1.6021 \times 10^{-12} \text{ erg} \\ 23.061 \text{ kcal mole}^{-1} \end{array}\right\}$	Energy associated with 1 electron volt	ev
13.60 electron volts	Energy associated with 1 rydberg	
1.98×10^{-16} erg	Energy associated with unit wave number	hc
1.24×10^{-4} cm	Wavelength associated with 1 electron volt	
8066 cm^{-1}	Photon wave number associated with 1 electron volt	
2.42×10^{14} sec^{-1}	Photon frequency associated with 1 electron volt	
11605 K	Temperature associated with 1 electron volt	
0.529×10^{-8} cm	Bohr radius of the ground state of hydrogen $\hbar^2/me^2$	a_0
0.927×10^{-20} erg gauss^{-1}	Bohr magneton $e\hbar/2mc$	μ_B
0.505×10^{-23} erg gauss^{-1}	Nuclear magneton $e\hbar/2M_pc$	μ_n
137.04	Reciprocal of fine-structure constant	$\hbar c/e^2$
Particles		
0.911×10^{-27} gm	Electron rest mass	m
1.6725×10^{-24} gm	Proton rest mass	M_p
1.6747×10^{-24} gm	Neutron rest mass	M_n
1.66042×10^{-24} gm	One unified atomic mass unit ($\equiv \frac{1}{12}$ mass of C^{12})	u
1836	Proton mass/electron mass	M_p/m
2.82×10^{-13} cm	Classical radius of the electron e^2/mc^2	r_0
4.803×10^{-10} esu	Charge on the proton	e
3.86×10^{-11} cm	Electron Compton wavelength $\hbar/mc$	$\lambdabar_C$

THERMAL PHYSICS

CHARLES KITTEL

Thermal Physics

John Wiley & Sons, Inc.   *New York, London, Sydney, Toronto*

Copyright © 1969, by John Wiley & Sons, Inc.

All rights reserved. No part of this book may be reproduced
by any means, nor transmitted, nor translated into a machine
language without the written permission of the publisher.

10 9 8 7 6 5 4 3 2 1

Library of Congress Catalogue Card Number: 72-88609

SBN 471 49030 X

Printed in the United States of America

QC
311
.5
K52
Copy 3
Phys

Preface

Thermal physics unites thermodynamics and statistical mechanics. The subject is simple, with few assumptions, and the results are broad and powerful.

The object of this book is to give undergraduate students a clear account of the concepts of thermal physics and a selection of the applications of these concepts to physics, chemistry, biology, and engineering. Readers should be familiar with several of the historical ideas of quantum physics, such as the de Broglie waves of a free particle and the nature of the Bohr atom, and with calorimetry as taught in elementary chemistry. No other knowledge of thermodynamics is assumed. I have tried to equip the readers for graduate courses in statistical mechanics and irreversible thermodynamics, and for work in other fields in which an understanding of entropy and free energy is essential.

The approach to the subject is that of Gibbs, so that all results follow as the logical consequence of one or two plain assumptions. What will be new to most readers is the simple way the results appear in the quantum language. Thermal physics is a remarkably easy subject if taught from a consistent quantum viewpoint in which we think of states of an entire system, however large or small. The classical approach has persisted so long because it leads quickly to the ideal gas law and to the heat capacity of the ideal gas; but this ease is deceptive, for the correct entropy is then obtained only by patchwork and jury-rigging. The advantages of doing things correctly the first time are nowhere more apparent than in thermal physics, for we can quickly obtain the quantum distributions and then pass to the ideal gas limit to find correct expressions for the entropy, the gas law, and the equilibrium constants. There is perhaps some pedagogical invention here, but the substance has been part of physics for two generations. In particular the present approach leads to a clear understanding of entropy, to simple derivations of the distribution laws for identical particles, and, through emphasis on the chemical potential, to a synthesis of the physical and chemical methods.

Several standard topics in thermodynamics have not been included, but they may be retrieved from existing texts, such as those by Zemansky and by Callen. For accounts of modern experiments my students have enjoyed the paperback by Mark Zemansky, *Temperatures very low and very high* (Van Nostrand, 1964). If the problems given are too easy, the reader may supplement them from the splendid collections by R. Kubo. In a ten-week course I usually treat Chapters 1 to 15, followed by Chapters 18 to 21.

The Contents is followed by tabular guides to summaries, definitions, and applications. A chart of the logical connections of the chapters forms one of the front endpapers. The appendices introduce at a more advanced level several additional topics, including the Nyquist theorem and the Boltzmann transport equation.

The book is derived from course notes developed in sophomore and upper-division classes at Berkeley with the assistance of two grants by the University of California. Many helpful suggestions and original ideas were offered by Edward M. Purcell in a review of the manuscript. It is a pleasure to remember the sympathetic help offered by students, teaching assistants, and others in the early development of the course. Among those who helped generously were Samuel P. Bowen, Bernard Feldman, Randolph Levine, Roger Knacke, Robert Gray, and Robert Cahn. I am grateful to H. Eugene Stanley for his careful criticism of the early notes. I have benefitted from much good advice given by Norman E. Phillips on experimental matters. My interest in a simple unified approach to thermal physics arose from a problem in molecular biology discussed with me by J.-P. Changeux, and the related work was supported by the Office of Naval Research. Mrs. Madeline Moore typed and organized the manuscript with unfailing skill and humor.

C. Kittel

Berkeley, California
August 1969

Contents

Guide to Summaries

Guide to Fundamental Definitions

Guide to Applications

Selected References for Further Study

Thermodynamics

H. B. Callen, *Thermodynamics*, Wiley, 1960. Excellent development of the basic principles. Referred to as Callen.

A. B. Pippard, *Elements of classical thermodynamics*, Cambridge University Press, 1957. Very careful discussion.

M. W. Zemansky, *Heat and thermodynamics*, 5th ed., McGraw-Hill, 1968. Especially good for experimental matter; very thorough. Referred to as Zemansky.

R. Kubo, *Thermodynamics*, Wiley, 1968. Consists in large part of problems with solutions; a very useful collection.

D. C. Spanner, *Introduction to thermodynamics*, Academic Press, 1964. Good treatment of biological applications.

Statistical mechanics

R. Kubo, *Statistical mechanics*, Wiley, 1965. See comment above.

R. C. Tolman, *Principles of statistical mechanics*, Oxford University Press, 1938. Full and careful discussion of the principles of quantum statistical mechanics.

K. Huang, *Statistical mechanics*, Wiley, 1963. Good discussion of imperfect gases and of phase transitions.

C. Kittel, *Elementary statistical physics*, Wiley, 1958. Parts 2 and 3 treat applications to noise and to elementary transport theory. Part 1 is made obsolete by the present text.

T. L. Hill, *An introduction to statistical thermodynamics*, Addison-Wesley, 1960. Particularly good treatment of the theory of real gases, solutions, and polymers.

L. D. Landau and E. M. Lifshitz, *Statistical physics*, Addison-Wesley, 1958. Discussion of many topics of interest to physicists. The field lacks a real handbook, but this text is often used as such.

Historical

J. R. Partington, *An advanced treatise on physical chemistry*, Longmans, Green (1949), Vol. I. Partington has studied and cited a fantastic number of papers in thermodynamics and statistical mechanics, from the earliest times through 1948. The volume is also an excellent textbook.

Applications

ASTROPHYSICS

D. H. Menzel, P. L. Bhatnagar, and H. K. Sen, *Stellar interiors*, Wiley, 1963.

BIOPHYSICS

C. Tanford, *Physical chemistry of macromolecules*, Wiley, 1961.

IRREVERSIBLE THERMODYNAMICS

I. Prigogine, *Introduction to the thermodynamics of irreversible processes*, Interscience, 1961.

A. Katchalsky and P. F. Curran, *Nonequilibrium thermodynamics in biophysics*, Harvard University Press, 1965.

NOISE AND RANDOM PROCESSES

N. Wax, ed., *Selected papers on noise and stochastic processes*, Dover, 1954, paperback. Collection of memorable basic papers.

D. K. C. MacDonald, *Noise and fluctuations*, Wiley, 1962.

PLASMA PHYSICS

L. Spitzer, *Physics of fully ionized gases*, 2nd ed., Interscience, 1962.

KINETIC THEORY

E. H. Kennard, *Kinetic theory of gases, with an introduction to statistical mechanics*, McGraw-Hill, 1938.

L. Loeb, *Kinetic theory of gases*, 2nd ed., Dover, 1934, paperback.

LOW TEMPERATURE PHYSICS

G. K. White, *Experimental techniques in low temperature physics*, 2nd ed., Oxford University Press, 1968.

SOLID STATE PHYSICS

C. Kittel, *Introduction to solid state physics*, 3rd ed., Wiley, 1966. Referred to as ISSP.

CHAPTER 1

Quantum States

"But although, as a matter of history, statistical mechanics owes its origin to investigations in thermodynamics, it seems eminently worthy of an independent development, both on account of the elegance and simplicity of its principles, and because it yields new results and places old truths in a new light in departments quite outside of thermodynamics."

(J. W. Gibbs, 1902)

"A theory is the more impressive the greater the simplicity of its premises are, the more different kinds of things it relates, and the more extended is its area of applicability. Therefore the deep impression that classical thermodynamics made upon me. It is the only physical theory of universal content concerning which I am convinced that, within the framework of applicability of its basic concepts, it will never be over-thrown."

(A. Einstein, 1949)

CHAPTER 1   QUANTUM STATES

Two works written in the first years of this century have determined the development of the subject of thermal physics. Planck[1] in Berlin in 1901 wrote a historic article on the distribution of energy in thermal radiation (the radiation emitted from a hot body). His article led to the theory of quanta, from which quantum mechanics developed. In the same year Gibbs[2] in New Haven wrote an extraordinarily perceptive, difficult, and hermetic treatise entitled *Elementary principles in statistical mechanics*. The theoretical physicist Lorentz said of this title that "the word 'elementary' rather indicates the modesty of the author than the simplicity of the subject."

We know now that the study of thermal physics is approached more easily from the viewpoint of quantum mechanics than from the viewpoint of classical mechanics available to Gibbs. This is not an unnatural circumstance, because quantum mechanics gives a valid description of nature, whereas classical mechanics gives a description that is incomplete at the atomic level. When we translate the principles of Gibbs into the language of quantum mechanics, then for the first time we find a clear, reasonable, and simple statement of the physical basis of all of thermodynamics and of statistical mechanics. The process of translation utilizes in an essential way only a single concept of quantum mechanics, the concept of a stationary quantum state of a system of particles.

When we can count the stationary quantum states accessible to a system, we know the entropy of the system, for the entropy is the logarithm of the number of accessible states. The entropy is the most important quantity of thermal physics: from the entropy we find the temperature, the pressure, the chemical potential, the magnetic moment, and the other functions of thermal physics.

[1] M. Planck, Annalen der Physik 4, 553–563 (1901). A summary of the theory was published in 1900. An English translation of the summary appears on pp. 82–90 of D. ter Haar, *Old quantum theory*, Pergamon, 1967, paperback.

[2] J. W. Gibbs, *Elementary principles in statistical mechanics, developed with especial reference to the rational foundation of thermodynamics*, Dover paperback reprint S707. The book was originally published in 1902 by Yale University Press.

The idea of a **stationary quantum state** was conceived by Niels Bohr[3] in his famous 1913 paper "On the constitution of atoms and molecules." A property of a stationary quantum state of a physical system of constant energy is that the probability of finding a particle in any element of volume is independent of the time. A stationary quantum state may be defined as a condition of a system such that all observable physical properties are independent of the time. The stationary quantum states of the systems we consider are usually denumerable, although infinite in number.

The system under consideration may be composed of a single particle or of many particles. We are usually concerned with the states of a system of many particles. Each stationary quantum state has a definite energy, but it may happen that several states have identical or nearly identical energies. For brevity we shall omit henceforth the word stationary; the quantum states that we treat are understood to be stationary unless otherwise indicated.

The **degeneracy** of an energy level is defined as the number of quantum states having the given energy or having an energy in a narrow range. It is the energy level and not the quantum state that is said to be degenerate. The practical definition of the degeneracy of an energy level depends on the resolving power of the particular method used to obtain and to display the results. With a finer pen many of the energy levels of Fig. 1 would be shown as split into multiple levels.

Let us look at the quantum states and energy levels of several atomic systems. The simplest atom is hydrogen, with one electron and one proton. The quantum states of the hydrogen atom are associated with the motion of the electron and the proton. The low-lying energy levels of hydrogen are shown in Fig. 1, where we have put in parentheses the number of quantum states[4] belonging to approximately the same energy level, within the resolution of the figure. The zero of energy is taken as the energy of the lowest energy level. The positions of the energy levels can be determined spectroscopically by measuring the wavelength λ of the quanta emitted from excited atoms. From the relation $\lambda \nu = c$, the speed of light, we determine the frequency ν. The energy is then given by $h\nu = \epsilon$, where h is Planck's constant. Here λ, ν, and ϵ are the Greek letters lambda, nu, and epsilon, respectively.

[3] Niels Bohr, Philosophical Magazine **26**, 1 (1913); reprinted in D. ter Haar, *Old quantum theory,* Pergamon, 1967, paperback.

[4] For the present we overlook the fact that the nucleus may possess a spin and a magnetic moment. We enumerate the degeneracies of the energy levels as if the spin of the nucleus were zero. The proton has a spin of $\frac{1}{2}$ in units of $h/2\pi$ or $\hbar$ and has two independent orientations. To take account of this we should double the values of the degeneracies shown for atomic hydrogen in Fig. 1.

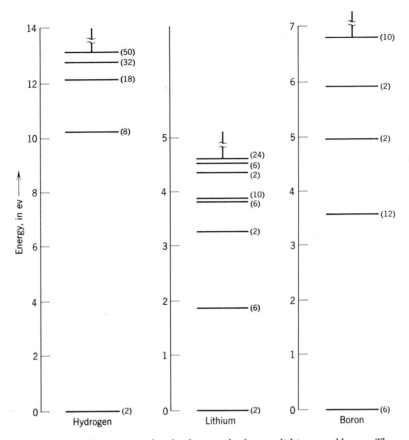

Figure 1   Low-lying energy levels of atomic hydrogen, lithium, and boron. The energies are given in electron volts, with 1 ev = 1.602×10^{-12} erg. The numbers in parentheses give the number of quantum states having approximately the same energy, with no account taken of the spin of the nucleus. The data are taken from *Atomic energy levels*, National Bureau of Standards Circular 467.

An atom of lithium has three electrons which move about the nucleus. Each electron interacts with the nucleus by an electrostatic interaction, and each electron also interacts electrostatically with all the other electrons. The energies of the levels of lithium shown in the figure are the collective energies of the entire system. The energy levels shown for boron, which has five electrons, are also energies of the entire system.

The **energy of the system** is the total energy of all particles, kinetic plus potential, with account taken of all mutual interactions. Thus the energy of a system of more than two particles cannot be described exactly as the excitation energy of an individual particle in the field of another particle, even though sometimes it may be a very good approximation to describe the low-

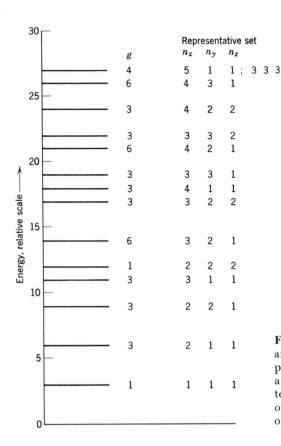

Figure 2   Energy levels, degeneracies, and quantum numbers n_x, n_y, n_z of a particle of spin zero which is confined to a cube. The quantum states of this system are discussed in Chapter 10 on the orbitals of a free particle. Quantum states of a one-particle system are called orbitals.

lying energy levels in this way. A quantum state of the system is a state of all particles of the system. When we have cause to speak of the state of one particle, we shall use the term orbital in place of state.

We shall be concerned with the properties of physical systems of many different types. To describe the statistical properties of a system of N particles it is essential to know the set of values of the energy $\epsilon_l(N)$, where the notation means that ϵ is the energy of the quantum state l of the N particle system. We use the Greek letter epsilon, ϵ, to denote energy. The indices such as l may be assigned to the quantum states in any convenient arbitrary way, but two states should not be assigned the same index.

The low-lying energy levels of a single particle confined to a cube of side L are shown in Fig. 2. The wavefunctions are derived in Chapter 10. We also derive the energies there, with the result

$$\epsilon = \frac{\hbar^2}{2M} \left(\frac{\pi}{L}\right)^2 (n_x^2 + n_y^2 + n_z^2) , \tag{1}$$

where n_x, n_y, n_z are integers. The degeneracies of the levels are indicated in the figure. The orbital with $n_x = 4$; $n_y = 1$; $n_z = 1$ has $n_x^2 + n_y^2 + n_z^2 = 18$.

The two orbitals with $n_x = 1$; $n_y = 1$; $n_z = 4$ and $n_x = 1$; $n_y = 4$; $n_z = 1$ also have $n_x^2 + n_y^2 + n_z^2 = 18$, so that the corresponding energy level has the degeneracy $g = 3$ as indicated in the figure.

It is a good idea to start by studying the properties of simple model systems for which the energies $\epsilon_l(N)$ can be calculated exactly in an elementary way. One such system is introduced in Chapter 2 and developed further in Chapter 4. The conclusions derived for the model system will be assumed to apply to all physical systems. This is a drastic step to take, but the consequences are in agreement with experiment. The importance of the model system lies precisely in the fact that its statistical and thermal properties can be treated exactly.

CHAPTER 2

An Elementary Soluble System

It is an encouragement to our imagination and physical intuition to have available a model many-body system, however trivial, which can be solved exactly for all statistical properties. We shall use one model system repeatedly to help us visualize what might otherwise be concealed by the hard shell of abstraction. The term model system means a system for which the states, degeneracies, and energies can be found explicitly and exactly. We shall assume throughout the book that the general statistical properties found for the model system apply equally well to any realistic physical system. This assumption leads to predictions that agree in every known instance with experiment.

STATES OF THE MODEL SYSTEM

The model system shown in Fig. 1 is a set of N distinct elementary magnets at N fixed points on a line. The magnitude of the magnetic moment of each elementary magnet is denoted by μ, the Greek letter mu.

We suppose that each moment can point only straight up or straight down. By up we mean in the direction of the $+z$ axis. If the magnet points up, we say that the magnetic moment is $+\mu$. If the magnet points down, the magnetic moment is $-\mu$. This system is much simpler to treat statistically than an ideal gas, and that is why we start with it. The ideal gas offers difficulties, but we shall learn how to avoid them in Chapter 11. The mathematical problems of the linear polymer (Appendix A) and the lattice gas (Appendix B) are related to our model system.

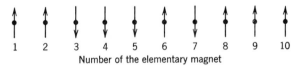

1   2   3   4   5   6   7   8   9   10
Number of the elementary magnet

Figure 1   Model system composed of elementary magnets at fixed points on a line, each having magnetic moments $\pm\mu$. There are no interactions among the magnets, and there is no external magnetic field. Each magnetic moment may be oriented in two ways, up or down, so that there are 2^{10} distinct arrangements of the ten magnetic moments shown in the figure. If the arrangements are selected in a random process, the probability of finding the particular arrangement shown is $1/2^{10}$.

A particle of spin angular momentum $\frac{1}{2}\hbar$, such as an electron, a neutron, or a proton, has two possible orientations of the spin, or of the magnetic moment, relative to any fixed direction. This is confirmed by the results of atomic beam experiments. In our model we choose a particle with two orientations as a matter of computational convenience. A system composed of one such particle has two different stationary quantum states, one with the spin pointing up and one with the spin pointing down. The four states of a system composed of two particles are shown in Fig. 2.

Now consider N different sites, each of which bears a moment that may assume the values of $\pm\mu$. Each moment may be oriented in two ways with a probability independent of the orientation of all other moments. The total number of arrangements of the N moments is 2^N. We specify a **state** of the system by giving the orientation of the moment on each site. There are 2^N states. We may use the following transparent notation to denote a single state of the system of N moments:

$$\uparrow\uparrow\downarrow\downarrow\downarrow\uparrow\downarrow\uparrow\uparrow\uparrow \cdots . \tag{1a}$$

The sites themselves are assumed to be arranged in a definite order, and we might number them in sequence from left to right, for example. On this convention the state (1a) could be written as

$$\uparrow_1\uparrow_2\downarrow_3\downarrow_4\downarrow_5\uparrow_6\uparrow_7\ \uparrow_8\uparrow_9\uparrow_{10} \cdots . \tag{1b}$$

Both sets of symbols (1a) and (1b) denote the same state of the system, the state in which the moment on site 1 is $+\mu$; on site 2, the moment is $+\mu$; on site 3, the moment is $-\mu$; and so forth.

Every different state of the system is contained in a symbolic development of the following product of N factors:

$$(\uparrow_1 + \downarrow_1)(\uparrow_2 + \downarrow_2)(\uparrow_3 + \downarrow_3) \cdots (\uparrow_N + \downarrow_N) . \tag{2a}$$

We define a multiplication rule for this symbol by the relation:

$$(\uparrow_1 + \downarrow_1)(\uparrow_2 + \downarrow_2) = \uparrow_1\uparrow_2 + \uparrow_1\downarrow_2 + \downarrow_1\uparrow_2 + \downarrow_1\downarrow_2 . \tag{2b}$$

The function (2a) on multiplication generates a sum of 2^N terms, one for each of the 2^N possible states. Each term is a product of N individual magnetic moment symbols, with one symbol for each elementary magnet on the line. Each term denotes an independent state of the system and is a simple product of the form, for example,

$$\uparrow_1\uparrow_2\downarrow_3 \cdots \uparrow_N . \tag{3}$$

For a system of two elementary magnets, we multiply $(\uparrow_1 + \downarrow_1)$ by $(\uparrow_2 + \downarrow_2)$ to obtain the four possible states:

$$(\uparrow_1 + \downarrow_1)(\uparrow_2 + \downarrow_2) = \uparrow_1\uparrow_2 + \uparrow_1\downarrow_2 + \downarrow_1\uparrow_2 + \downarrow_1\downarrow_2 , \tag{4}$$

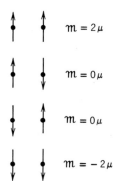

$$m = 2\mu$$

$$m = 0\mu$$

$$m = 0\mu$$

$$m = -2\mu$$

Figure 2   The four different states of a system of two elementary magnets ($N = 2$). These states may be constructed by inspection. For large values of N it is convenient to generate the states from the function displayed in (2a). The quantity m is the total magnetic moment of the state.

as in (2b). The sum here is not a state but a way of listing the four possible states of the system. The product on the left-hand side of the equation is called a generating function: it generates the states of the system.

The generating function for the states of a system of three spins is

$$(\uparrow_1 + \downarrow_1)(\uparrow_2 + \downarrow_2)(\uparrow_3 + \downarrow_3) \ .$$

This expression on multiplication generates $2^3 = 8$ different states:

$$\uparrow_1\uparrow_2\uparrow_3$$
$$\uparrow_1\uparrow_2\downarrow_3 \qquad \uparrow_1\downarrow_2\uparrow_3 \qquad \downarrow_1\uparrow_2\uparrow_3$$
$$\uparrow_1\downarrow_2\downarrow_3 \qquad \downarrow_1\uparrow_2\downarrow_3 \qquad \downarrow_1\downarrow_2\uparrow_3$$
$$\downarrow_1\downarrow_2\downarrow_3 \ .$$

The **total magnetic moment** of our model system will be denoted by the script capital letter m. The value of m varies from $N\mu$ to $-N\mu$. The set of possible values of m is given by

$$m = N\mu, (N - 2)\mu, (N - 4)\mu, (N - 6)\mu, \cdots, -N\mu \ . \tag{5}$$

The set of possible values of m is obtained if we start with the state for which all spins are up ($m = N\mu$) and reverse one spin at a time. We may reverse N spins to obtain the ultimate state for which all spins are down ($m = -N\mu$).

There are $N + 1$ possible values of the total moment, whereas there are 2^N states. If $N > 1$, then $2^N > N + 1$. There are more states than values of the total moment. For example, if $N = 10$, there are $2^{10} = 1024$ states distributed among eleven different values of the total magnetic moment. For large N many different states of the system may have the same value of the total moment m.

The system with $N = 2$ described by the function (4) has one state with $m = 2\mu$; two states with $m = 0\mu$; and one state with $m = -2\mu$. These states were shown in Fig. 2.

Only one state of a system has the moment $m = N\mu$; that state is

$$\uparrow\uparrow\uparrow\uparrow \cdots \uparrow\uparrow\uparrow\uparrow \ . \tag{6}$$

But there are N ways to form a state with one spin[1] down:

$$\downarrow\uparrow\uparrow\uparrow \cdots \uparrow\uparrow\uparrow\uparrow \qquad (7)$$

is one such state; another state is

$$\uparrow\downarrow\uparrow\uparrow \cdots \uparrow\uparrow\uparrow\uparrow , \qquad (8)$$

and the other states are formed from (6) by reversing any single spin. The states (7) and (8) have total moment $\mathcal{m} = N\mu - 2\mu$.

ENUMERATION OF STATES:
THE DEGENERACY FUNCTION $g(N, m)$

We can easily find an analytic expression for the number of states with $\frac{1}{2}N + m$ spins up and $\frac{1}{2}N - m$ spins down, where m is an integer.[2] It is convenient to assume that N is an even number. Some of our results in this chapter would be slightly different if N were an odd number, but it is not worth our while to give the results separately. We are generally interested in very large values of N, and it cannot make any significant difference whether N is even or odd. The difference is

$$(\text{number up}) - (\text{number down}) = 2m .$$

We call $2m$ the **spin excess**. We illustrate the definition of the spin excess in Fig. 3. It will turn out to be quite convenient to keep the factor of 2 in the definition of the spin excess.

The N-factor product in (2a) may be written symbolically as

$$(\uparrow + \downarrow)^N . \qquad (9)$$

We are allowed to drop the site labels (the subscripts) from (2a) if we are interested only in how many spins are up or down, and not in which particular spins are up or down. If we drop the labels and neglect the order in which the arrows appear in a given product, then (2b) becomes

$$(\uparrow + \downarrow)^2 = \uparrow\uparrow + 2\uparrow\downarrow + \downarrow\downarrow ,$$

and

$$(\uparrow + \downarrow)^3 = \uparrow\uparrow\uparrow + 3\uparrow\uparrow\downarrow + 3\uparrow\downarrow\downarrow + \downarrow\downarrow\downarrow ,$$

where the right-hand side may be written in condensed form as

$$\uparrow^3 + 3\uparrow^2\downarrow + 3\uparrow\downarrow^2 + \downarrow^3 .$$

---

[1] Here the word **spin** is used as a shorthand for elementary magnet.

[2] When we turn one elementary magnet from the up to the down orientation, $\frac{1}{2}N + m$ goes to $\frac{1}{2}N + m - 1$ and $\frac{1}{2}N - m$ goes to $\frac{1}{2}N - m + 1$. The difference (number up) $-$ (number down) changes from $2m$ to $2m - 2$.

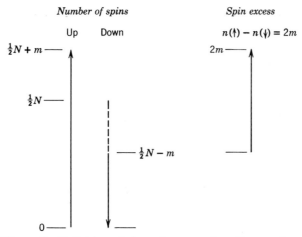

Figure 3 Definition of the spin excess $2m$. The number $n(\uparrow)$ of up spins is $\frac{1}{2}N + m$; the number $n(\downarrow)$ of down spins is $\frac{1}{2}N - m$. The total number of spins is $n(\uparrow) + n(\downarrow) = (\frac{1}{2}N + m) + (\frac{1}{2}N - m) = N$. The spin excess $n(\uparrow) - n(\downarrow) = (\frac{1}{2}N + m) - (\frac{1}{2}N - m) = 2m$. If each spin has magnetic moment μ, the magnetic moment of the up spins alone is $\mu n(\uparrow)$ and that of the down spins alone is $-\mu n(\downarrow)$. The total magnetic moment is $\mathfrak{M} = \mu[n(\uparrow) - n(\downarrow)] = 2\mu m$. If N is an even number, m is an integer; if N is an odd number, m is a half-integer. We shall always assume that N is even. The integer m assumes all values between $\frac{1}{2}N$ and $-\frac{1}{2}N$.

By the binomial theorem of algebra we know that

$$(x + y)^N = x^N + Nx^{N-1}y + \frac{1}{2}N(N-1)x^{N-2}y^2 + \cdots + y^N =$$
$$\sum_{s=0}^{N} \frac{N!}{(N-s)!\,s!} x^{N-s}y^s . \tag{10a}$$

Here $s! \equiv 1 \cdot 2 \cdot 3 \cdots s$ and is called "s factorial." We may write the exponents of x and y in a slightly different, but equivalent, form:

$$(x + y)^N = \sum_{m=-\frac{1}{2}N}^{\frac{1}{2}N} \frac{N!}{(\frac{1}{2}N + m)!\,(\frac{1}{2}N - m)!} x^{\frac{1}{2}N + m}y^{\frac{1}{2}N - m} . \tag{10b}$$

With this result the symbolic expression $(\uparrow + \downarrow)^N$ becomes

$$(\uparrow + \downarrow)^N \equiv \sum_{m} \frac{N!}{(\frac{1}{2}N + m)!\,(\frac{1}{2}N - m)!} \uparrow^{\frac{1}{2}N + m}\downarrow^{\frac{1}{2}N - m} . \tag{11}$$

The notation $\uparrow^{\frac{1}{2}N + m}\downarrow^{\frac{1}{2}N - m}$ does not actually denote a single specific state because we have dropped the site labels. However, the coefficient of the term $\uparrow^{\frac{1}{2}N + m}\downarrow^{\frac{1}{2}N - m}$ is the number of distinct states having $\frac{1}{2}N + m$ spins up and $\frac{1}{2}N - m$ spins down. Such states have total moment $\mathfrak{M} = 2m\mu$ and have spin excess $2m$.

We denote the coefficient of $\uparrow^{\frac{1}{2}N + m}\downarrow^{\frac{1}{2}N - m}$ in (11) by $g(N, m)$, where

$$g(N, m) = \frac{N!}{(\frac{1}{2}N + m)!\,(\frac{1}{2}N - m)!} = \text{number of states with a spin ex-} \quad (12)$$
$$\text{cess } 2m, \text{ for a system of } N \text{ spins.}$$

Thus

$$(\uparrow + \downarrow)^N = \sum_{m = -\frac{1}{2}N}^{\frac{1}{2}N} g(N, m)\uparrow^{\frac{1}{2}N + m}\downarrow^{\frac{1}{2}N - m} . \quad (13)$$

The quantity $g(N, m)$ is a binomial coefficient, where m is any integer (or, if N is odd, any half-integer) between $-\frac{1}{2}N$ and $\frac{1}{2}N$. Most collections of mathematical tables include a table of binomial coefficients.

We shall call $g(N, m)$ the **degeneracy function**; it is the number of states having the same value of m (or of $\mathcal{M}$). This usage appears to differ slightly from that of the degeneracy in Chapter 1. The reason for our present definition will emerge later in this chapter when a magnetic field is applied to the system. In a magnetic field states of different m will have different values of the energy, so that our g is equal to the usual degeneracy in a magnetic field. At present we have not introduced a magnetic field; until we do so all states of the model system may be supposed to have the same energy. Notice from (10) that the total number of states is given by

$$(1 + 1)^N = 2^N = \sum_{m = -\frac{1}{2}N}^{m = \frac{1}{2}N} g(N, m) .$$

[The quantity $g(N, m)$ is often derived in probability theory as the number of ways of choosing $\frac{1}{2}N + m$ up spins and $\frac{1}{2}N - m$ down spins from a group of N spins. Our argument from the generating function is equivalent, but is more direct. There is no mystique about how to count correctly.]

Examples related to $g(N, m)$ for $N = 10$ are given in Figs. 4 and 5. For a coin, "heads" could stand for "spin up" and "tails" could stand for "spin down."

EXAMPLE. *Value of $\langle \mathcal{M} \rangle$ and $\langle \mathcal{M}^2 \rangle$.* Here the symbol $\langle \cdots \rangle$ denotes the average value over all the states of the model system. We assume if up and down spin orientations are selected at random that all 2^N states of the system are equally likely to occur. This is a basic assumption to which we shall return in Chapter 3.

By the definition of $\mathcal{M}$, the average value of the total magnetic moment is

$$\langle \mathcal{M} \rangle = \left\langle \sum_{s=1}^{N} \mu_s \right\rangle = \sum_{s=1}^{N} \langle \mu_s \rangle . \quad (14)$$

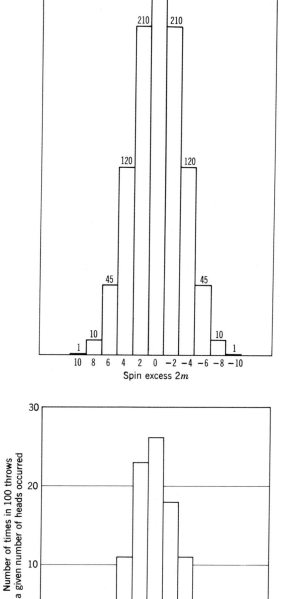

Figure 4   Number of distinct arrangements of $5 + m$ spins up and $5 - m$ spins down. Values of $g(N, m)$ for $N = 10$, where $2m$ is the spin excess, or $n(\uparrow) - n(\downarrow)$. The total number of states is

$$2^{10} = \sum_{m = -5}^{5} g(10, m) \ .$$

The values of the g's were taken from a table of the binomial coefficients.

Figure 5   An experiment was done in which 10 pennies were thrown 100 times. The number of heads in each throw was recorded. (Courtesy of Tim Kittel.)

But $\langle \mu_s \rangle = 0$, for the moment at s is as likely to be $+\mu$ as $-\mu$. To say this another way, the average moment on any site s is

$$\langle \mu_s \rangle = \tfrac{1}{2}[(+\mu) + (-\mu)] = 0 \ , \tag{15}$$

whence $\langle \mathcal{M} \rangle = 0$.

Now consider the average value of the square of the total magnetic moment:

$$\langle \mathcal{M}^2 \rangle = \left\langle \left(\sum_{s=1}^{N} \mu_s \right)^2 \right\rangle = \sum_r \sum_s \langle \mu_r \mu_s \rangle \ , \tag{16}$$

where r and s run independently from 1 to N. In the double summation there appear terms for which $r = s$; for such a term the contribution to $\langle \mathcal{M}^2 \rangle$ is

$$\langle \mu_r \mu_r \rangle = \langle \mu_r^2 \rangle = \tfrac{1}{2}[(\mu^2) + (-\mu)^2] = \mu^2 \ . \tag{17}$$

There are N such terms in the double summation.

If $r \neq s$, then $\langle \mu_r \mu_s \rangle = 0$. We see this by considering the average over states of two spins, μ_1 and μ_2, for which the four possible states were shown in Fig. 2:

$$\langle \mu_1 \mu_2 \rangle = \tfrac{1}{4}[(+\mu)(+\mu) + (+\mu)(-\mu) + (-\mu)(+\mu) + (-\mu)(-\mu)] = 0 \ . \tag{18}$$

Terms with $r \neq s$ make no contribution to $\langle \mathcal{M}^2 \rangle$. Therefore $\langle \mathcal{M}^2 \rangle$ contains only the N terms of value μ^2:

$$\langle \mathcal{M}^2 \rangle = N\mu^2 \ . \tag{19}$$

The root mean square value of the total moment is defined as $\langle \mathcal{M}^2 \rangle^{\frac{1}{2}}$ and is denoted by $\mathcal{M}_{\mathrm{rms}}$. Thus, by (19),

$$\mathcal{M}_{\mathrm{rms}} = \sqrt{N}\mu \ . \tag{20}$$

We can show that the distribution of values of $\mathcal{M}$ must have a narrow peak. Suppose that we divide $\mathcal{M}_{\mathrm{rms}}$ by the maximum value of $\mathcal{M}$, which is $N\mu$:

$$\frac{\mathcal{M}_{\mathrm{rms}}}{\mathcal{M}_{\mathrm{max}}} = \frac{\sqrt{N}\mu}{N\mu} = \frac{1}{\sqrt{N}} \ . \tag{21}$$

The ratio can be very small if N is a macroscopic number, and we assert that then the peak must be very narrow and centered at $\mathcal{M} = 0$. A broad distribution would have a large value of $\mathcal{M}_{\mathrm{rms}}$. (We shall call N a **macroscopic number** if it is of the order of the number of atoms in a tangible specimen; if $N = 10^{20}$, then $1/\sqrt{N} = 10^{-10}$.)

SHARP PEAK OF $g(N, m)$

We now show explicitly that for a very large system ($N \gg 1$) the function $g(N, m)$ defined by (12) is peaked very sharply about the value $m = 0$. We first look for an approximation that allows us to examine the form of $g(N, m)$ versus m when $N \gg 1$ and $|m| \ll N$. Common tables of factorials do not go above $N = 100$, and we may be interested in $N \approx 10^{20}$, so that an approximation is clearly needed. A good one is available; the result is given in (37).

It is convenient to work with $\log g$. When you are confronted with a very large number, it is a good rule to consider instead the logarithm of the number. On taking the logarithm[3] of both sides of (12) we have

$$\log g(N, m) = \log N! - \log (\tfrac{1}{2}N + m)! - \log (\tfrac{1}{2}N - m)! , \qquad (22)$$

by virtue of the characteristic property of the logarithm of a product:

$$\log xy = \log x + \log y ; \qquad \log \frac{x}{y} = \log x - \log y . \qquad (23)$$

We may construct a simple identity:

$$n! = 1 \cdot 2 \cdot 3 \cdots n = 1 \cdot 2 \cdot 3 \cdots (k - 1)(k)(k + 1)(k + 2) \cdots n$$
$$= k!(k + 1)(k + 2) \cdots n . \qquad (24)$$

We use this identity to evaluate one of the terms of (22):

$$(\tfrac{1}{2}N + m)! = (\tfrac{1}{2}N)! \, (\tfrac{1}{2}N + 1)(\tfrac{1}{2}N + 2) \cdots (\tfrac{1}{2}N + m) ;$$

$$\log (\tfrac{1}{2}N + m)! = \log (\tfrac{1}{2}N)! + \sum_{s=1}^{m} \log (\tfrac{1}{2}N + s) . \qquad (25)$$

It is convenient to phrase our discussion for positive m. This is no restriction, because $g(N, m)$ is an even function of m.

By a similar argument

$$\log (\tfrac{1}{2}N - m)! = \log (\tfrac{1}{2}N)! - \sum_{s=1}^{m} \log (\tfrac{1}{2}N - s + 1) . \qquad (26)$$

We combine (25) and (26) to obtain

$$\log (\tfrac{1}{2}N + m)! + \log (\tfrac{1}{2}N - m)! \cong 2 \log (\tfrac{1}{2}N)! + \sum_{s=1}^{m} \log \frac{1 + (2s/N)}{1 - (2s/N)} , \qquad (27)$$

where in the argument of the logarithm in (26) we have approximated $\tfrac{1}{2}N - s + 1$ by $\tfrac{1}{2}N - s$.

[3] Except where otherwise specified, all logarithms are understood to be log base e, written here as log.

By a famous power series expansion

$$e^{\pm x} = 1 \pm x + \tfrac{1}{2}x^2 \pm \cdots . \tag{28}$$

We assume that $x^2 \ll 1$ and take the log of both sides to obtain

$$\pm x \cong \log (1 \pm x) , \tag{29}$$

to order x. Then

$$\log (1 + x) - \log (1 - x) \cong 2x , \tag{30}$$

or

$$\log \frac{1 + x}{1 - x} \cong 2x ; \qquad \log \frac{1 + (2s/N)}{1 - (2s/N)} \cong \frac{4s}{N} , \tag{31}$$

where x denotes $2s/N$. We have dropped terms of order s^3/N^3 and higher from this expansion, because $s \le m$, and we have assumed $m/N \ll 1$. Although m may be a large number, N is assumed to be much larger.

Now carry out the sum over s in (27) with the approximation (31). We sum

$$\sum_{s=1}^{m} \log \frac{1 + (2s/N)}{1 - (2s/N)} \cong \frac{4}{N} \sum_{s=1}^{m} s . \tag{32}$$

By the standard arithmetical series

$$1 + 2 + 3 + \cdots + m = \tfrac{1}{2}m(m + 1) , \tag{33}$$

whence (32) has the value

$$\sum_{s=1}^{m} \log \frac{1 + (2s/N)}{1 - (2s/N)} \cong \frac{4}{N} \sum_{s=1}^{m} s = \frac{4}{N} \cdot \tfrac{1}{2}m(m + 1) \cong \frac{2m^2}{N} , \tag{34}$$

in the approximation $1 \ll m \ll N$.

Thus the expression (22) for $\log g(N, m)$ becomes

$$\log g(N, m) \cong \log N! - 2 \log (\tfrac{1}{2}N)! - \frac{2m^2}{N} . \tag{35}$$

We raise e to powers of both sides of (35) to find

$$g(N, m) \cong \frac{N!}{(\tfrac{1}{2}N!) (\tfrac{1}{2}N!)} e^{-2m^2/N} , \tag{36}$$

for $1 \ll |m| \ll N$. We may write this result as

$$\boxed{g(N, m) \cong g(N, 0) \, e^{-2m^2/N} ,} \tag{37}$$

where

$$g(N, 0) = \frac{N!}{(\tfrac{1}{2}N)! (\tfrac{1}{2}N)!} . \tag{38}$$

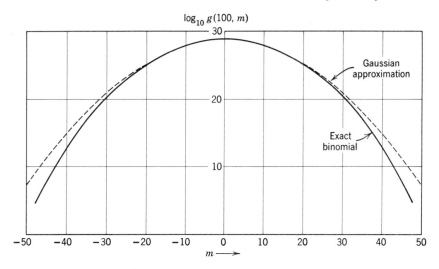

Figure 6 (above) Comparison of exact (12) and approximate (37) expressions for the binomial coefficients $g(N, m)$ for $N = 100$. The range of the index m is from -50 to $+50$. Values of $\log_{10} g(100, m)$ are plotted, rather than of $g(100, m)$, in order to emphasize the regions of m in which the approximation differs significantly from the exact values. (Courtesy of Peter Kittel.)

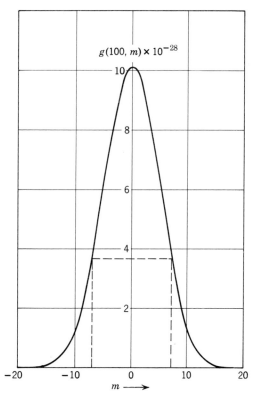

Figure 7 (right) The Gaussian approximation to the binomial coefficients $g(100, m)$ plotted on a linear scale. On this scale it is not possible on the drawing to distinguish the approximation from the exact values over the range of m plotted. The entire range of m is from -50 to $+50$. The dashed lines are drawn from the points at $1/e$ of the maximum of g.

The exact binomial coefficient (12) and the approximate expression (37) for $g(N, m)$ are plotted in Fig. 6 for $N = 100$. To exhibit the deviation of (37) from (12) at large m we found it convenient to plot $\log_{10} g$ versus m, rather than g versus m.

The distribution defined by the right-hand side of (37) is called **Gaussian**, and it is plotted in Fig. 7. The distribution is centered in a maximum at

$m = 0$. When $m^2 = \frac{1}{2}N$, the value of g is reduced to e^{-1} of the maximum value.[4] That is, when

$$\frac{m}{N} = \left(\frac{1}{2N}\right)^{\frac{1}{2}} , \tag{39}$$

the value of g is e^{-1} of $g(N, 0)$. The quantity $(1/2N)^{\frac{1}{2}}$ is thus a reasonable measure of the fractional width of the distribution. For $N \approx 10^{22}$, the fractional width is of the order of 10^{-11}. When N is very large the distribution is exceedingly sharply defined. This result confirms our inference from the example in which we calculated $\langle \mathfrak{M}^2 \rangle$. We conclude that the distribution of the total magnetic moment $\mathfrak{M}$ for a macroscopic number of moments oriented at random is sharply defined about the mean value $\mathfrak{M} = 0$.

EXAMPLE. *Stirling approximation.* The Stirling approximation to the value of $n!$ for $n \gg 1$ is

$$n! \cong (2\pi n)^{\frac{1}{2}} n^n \exp\left(-n + \frac{1}{12n} + \cdots\right) . \tag{40}$$

This useful result is stated in almost every collection of mathematical formulae, and it is derived in numerous texts on advanced calculus. The terms $1/12n + \cdots$ in the argument of the exponential may be neglected for sufficiently large n. From (38) and (40) we have

$$g(N, 0) \cong 2^N \left(\frac{2}{\pi N}\right)^{\frac{1}{2}} . \tag{41}$$

For $N = 50$, the exact value of $g(50, 0)$ is 1.264×10^{14}, from (38). The approximate value (41) from the Stirling approximation is 1.255×10^{14}.

We may combine (41) with (37) to give

$$g(N, m) \cong 2^N \left(\frac{2}{\pi N}\right)^{\frac{1}{2}} e^{-2m^2/N} . \tag{42}$$

This is *not* as good an approximation as (37) with (38) for $g(N, 0)$. However, it has the advantage that the integral over a range of m from $-\infty$ to $+\infty$ gives the correct value, 2^N, for the total number of states. The integral is carried out in the example below. We recall from (13) that the exact distribution (12) when summed over m from $-\frac{1}{2}N$ to $\frac{1}{2}N$ must give the total number of states, 2^N, correctly.

---

[4] Useful constants are $e = 2.71828 \cdots$; $e^{-1} = 0.36787 \cdots$; $\log_{10} e = 0.43429 \cdots$; and $\log_e 10 = 2.30528 \cdots$.

EXAMPLE. *Gauss integral.* Verify by integration of (42) that

$$\int_{-\infty}^{\infty} dm \, g(N, m) = 2^N \; . \tag{43}$$

The required definite integral is the Gauss integral:

$$\mathcal{I} = \int_{-\infty}^{\infty} dx \, e^{-x^2} = \pi^{\frac{1}{2}} \; , \tag{44}$$

where we have set $x^2 \equiv 2m^2/N$ or $m^2 = Nx^2/2$. Thus

$$\int dm \, g(N, m) \cong 2^N \left(\frac{2}{\pi N}\right)^{\frac{1}{2}} \left(\frac{N}{2}\right)^{\frac{1}{2}} \mathcal{I} \; . \tag{45}$$

To evaluate $\mathcal{I}$, form $\mathcal{I}^2$:

$$\mathcal{I}^2 = \int_{-\infty}^{\infty}\int_{-\infty}^{\infty} dx \, dy \, e^{-(x^2 + y^2)} = \int_{0}^{2\pi} d\varphi \int_{0}^{\infty} e^{-\rho^2}\rho \, d\rho = \pi \; . \tag{46}$$

We have changed from a surface integral in Cartesian coordinates to a surface integral in polar coordinates with $\rho^2 = x^2 + y^2$. The area element $dx \, dy$ becomes $\rho \, d\varphi \, d\rho$. The integral over $d\rho$ is simple.

With the result (46), we have for (45):

$$\int_{-\infty}^{\infty} dm \, g(N, m) = 2^N \left(\frac{2}{\pi N}\right)^{\frac{1}{2}} \left(\frac{N}{2}\right)^{\frac{1}{2}} \pi^{\frac{1}{2}} = 2^N \; . \tag{47}$$

We should really have taken the limits of integration as $\pm\frac{1}{2}N$, but for $N \gg 1$ the outer wings of the integrand do not contribute significantly to the integral, as we see from Fig. 6. The approximation (42) gives slightly too low a value at small m and too high a value at large m; the two errors exactly compensate each other in the integral.

Problem 1. *Evaluation of definite integrals.* Show that

$$\int_{-\infty}^{\infty} dx \, x^2 \, e^{-x^2} = \tfrac{1}{2}\sqrt{\pi} \; ; \tag{48}$$

$$\int_{-\infty}^{\infty} dx \, x^4 \, e^{-x^2} = \tfrac{3}{4}\sqrt{\pi} \; . \tag{49}$$

Hint. From (44) evaluate

$$\int_{-\infty}^{\infty} dx \, e^{-\alpha x^2} \; ; \tag{50}$$

then take $-(d/d\alpha)$ of the result to obtain the first integral. Repeat the operation to obtain the second integral.

Problem 2. *Lattice gas.* Consider as a mathematical model N_0 lattice sites, each of which may be occupied by 0 or 1 atom. Suppose that N atoms are distributed at random among the N_0 sites. A vacant site is represented by a hollow circle o, and an occupied site is represented by a solid circle ●. By consideration of the quantity

$$(● + o)^{N_0} ,\tag{51}$$

which is identical to $(\uparrow + \downarrow)^{N_0}$ in (11), show that the number of different arrangements of the N atoms among the N_0 sites is

$$\frac{N_0!}{(N_0 - N)!\, N!} .\tag{52}$$

We may denote this quantity by $g(N_0, N)$. It is the number of distinct states that may be formed by N atoms on N_0 sites, with 0 or 1 atom per site. Be careful: the meaning of N is not quite the same as the meaning of m in (11) or (12).

ENERGY OF THE MAGNETIC MODEL SYSTEM

The thermodynamic properties of the model system of free elementary magnets treated above are not especially interesting, for the reason that all states were assumed to have the same energy. This assumption is made also for the model systems of polymeric chains in Appendix A. However, we have worked out some significant statistical properties of the systems, such as the mean square magnetic moment $\langle \mathfrak{m}^2 \rangle$ and the mean square polymer length $\langle r^2 \rangle$, by use of the assumption that all states are equally likely to occur in a random sample of the states of the system.

The thermodynamic properties such as the energy become physically relevant if the system of elementary magnets is placed in a magnetic field, for then the energies of the different states are no longer all equal. If the energy of the system is specified, then only the states having this energy may occur in the sampling process. The energy of interaction of a single magnetic moment μ_s with a fixed external magnetic field $\mathbf{H}$ as derived in Fig. 8 is

$$U_s = -\boldsymbol{\mu}_s \cdot \mathbf{H} .\tag{53}$$

This is the potential energy of the magnet $\boldsymbol{\mu}_s$ in the field $\mathbf{H}$.

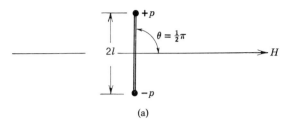

Figure 8   A magnetic dipole of strength μ is represented by magnetic monopoles of strength $\pm p$ separated by a distance $2l$, so that $\mu = 2pl$. The reference position of zero potential energy is shown in (a), with $\theta = \frac{1}{2}\pi$. To attain the position shown in (b) the dipole releases energy $2plH \cos \theta$, or $\mu H \cos \theta$. The potential energy in (b) referred to (a) as zero is $-\mu H \cos \theta$, or $-\boldsymbol{\mu} \cdot \mathbf{H}$.

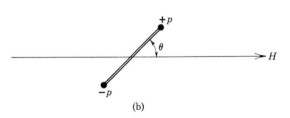

In Chapters 22 and 23 we give a careful discussion of the energy of electric and magnetic systems. But we may derive (53) by considering a dipole of magnetic moment

$$\mu = 2pl \tag{54}$$

as made up of magnetic poles of strength $+p$ and $-p$ separated by a distance $2l$, as in Fig. 8. If we rotate the dipole from a direction perpendicular to the field to a direction making an angle θ with the field, we move the pole $+p$ by a distance $l \cos \theta$ in the direction of the field, and we move the pole $-p$ by $-l \cos \theta$. The work done on the dipole in this rotation is given by the force times the displacement. The force on the pole $+p$ is $+pH$; the force on $-p$ is $-pH$. Thus the work done on the dipole in a rotation from an angle $\frac{1}{2}\pi$ to an angle θ with the field is

$$(pH)(l \cos \theta) + (-pH)(-l \cos \theta) = 2plH \cos \theta = \boldsymbol{\mu} \cdot \mathbf{H} . \tag{55}$$

The potential energy of the magnet is decreased by the rotation from $\frac{1}{2}\pi$ to θ: we have to go from position b to position a in Fig. 8. We change the sign in (55) to obtain the potential energy of the magnetic moment μ in the field $\mathbf{H}$. We thereby obtain the result (53) for the potential energy.

For the model system of N elementary magnets, each with two allowed orientations in a uniform magnetic field H, the total potential energy U is

$$U = \sum_{s=1}^{N} U_s = -H \sum_{s=1}^{N} \mu_s = -H \mathfrak{M} = -2m\mu H , \tag{56}$$

using the expression $2m\mu$ for the total magnetic moment, where the spin excess $2m$ is defined to be $n(\uparrow) - n(\downarrow)$.

m		$U(m)/\mu H$	$g(m)$	$\log g(m)$
-5	————————	$+10$	1	0
-4	————————	$+8$	10	2.30
-3	————————	$+6$	45	3.80
-2	————————	$+4$	120	4.78
-1	————————	$+2$	210	5.35
0	————————	0	252	5.53
$+1$	————————	-2	210	5.35
$+2$	————————	-4	120	4.78
$+3$	————————	-6	45	3.80
$+4$	————————	-8	10	2.30
$+5$	————————	-10	1	0

Figure 9 Energy levels of the model system of ten magnetic moments μ in a magnetic field H. The levels are labeled by their m values, where $2m$ is the spin excess and $\frac{1}{2}N + m = 5 + m$ is the number of up spins. The energies $U(m)$ and degeneracies $g(m)$ are shown. For this problem the energy levels are spaced equally, with separation $\Delta\epsilon = 2\mu H$ between adjacent levels. The values of the $g(m)$ are taken from Fig. 4.

In this example the spectrum of values of the energy U is discrete. We shall see later that a continuous or quasicontinuous spectrum will create no difficulty. Furthermore, the spacing between adjacent energy levels of this model is constant, as in Fig. 9. Constant spacing is a special feature of the particular model, but this feature will not restrict the generality of the argument that is developed in the following sections.

The value of U for spins that interact only with the external magnetic field is completely determined by the value of m, and we indicate this functional dependence by writing $U(m)$. Reversing a single moment away from the direction of the field lowers $2m$ by -2, lowers the total magnetic moment by -2μ, and raises the energy by $2\mu H$. The energy difference between adjacent levels is denoted by $\Delta\epsilon$, where

$$\Delta\epsilon = U(m) - U(m + 1) = 2\mu H \ . \tag{57}$$

The Fundamental Assumption

The fundamental assumption of thermal physics is that a **closed system is equally likely to be in any of the stationary quantum states accessible to it.** The fundamental assumption is used, for example, in the definition of the probability of a state in (1) below and in the definition of average value of a physical property in (3) below. It is also used in the analysis of what happens when two systems are placed in contact, as in (4.5). Let us consider what this assumption means.

CLOSED SYSTEM

A closed system is defined as a system with constant energy, constant number of particles, and constant volume.

ACCESSIBLE STATE

A state is accessible if its properties are compatible with the specification of the system. This means that the energy of the state must be in the range within which the energy of the system is specified, and the number of particles represented by the state must be equal to the number of particles in the specification of the system. There may sometimes be unusual properties of the system that make it impossible for certain quantum states to be accessible within the time the system is under observation. The states of the crystalline form of SiO_2 is essentially inaccessible at low temperatures if we start from the glassy form: fused silica will not convert to quartz in our lifetime in a low temperature experiment. You will recognize most situations of this type by common sense. To summarize, we treat all quantum states as accessible unless they are excluded by the specification of the system (Fig. 1) and the time scale of the measurement.

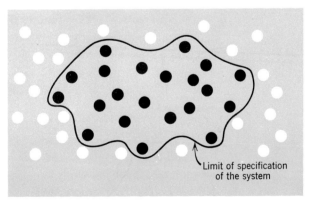

Figure 1 A purely symbolic diagram: each solid spot represents an accessible quantum state of a closed system. The fundamental assumption of statistical physics is that a closed system is equally likely to be in any of the quantum states accessible to it. The empty circles represent some of the states that are not accessible because their properties do not satisfy the specification of the system.

Limit of specification of the system

Of course, we can overspecify the configuration of a closed system to a point that its statistical properties are of no interest. If we can ascertain that the system is exactly in a stationary quantum state l, then the system will always remain in the state l, and no other state is accessible. No statistical aspect is left in the problem. We can usually recognize such an extreme condition.

PROBABILITY

Imagine that at a large number of successive times $t_1, t_2, t_3, \cdots, t_q$ we make observations, q in all, each of which reveals the state of the system. Let $n(l)$ denote the number of times in this series of observations that the system is found to be in the state labeled l. Then the probability $P(l)$ of finding the system in the state l is defined by

$$P(l) = \frac{n(l)}{q} .$$ (1)

We assume that $P(l)$ approaches a limit as the number of observations q is increased. We will take q to be large enough so that $P(l)$ is not likely to change significantly on doubling or tripling the number of observations. The value of q at which we stop taking observations is again a matter of common sense.

We note that the definition of the probability $P(l)$ ensures that

$$\sum_l P(l) = 1 .$$ (2)

That is, the total probability that the system is in *some* state is unity. We say that the probability is normalized to unity.

The probabilities defined by (1) lead naturally to a definition of the **average value** of any physical property. Suppose that A, the physical property of interest, has the value $A(l)$ when the system is in the state l. Here A might denote magnetic moment, energy, square of the energy, charge density near a point $\mathbf{r}$, or any other property that can be observed when the system is in a quantum state. Then the average $\langle A \rangle$ of our observations of the quantity A taken over a system described by the probabilities (1) is defined by

$$\langle A \rangle = \sum_l A(l)P(l) = \frac{1}{q} \sum_l A(l)n(l) .$$ (3)

This is the natural definition of **average value** of A. Here $P(l)$ is the probability the system is in the state l, and $n(l)$ is the number of times in a series of q observations that the system is found in the state l.

This average is a time average over a single system, because the values

(a) Ideal plane boundaries

(b) Rough plane boundaries

Figure 2   It is relatively easy to find the exact states ψ_l of a system of one particle confined in a cube with ideal plane boundaries, as in (a). It is difficult to find the exact states φ_r for one particle confined within rough boundaries, as in (b), for the precise shape of the boundary may not even be known. We may be able to approximate a state φ_r by a solution $\psi_{l\,=\,r}$ of the idealized problem, but ψ_r will not be a stationary time-independent exact solution of the real problem. Suppose that the real system is prepared in the state ψ_r at $t = 0$. Other states of the set ψ_l will appear and fade in and out in the course of time, particularly those states with energy close to the energy of ψ_l. These other states are said to be accessible to the real system.

of the $n(l)$ were determined by observations at successive times. It is important to our definition of probability that the elapsed time between the initial and final observations be "sufficiently long." It is an experimental fact that the complex systems with which we deal appear to randomize themselves over a sufficiently long time. The necessary time interval is called the **relaxation time**. (A system will have many different relaxation times, according to the property under study.) The time of observation must be much longer than a relaxation time. The randomization of a simple system of one particle is discussed in Fig. 2. We speak of the quantum states as stationary, but we always assume in the problems of thermal physics that the quantum states are not absolutely stationary. We suppose that weak perturbations always occur which do not sensibly affect the energy of the system, but which cause a system in course of time to explore all the quantum states that are compatible with the original specification.

The relaxation time describes approximately the time required for a fluctuation in the properties of the system to damp out. The value of the relaxation time may depend on the particular property being observed: it may require a year for a crystal of copper sulfate in a beaker of water to diffuse to produce a reasonably uniform solution, yet the pressure fluctuation produced when the crystal is dropped in the beaker may damp out in a second.

There may be properties that do not become random over any practical time interval. Common sense will exclude these properties from a statistical theory.

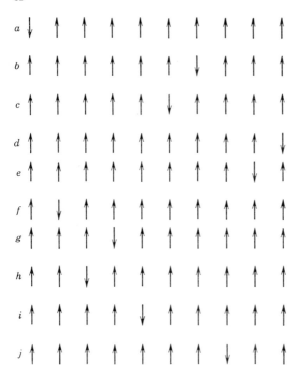

Figure 3 This ensemble represents a system of 10 spins with energy $-8\mu H$ and spin excess $2m = 8$, as in the second lowest energy level of Fig. 2.9. The degeneracy $g(N, m)$ is $g(10, 4) = 10$, so that the representative ensemble must contain 10 systems. The order in which the various systems in the ensemble are listed makes no difference.

ENSEMBLE AVERAGE

Boltzmann and Gibbs made a conceptual advance in the problem of calculating average values of physical quantities. Instead of taking time averages over a single system, they imagined a group of a large number of similar systems, suitably randomized. Averages at a single time are taken over this group of systems. The group of similar systems is called an **ensemble of systems.** The average is called the ensemble average or the thermal average.

An ensemble is an intellectual construction that represents at one time the properties of the actual system as they develop in the course of time. The word ensemble is thus used in a special sense in thermal physics, a sense unrecognized by most lexicographers.

An ensemble of systems is composed of very many systems, all constructed alike. Each system in the ensemble is a replica of the actual system in one of the quantum states accessible to the system. If there are g accessible states, then there are g systems in the ensemble. Each system in the ensemble is equivalent for all practical purposes to the actual system. Each system satisfies all external requirements placed on the original system and is in this sense just as good as the actual system.

The ensemble is randomized suitably: every quantum state accessible to the actual system is represented in the ensemble by one system in a stationary quantum state, as in Fig. 3. Our assumption is that this ensemble represents the system in the sense that an average over the ensemble gives correctly the value of the average for the system.

The Gibbs scheme replaces time averages over a single system by ensemble averages, which are averages over all systems in an ensemble. The demonstration of the equivalence of the ensemble and time averages is difficult and has challenged many mathematicians. The book by Tolman gives an excellent and readable discussion of the general question. It is certainly plausible that the two averages might be equivalent, but one does not know how to state the necessary and sufficient conditions that they are exactly equivalent. Our averages will be ensemble averages, unless otherwise stated.

EQUAL PROBABILITIES

We have constructed an ensemble by establishing a one-to-one correspondence between a system of the ensemble and an accessible state of the system of interest. Our fundamental assumption now means that we accept any system of the ensemble as just as good—that is, just as likely—as any other system of the ensemble. The assumption is not unreasonable because of our real ignorance of the detailed motion of the system, but it is difficult to justify rigorously. We treat the procedure as part of the basic assumption.

This completes our analysis of the fundamental assumption that a closed system is equally likely to be in any of the stationary quantum states accessible to it. The fundamental assumption applies only to **closed systems.** An ensemble that contains one system for each accessible state is assumed to represent the actual system of interest. Other situations are discussed in Chapters 4, 5, and 6, where we derive results for systems in contact with a reservoir. Such systems are not closed.

It would be difficult to develop a science of statistical mechanics without making this or an equivalent postulate. One modern viewpoint takes the fundamental assumption as a postulate that is justified because its consequences have always agreed with experimental results. The methods that follow from the postulate are so simple and so powerful that we would be tempted to develop them even if the consequences only related to experiment once or twice in a generation. Fortunately, our motivation for accepting the assumption is better grounded: the consequences have always agreed with experiment.

It is useful at this point to quote at length from Tolman's summary of what is regarded as a common modern attitude toward the validity of statistical mechanics:

"In the first place, it is to be emphasized, in accordance with the viewpoint here chosen, that the proposed methods are to be regarded as *really statistical* in character, and that the results which they provide are to be regarded as true *on the average* for the systems in an appropriately chosen ensemble, rather than as necessarily precisely true in any individual case. In the second place, it is to be emphasized that the representative ensembles chosen as appropriate are to be constructed with the help of an hypothesis, as to equal *a priori* probabilities, which is introduced at the start, *without proof*, as a necessary postulate.

"Concerning the first of these apparent limitations, it is to be remarked that we have, of course, no just grounds for objecting to the fact that our methods provide us with average rather than precise results. This is merely an inevitable consequence of the statistical nature of our attack, and we have committed ourselves to statistical rather than precise methods, either because we are forced thereto by lack of precise initial knowledge or because the practical problems which we have in mind are otherwise too complicated for treatment. Moreover, it is to be noted that the proposed methods make it possible to compute not only the average values of quantities but also the average *fluctuations* around those values. This, then, makes it possible to draw conclusions also as to the frequency with which we may expect to find systems with properties differing from the average to any specified extent. In the case of typical applications the computed fluctuations are extremely small. In the special cases where they are large enough they may be compared with what is found experimentally.

"Concerning the second of the above-mentioned limitations on the character of the proposed methods, two remarks already made in the preceding section may again be emphasized. In the first place, it is to be appreciated that *some* postulate as to the *a priori* probabilities . . . has in any case to be chosen. This again is merely a consequence of our commitment to statistical methods. It is analogous to the necessity of making some preliminary assumption as to the probabilities for heads or tails in order to predict the results to be expected on flipping a coin. In the second place, it is to be emphasized that the actual assumption, of equal *a priori* probabilities . . . is the only general hypothesis that can reasonably be chosen. . . . In the absence of any knowledge except that our systems do obey the laws of mechanics, it would be arbitrary to make any assumption other than that of equal *a priori* probabilities. . . . The procedure may be regarded as roughly analogous to the assumption of equal probabilities for heads and tails, after a preliminary investigation has shown that the coin has not been loaded.

"In further support of the validity of the proposed methods it may, of course, again be emphasized that they have the *a posteriori* justification of leading to conclusions which do agree with empirical facts. This includes agreement with conclusions not only as to average values but also as to fluctuations.

"Hence the present point of view as to the validity of the methods of statistical mechanics may be summarized as follows. The methods are essentially statistical in character and only purport to give results that may be expected on the average rather than precisely expected for any particular system. The methods lead to calculated fluctuations around the averages which are exceedingly small in the case of the usual typical applications, and in other cases can be compared with empirical findings. The methods being statistical in character have to be based on some hypothesis as to *a priori* probabilities, and the hypothesis chosen is the only postulate that can be introduced without proceeding in an arbitrary manner. The methods lead to results which do agree with empirical findings."

REFERENCE

R. C. Tolman, *Principles of statistical mechanics*, Oxford University Press, 1938.

CHAPTER 4

Two Systems in Thermal Contact: Definition of Entropy and Temperature

"If we wish to find in rational mechanics an *a priori* foundation for the principles of thermodynamics, we must seek mechanical definitions of temperature and entropy." (J. W. Gibbs)

"The general connection between energy and temperature may only be established by probability considerations. [Two systems] are in statistical equilibrium when a transfer of energy does not increase the probability." (M. Planck)

The purpose of this chapter is to define the temperature and entropy of a system. To do this we need to know the number of accessible states of the system. The logarithm of this number is called the entropy and is the key to the thermal properties of the system. Interesting questions arise when contact is established between two systems to allow the exchange of energy or the exchange of energy and particles, as in Fig. 1. In this chapter we consider energy exchange between two systems; in Chapter 5 we consider exchange of both energy and particles. In **thermal contact** the two systems have been brought together and allowed to exchange energy, but not particles.

What determines when there will be a net flow of energy from one system to another? The answer to this question is the origin of the concept of temperature. The direction of energy flow is not simply a matter of whether the energy of one system is greater than the energy of the other, for the systems can be quite different in size and constitution. The total energy $U = U_1 + U_2$ can be shared in many ways between the two systems, while

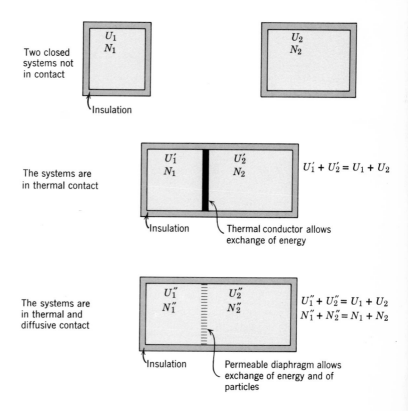

Figure 1   Modes of contact between two systems.

keeping the total energy constant. The first task of thermal physics is to discuss the most probable division of the energy between two systems.

The most probable division of the energy is defined as that for which the combined system has the maximum number of accessible states. Every accessible state of the combined system is equally probable in the sense of Chapter 3. We enumerate below the accessible states of two model systems, and we shall find the most probable configuration of the systems when in thermal contact. We then generalize the result to two arbitrary systems in thermal contact.

ENERGY EXCHANGE AND THE
MOST PROBABLE CONFIGURATION

We solve in detail the problem of thermal contact between two model spin systems, 1 and 2, in a magnetic field. The numbers of spins N_1, N_2 may be different, and the values of the spin excess $2m_1$, $2m_2$ may be different. The actual exchange of energy might take place via a weak magnetic coupling between the spins near the interface between the two systems. We keep N_1, N_2 constant, but the values of the spin excess are allowed to change.

The spin excess of a state of the combined system will be denoted by $2m$. We have

$$m = m_1 + m_2 . \tag{1}$$

The energy of the combined system is

$$U(m) = U_1(m_1) + U_2(m_2) ; \tag{2}$$

the number of particles is

$$N = N_1 + N_2 . \tag{3}$$

We assume that the level splittings $2\mu H$ are equal in both systems, hence that the energy given up by system 1 when one spin is reversed can be taken up by the reversal of one spin of system 2 in the opposite sense. Any large physical system will have enough diverse means of energy storage, so that energy exchange with another system is always possible. The value of $m = m_1 + m_2$ is constant because the total energy is constant, but a redistribution may occur in the values of m_1, m_2 and thus in the energy when the two systems are brought into thermal contact.

We show below that the degeneracy function $g(N, m)$ of the combined system is related to the product of the degeneracy functions of the individual systems 1 and 2:

$$g(N, m) = \sum_{m_1} g_1(N_1, m_1)g_2(N_2, m - m_1) . \tag{4}$$

The range of m_1 in the summation is from $-\frac{1}{2}N_1$ to $\frac{1}{2}N_1$ if $N_1 < N_2$, as we assume for convenience.

We consider first the configuration of the combined system for which the first system has spin excess $2m_1$ and the second system has spin excess $2m_2$. **A configuration consists of the set of states specified by fixed values of m_1 and m_2.** The first system has $g_1(N_1, m_1)$ accessible states, and each of these states may occur together with any of the $g_2(N_2, m_2)$ accessible states of the second system. The total number of states in a configuration of the combined system is given by the product $g_1(N_1, m_1)g_2(N_2, m_2)$. Because $m = m_1 + m_2$ is constant, we have $m_2 = m - m_1$, and the product may be written as

$$g_1(N_1, m_1)g_2(N_2, m - m_1) .$$

Other accessible configurations of the combined system are characterized by different values of m_1. We sum over all possible values of m_1 to obtain the total degeneracy of all the accessible configurations:

$$g(N, m) = \sum_{m_1} g_1(N_1, m_1)g_2(N_2, m - m_1) . \qquad (5)$$

Here $g(N, m)$ is the number of accessible states of the combined system. In the sum m, N_1, and N_2 have been held constant, as part of the specification of thermal contact.

We now come to a matter of great importance in thermal physics. The result (5) is a sum of products of the form $g_1(N_1, m_1)g_2(N_2, m - m_1)$. Such a product will have a maximum for some value of m_1, say $\hat{m}_1$. The number of states in the **most probable configuration** is

$$g_1(N_1, \hat{m}_1)g_2(N_2, m - \hat{m}_1) . \qquad (5a)$$

If the number of particles in at least one of the two systems is very large, then we can show that the maximum will be extremely sharp with respect to changes in m_1. A sharp maximum means that a relatively small number of configurations will dominate the statistical properties of the combined system. This is a property of every type of large system for which exact solutions are available, and we suppose that it is a general property of all large systems. We use the sharpness property whenever we assume that fluctuations about the most probable configuration are small, as whenever we assume that the average properties of a system in thermal contact with a reservoir are accurately described by the properties of the most probable configuration.

We shall always assume that at least one of the systems in contact is composed of an arbitrarily large number of particles, and this system will be called a **reservoir** (Fig. 2). Under these conditions we shall often replace the average of a physical quantity over all accessible configurations (5) by an average over the most probable configuration (5a). We now show what is involved in such an approximation.

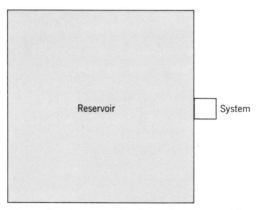

Figure 2 A system in thermal contact with a reservoir. A reservoir is always assumed to be composed of an arbitrarily large number of particles, certainly much larger than the number of particles in the system. The totality of system + reservoir is closed.

EXAMPLE. *Two spin systems in thermal contact.* We investigate for our model system the important question of the sharpness of the product near the maximum term written in (5a). A little tedious calculation will give us essentially exact answers. We use the distribution functions for $g_1(N_1, m_1)$ and $g_2(N_2, m_2)$ as given by (2.37). We form the product

$$g_1(N_1, m_1)g_2(N_2, m_2) = g_1(0)g_2(0) \exp\left(-\frac{2m_1{}^2}{N_1} - \frac{2m_2{}^2}{N_2}\right), \qquad (6)$$

where $g_1(0)$ denotes $g_1(N_1, 0)$ and $g_2(0)$ denotes $g_2(N_2, 0)$. Because $m_1 + m_2 = m$, we may replace m_2 by $m - m_1$:

$$g_1(N_1, m_1)g_2(N_2, m - m_1) = g_1(0)g_2(0) \exp\left(-\frac{2m_1{}^2}{N_1} - \frac{2(m - m_1)^2}{N_2}\right). \qquad (7)$$

This product gives the number of states accessible to the combined system when the spin excess of the first system is $2m_1$ and the spin excess of the combined system is $2m$. The schematic plot in Fig. 3 may help to give a feeling for the product, although the plot is only for a small system.

We wish to find the maximum value of (7) as a function of m_1. Now the maximum of $\log y(x)$ occurs at the same value of x as the maximum of $y(x)$. From (7),

$$\log g_1(N_1, m_1)g_2(N_2, m - m_1) = \log g_1(0)g_2(0) - \frac{2m_1{}^2}{N_1} - \frac{2(m - m_1)^2}{N_2}. \qquad (8)$$

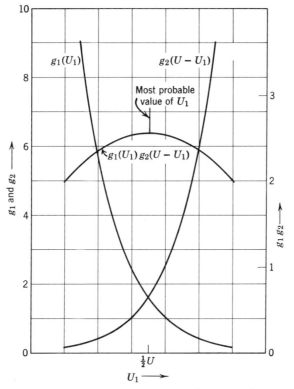

Figure 3 Schematic plot for two small systems of g_1, g_2 and g_1g_2. The function plotted as g_1 is $\dfrac{2}{\sqrt{\pi}}e^{-x^2}$; that plotted as g_2 is $\dfrac{2}{\sqrt{\pi}}e^{-(8 - x)^2}$. The product g_1g_2 as plotted has been multiplied by 5×10^{13} in order to make it visible. (The factor $2/\sqrt{\pi}$ is usually included in standard tables of the Gauss function.)

This quantity is an extremum when the derivative with respect to m_1 is zero. An extremum may be a maximum, a minimum, or a point of inflection. The extremum is a maximum if the second derivative of the function is negative, so that the curve bends downward.

The first derivative is

$$\frac{\partial}{\partial m_1}\{\log g_1(N_1, m_1)g_2(N_2, m - m_1)\} = -\frac{4m_1}{N_1} + \frac{4(m - m_1)}{N_2} = 0 , \quad (9)$$

where N_1, N_2, and m are held constant as m_1 is varied. The second derivative $\partial^2/\partial m_1^2$ of Eq. (8) is found to be

$$-4\left(\frac{1}{N_1} + \frac{1}{N_2}\right)$$

and is negative, so that the extremum is indeed a maximum. Thus the most probable configuration of the combined system is that for which (9) is satisfied. The condition (9) may be written

$$\frac{m_1}{N_1} = \frac{m - m_1}{N_2} = \frac{m_2}{N_2} . \tag{10}$$

That is, the two systems are in equilibrium with respect to interchange of energy when the fractional spin excess of system 1 is equal to the fractional spin excess of system 2. We shall see that nearly all the accessible states satisfy (10), or very nearly satisfy it.

If $\hat{m}_1$ and $\hat{m}_2$ denote the values of m_1 and m_2 at the maximum, then (10) appears as

$$\frac{\hat{m}_1}{N_1} = \frac{\hat{m}_2}{N_2} = \frac{m}{N} . \tag{11}$$

The symbol $\hat{m}$ is read "m hat" or "m caret." To find the value of the product $g_1 g_2$ at the maximum, we simply substitute (11) in (6) to obtain

$$(g_1 g_2)_{\max} \equiv g_1(\hat{m}_1) g_2(m - \hat{m}_1) = g_1(0) g_2(0) \, e^{-2m^2/N} . \tag{12}$$

How sharp is the maximum of $g_1 g_2$ at a given value of m? Let

$$m_1 = \hat{m}_1 + \delta ; \qquad m_2 = \hat{m}_2 - \delta . \tag{13}$$

Here δ measures the deviation of m_1, m_2 from their values $\hat{m}_1, \hat{m}_2$ at the maximum of $g_1 g_2$.

We square the relations in (13) to form

$$m_1^2 = \hat{m}_1^2 + 2\hat{m}_1\delta + \delta^2 ; \qquad m_2^2 = \hat{m}_2^2 - 2\hat{m}_2\delta + \delta^2 ,$$

which we substitute in (6) and then use (12) to obtain the number of states

$$g_1(N_1, m_1) g_2(N_2, m_2) = (g_1 g_2)_{\max} \exp\left(-\frac{4\hat{m}_1\delta}{N_1} - \frac{2\delta^2}{N_1} + \frac{4\hat{m}_2\delta}{N_2} - \frac{2\delta^2}{N_2}\right) .$$

Now by (11) we know that $\hat{m}_1/N_1 = \hat{m}_2/N_2$, so that the number of states in a configuration of deviation δ is

$$g_1(N_1, \hat{m}_1 + \delta) g_2(N_2, \hat{m}_2 - \delta) = (g_1 g_2)_{\max} \exp\left(-\frac{2\delta^2}{N_1} - \frac{2\delta^2}{N_2}\right) . \tag{14}$$

As a numerical example, let $N_1 = N_2 = 10^{22}$ and $\delta = 10^{12}$; that is, the fractional deviation is $\delta/N_1 = 10^{-10}$. For this small fractional deviation from equilibrium we have $2\delta^2/N_1 = 200$, and the product $g_1 g_2$ is reduced to $e^{-400} \approx 10^{-173}$ of its maximum value. This is a large reduction indeed, and we see that $g_1 g_2$ must be a very, very sharply peaked function of m_1.

When two systems are in thermal contact, the values of m_1, m_2 that occur most often will be very close to the values $\hat{m}_1$, $\hat{m}_2$ for which the product $g_1 g_2$ is a maximum. It will be extremely rare to find the systems with values of m_1, m_2 widely different from $\hat{m}_1$, $\hat{m}_2$.

What does it really mean to say that the probability of finding the system with a fractional deviation $\delta/N_1 = 10^{-10}$ is only 10^{-173} of the probability of finding the system at $\delta/N_1 = 0$? We mean that the system will *never* be found with as large a deviation as 1 part in 10^{10}, minute as this deviation seems. Suppose that each spin changes from one orientation to another once every 10^{-12} sec, under the action of some unspecified interaction.[1] There are 10^{22} spins, so that the system changes from one quantum state to another $10^{12} \times 10^{22} = 10^{34}$ times per sec. Then if we wait

$$(10^{-34} \text{ sec}) \times 10^{173} = 10^{139} \text{ sec}$$

we might expect to catch the system with $\delta/N_1 = 10^{-10}$. But the age of the universe is only about 10^{18} sec! So we can say with great surety that **the event will never be observed.**[2] The estimate was rough, but the message is correct.[3]

We can expect to observe substantial fractional deviations in the properties of a small system. A small system in thermal contact with a large system poses no difficulty for the theory. We shall see, for example, that the temperature of the small system is defined as that of the system with which it is in contact. The energy of the small system may undergo fluctuations that are large in a fractional sense, as have been observed in experiments on the Brownian motion of small particles in suspension and on the spontaneous fluctuations of galvanometer mirrors. But the average energy of a small system can always be determined accurately by observations over a long period of time or by observations on a large number of identical small systems.

The result (5) for the number of accessible states of two model systems in thermal contact may be generalized to any two arbitrary systems. By an obvious extension of the earlier argument we have the result for the degeneracy $g(N, U)$ of the combined system:

$$g(N, U) = \sum_{U_1} g_1(N_1, U_1) g_2(N_2, U - U_1) , \tag{15}$$

where the sum is taken over all values of U_1 which are $\leq U$. Here $g_1(N_1, U_1)$ is the number of accessible states of system 1 at the energy U_1. A configura-

---

[1] This is not a bad choice for the relaxation time of one spin, because many single-particle relaxation times of importance in solids and liquids at room temperature have roughly this value.

[2] We can also ask whether we will observe the system with δ/N_1 equal to *or larger than* 10^{-10}. The answer to this question is *never*.

[3] A quotation from L. Boltzmann (1898) is relevant: "One should not imagine that two gases in a 0.1 liter container, initially unmixed, will mix, then again after a few days separate, then mix again, and so forth. On the contrary, one finds . . . that not until a time enormously long compared to $10^{(10^{10})}$ years will there be any noticeable unmixing of the gases. One may recognize that this is practically equivalent to never . . . ." This example is discussed in Problem 12.3.

tion of the combined system is specified by the values U_1 and U_2. The number of accessible states in a configuration is the product $g_1(N_1, U_1)g_2(N_2, U - U_1)$. The sum over all configurations gives $g(N, U)$.

We wish to find the largest term in the sum in (15). For an extremum it is necessary that the differential be zero for an infinitesimal exchange of energy:

$$dg = \left(\frac{\partial g_1}{\partial U_1}\right)_{N_1} g_2 \, dU_1 + g_1\left(\frac{\partial g_2}{\partial U_2}\right)_{N_2} dU_2 = 0 \; ; \qquad dU_1 + dU_2 = 0 \; . \quad (16)$$

Whenever it is of special importance to us, we shall investigate the nature of the extremum, but we suppose that the extremum here is a maximum.[4]

The **most probable configuration** of the combined system satisfies (16). We divide (16) by g_1g_2 and use the result $dU_2 = -dU_1$ to obtain

$$\frac{1}{g_1}\left(\frac{\partial g_1}{\partial U_1}\right)_{N_1} = \frac{1}{g_2}\left(\frac{\partial g_2}{\partial U_2}\right)_{N_2} \; . \qquad (17a)$$

Because
$$\frac{d}{dx}\log y = \frac{1}{y}\frac{dy}{dx} \; ,$$

we may write (17a) as

$$\left(\frac{\partial \log g_1}{\partial U_1}\right)_{N_1} = \left(\frac{\partial \log g_2}{\partial U_2}\right)_{N_2} \; . \qquad (17b)$$

The energy dependence of the number of accessible states of each system is an important physical property: it determines the most probable configuration of the combined system.

DEFINITION OF ENTROPY

We say that the two systems are in **thermal equilibrium** with each other when the combined system is in its most probable configuration; that is, the configuration for which the number of accessible states is a maximum.

The values of the degeneracies of g are usually very large numbers. It is convenient (Fig. 4) to work with a smaller number σ defined as the natural logarithm of g and called the entropy:[5]

$$\boxed{\sigma(N, U) \equiv \log g(N, U) \; .} \qquad (18)$$

This is a definition whose simplicity leaves us breathless: **the entropy is the**

[4] The notation
$$\left(\frac{\partial g_1}{\partial U_1}\right)_{N_1}$$
means that N_1 is held constant in the differentiation of $g_1(N_1, U_1)$ with respect to U_1. That is, the partial derivative with respect to U_1 is defined as

$$\left(\frac{\partial g_1}{\partial U_1}\right)_{N_1} = \lim_{\Delta U_1 \to 0} \frac{g_1(N_1, U_1 + \Delta U_1) - g_1(N_1, U_1)}{\Delta U_1} \; .$$

For example, if $g(x, y) = 3x^4y$, then $(\partial g/\partial x)_y = 12x^3y$ and $(\partial g/\partial y)_x = 3x^4$.

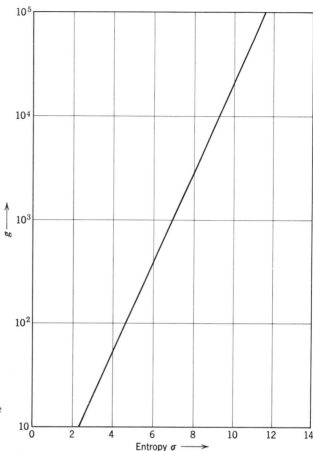

Figure 4   Plot of number of accessible states g versus the entropy σ.

logarithm of the number of states accessible to the system. The entropy is a dimensionless number, for the logarithm of a number is dimensionless.

It is said that entropy is a measure of the randomness of a system; this statement is made precise by the definition $\sigma \equiv \log g$. The more states that are accessible, the greater the entropy. In the definition (18) we have indicated a functional dependence of $g(N, U)$ on the number of particles in the system and on the energy of the system. The entropy may depend on additional independent variables: the entropy of a gas depends on the volume, as we shall see in Chapters 11 and 12.

---

[5] Here σ is the Greek letter sigma. The conventional thermodynamic entropy S is defined by $S = k_B \sigma$, where k_B is the Boltzmann constant 1.381×10^{-16} erg deg^{-1}. The conventional absolute thermodynamic temperature T is given by $k_B T \equiv \mathcal{T}$, where $\mathcal{T}$ is defined by (21) below. This relation defines k_B. The experimental determination of k_B and T is discussed in Chapter 8.

THIRD LAW OF THERMODYNAMICS

The definition (18) leads to the statement below of what is called the third law of thermodynamics. The dignity of a law of nature should not be conferred on what is essentially a definition. However, in the early history of thermal physics the physical significance of the entropy was not known. In that period, for example, the author of the article on thermodynamics in the *Encyclopaedia Britannica*, 11th ed., wrote that "The utility of the conception of entropy . . . is limited by the fact that it does not correspond directly to any directly measurable physical property, but is merely a mathematical function of the definition of absolute temperature." But we know now what physical property the entropy measures, and we know also that entropy is of central importance in thermal physics.

One statement of the **third law of thermodynamics** is that the entropy is zero when the system is in its lowest energy level, as at the absolute zero of temperature. This result follows directly from the definition of σ if the lowest energy level corresponds to only a single state of the system with $g = 1$ and $\sigma = \log g = 0$. But for many systems the lowest energy level may be degenerate, so that g is not equal to one and σ is not equal to zero. Another statement of the third law is that as the temperature approaches absolute zero the entropy becomes independent of the external parameters (such as volume and magnetic field intensity) involved in the specification of the system.

In many experiments on entropy the lowest temperature reached, perhaps one degree Kelvin, is still much too high to remove the entropy associated with the disorder of the orientation of the spins of the nuclei. If the nuclear spin entropy does not change at all throughout the experimental range of temperatures, it will often be omitted from tables of entropy values.

EXAMPLE. *Additivity of the entropy.* Consider two closed systems, not in contact. The first system has g_1 accessible states; the second system has g_2 accessible states. The combined system (with the parts still not in contact) has $g_1 g_2$ accessible states, for any accessible state of system 1 may be found with any accessible state of system 2. The entropy of the combined system is

$$\sigma = \log (g_1 g_2) = \log g_1 + \log g_2 = \sigma_1 + \sigma_2 . \tag{19}$$

Thus the total entropy is the sum of the entropies of the separate systems.

TEMPERATURE

We define a quantity called the temperature in such a way that two systems in thermal equilibrium with each other will have the same value for this quantity. We found that two systems are in equilibrium with respect to energy exchange when

$$\left(\frac{\partial \log g_1}{\partial U_1}\right)_{N_1} = \left(\frac{\partial \log g_2}{\partial U_2}\right)_{N_2} . \tag{20a}$$

Two systems are at the same temperature when this condition is satisfied. We can now give a general definition of the temperature of a system. In terms of the entropy the equilibrium condition (20a) becomes

$$\left(\frac{\partial \sigma_1}{\partial U_1}\right)_{N_1} = \left(\frac{\partial \sigma_2}{\partial U_2}\right)_{N_2} . \tag{20b}$$

We are led to define the **temperature T** by

$$\boxed{\frac{1}{T} \equiv \left(\frac{\partial \sigma}{\partial U}\right)_N .} \tag{21}$$

The reciprocal of the temperature is equal to the derivative of the entropy with respect to the energy. We shall refer to T, the Greek letter tau, as the fundamental temperature or simply as the temperature. We show in Chapter 8 that T is proportional to the conventional absolute temperature which is measured in degrees Kelvin. All external parameters such as the volume are held constant in the partial derivative in (21).

The temperatures of two bodies in thermal contact are exactly the same, but the contact allows the spontaneous exchange of energy between the two bodies. Thus small fluctuations in energy of one of the bodies will always occur. Any measurement of the temperature of a system involves the establishment of some contact or interaction of a device with the system, and it follows that exchange of energy between the system and the thermometer may always occur.

Because σ is dimensionless, T has the dimensions of energy. We define $1/T$, rather than T, as equal to $(\partial \sigma / \partial U)_N$ because we want our definition to be consistent with the idea that energy flows from a system at a high temperature to a system at a lower temperature. This point will now be demonstrated.

TENDENCY OF THE ENTROPY TO INCREASE

When we establish thermal contact between two arbitrary systems we may expect the total entropy to increase. This will occur if the product $g_1(U_1)g_2(U_2)$ evaluated at the initial values of the energies U_1 and U_2 is smaller than the maximum value of the product $g_1(\hat{U}_1)g_2(\hat{U}_2)$ which may occur at some other partition[6] $\hat{U}_1$, $\hat{U}_2$ of the same total energy U. The most probable condition of the combined system is that for which g_1g_2 is a maximum. Thus if the combined system attains the most probable condition after contact is established, we have

$$g_1(\text{final})g_2(\text{final}) \geq g_1(\text{initial})g_2(\text{initial}) \ .$$

Because $\log x$ increases as x increases, the inequality is preserved on taking the logarithm of each side:

$$\sigma_1(\text{final}) + \sigma_2(\text{final}) \geq \sigma_1(\text{initial}) + \sigma_2(\text{initial}) \ . \tag{22}$$

The final entropy is greater than (or equal to) the initial entropy. **The total entropy tends to increase when two systems are brought into thermal contact.** The equality applies only when the systems are initially at the same temperature.

We can understand intuitively the tendency of the entropy to increase: two separated systems are subject to individual constraints on their energies U_1 and U_2, but the combined system has only the single constraint $U = U_1 + U_2$ on the energy. There is now one constraint in place of two. The removal of a constraint can only increase the total number of accessible states.

When two bodies are brought into contact, there is a transfer of energy from body 1 at the higher temperature T_1 to body 2 at the lower temperature T_2. To prove this, we consider the total entropy change $\delta\sigma$ when we remove a positive amount of energy δU from 1 and add the same amount of energy to 2, as in Fig. 5. The total entropy change is

$$\delta\sigma = \left(\frac{\partial\sigma_1}{\partial U_1}\right)_{N_1}(-\delta U) + \left(\frac{\partial\sigma_2}{\partial U_2}\right)_{N_2}(\delta U) = \left(-\frac{1}{T_1} + \frac{1}{T_2}\right)\delta U \ . \tag{23}$$

When $T_1 > T_2$ the quantity in parentheses on the right-hand side is positive, so that the total change of entropy is positive as required by (22). The direction of the flow of energy agrees with the conventional sense of high and low temperature: energy flows from a body at a high temperature to a body at a low temperature.

---

[6] A **partition** of something is a sharing or allocation of the quantity into two or more parts.

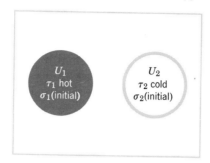

Figure 5 If the temperature T_1 is higher than T_2, the transfer of a positive amount of energy δU from system 1 to system 2 will increase the total entropy $\sigma_1 + \sigma_2$ of the combined systems over the initial value $\sigma_1(\text{initial}) + \sigma_2(\text{initial})$. In other words, the final system will be in a more probable condition if energy flows from the warmer body to the cooler body when thermal contact is established. This is an example of the law of increasing entropy.

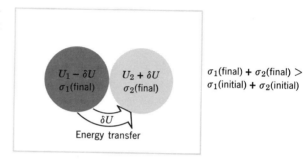

EXAMPLE. *Increase of energy with increase of temperature.* We can show that the energy of a system increases as the temperature increases. This result will follow from the requirement that σ be a maximum, rather than merely an extremum, in equilibrium.

We consider to second order in δU the spontaneous exchange of energy $\delta U_1 = -\delta U_2 = -\delta U$ between two bodies that are in thermal contact. We have seen that such exchanges are allowed between bodies maintained at the same temperature by means of thermal contact. We have

$$\delta \sigma_1 = \frac{\partial \sigma_1}{\partial U_1} \delta U_1 + \frac{1}{2} \frac{\partial^2 \sigma_1}{\partial U_1{}^2} (\delta U_1)^2 + \cdots \; ; \tag{24}$$

$$\delta \sigma_2 = \frac{\partial \sigma_2}{\partial U_2} \delta U_2 + \frac{1}{2} \frac{\partial^2 \sigma_2}{\partial U_2{}^2} (\delta U_2)^2 + \cdots \; , \tag{25}$$

so that the total entropy change[7] is

$$\delta \sigma = \delta \sigma_1 + \delta \sigma_2 = \tfrac{1}{2}(\delta U)^2 \left(\frac{\partial^2 \sigma_1}{\partial U_1{}^2} + \frac{\partial^2 \sigma_2}{\partial U_2{}^2} \right) , \tag{26}$$

to second order in δU. We have used the fact that $\partial \sigma_1 / \partial U_1 = \partial \sigma_2 / \partial U_2$ for two bodies at the same temperature. We now use the definition of temperature to express the second derivatives in another way:

---

[7] Strictly, this discussion might be written in terms of the generalized entropy of (50) below.

$$\frac{\partial^2 \sigma_1}{\partial U_1{}^2} = \frac{\partial}{\partial U_1}\left(\frac{\partial \sigma_1}{\partial U_1}\right) = \frac{\partial}{\partial U_1}\left(\frac{1}{T_1}\right) = -\frac{1}{T_1{}^2}\frac{\partial T_1}{\partial U_1} \;,$$

which gives, because $T_1 = T_2$,

$$\delta\sigma = -\frac{(\delta U)^2}{2T^2}\left(\frac{\partial T_1}{\partial U_1} + \frac{\partial T_2}{\partial U_2}\right) . \tag{27}$$

We know that σ is a maximum in equilibrium, so that the entropy change $\delta\sigma$ must be negative when the bodies exchange a finite amount of energy δU in a fluctuation from equilibrium. This will be satisfied if for each system $(\partial T/\partial U)_N > 0$, which is equivalent to the condition

$$\left(\frac{\partial U}{\partial T}\right)_N > 0 . \tag{28}$$

Thus the energy of a system increases as its temperature increases.

EXAMPLE.[8] *Entropy change on heat flow.* The specific heat of metallic copper over the temperature range 15 to 100 C is approximately 0.093 cal gm^{-1} deg^{-1}, according to a standard handbook. We suppose that thermal expansion may be neglected, so that no external work is done when a specimen is heated.

(a) What is the heat capacity of a 10 gm specimen, in ergs deg^{-1}?

One calorie is equal to 4.184×10^7 ergs. The term specific heat refers to the heat capacity of 1 gm of material; in ergs gm^{-1} deg^{-1} the specific heat of copper is

$(0.093$ cal gm^{-1} $deg^{-1})(4.184 \times 10^7$ ergs $cal^{-1})$
$$= 3.89 \times 10^6 \text{ ergs gm}^{-1} \text{ deg}^{-1} .$$

The heat capacity of the 10 gm specimen is 3.89×10^7 ergs deg^{-1}.

(b) A 10 gm specimen of copper at a temperature of 350 K is placed in thermal contact with a second 10 gm specimen of copper at a temperature of 290 K. What quantity of energy is transferred when the two specimens are placed in contact?

The energy increase of the second specimen is equal to the energy loss of the first; thus the energy increase of the second specimen is

$$\Delta U = (3.89 \times 10^7)(T_f - 290) = (3.89 \times 10^7)(350 - T_f) ,$$

so that the final temperature after contact is

$$T_f = \tfrac{1}{2}(350 + 290) = 320 \text{ K} .$$

Thus

$$\Delta U_1 = (3.89 \times 10^7 \text{ ergs deg}^{-1})(-30 \text{ K}) = -1.17 \times 10^8 \text{ ergs}$$

[8] This example uses the definition of heat capacity in (6.40) and the definition of the calorie in (12.6).

and

$$\Delta U_2 = -\Delta U_1 = 1.17 \times 10^8 \text{ ergs} .$$

(c) What is the change of entropy of the two specimens when a transfer of 1×10^6 ergs has taken place, immediately after initial contact? Notice that this is a small fraction of the final energy transfer as calculated above.

The energy transfer considered is small, so that we may suppose the specimens are approximately at their initial temperatures of 350 and 290 K. The entropy of the first body is changed by

$$\Delta S_1 = \frac{-1 \times 10^6 \text{ ergs}}{350 \text{ K}} = -2.86 \times 10^3 \text{ ergs deg}^{-1} .$$

The entropy of the second body is changed by

$$\Delta S_2 = \frac{1 \times 10^6 \text{ ergs}}{290 \text{ K}} = 3.45 \times 10^3 \text{ ergs deg}^{-1} .$$

The net entropy increases by

$$\Delta S_1 + \Delta S_2 = (-2.86 + 3.45) \times 10^3 = 0.59 \times 10^3 \text{ ergs deg}^{-1} .$$

In absolute units the increase of entropy is

$$\frac{0.59 \times 10^3}{k_B} = \frac{0.59 \times 10^3 \text{ ergs deg}^{-1}}{1.38 \times 10^{-16} \text{ ergs deg}^{-1}} = 0.43 \times 10^{19} ,$$

where k_B is the Boltzmann constant.

Problem 1. *Entropy and temperature.* Suppose $g = CU^N$, where C is a constant and N is the number of particles. (*a*) Show that $U = N\mathcal{T}$. (*b*) Show that $(\partial^2 \sigma / \partial U^2)_N$ is negative.

EXAMPLE. *Paramagnetism.*[9] Find the equilibrium value at temperature $\mathcal{T}$ of the fractional magnetization

$$\frac{\mathfrak{m}}{N\mu_0} = \frac{2m}{N} \tag{29}$$

of the model system of N spins in a magnetic field H. These results will be used in several subsequent examples.

The entropy is given by the logarithm of the expression (2.37) for the degeneracy $g(N, m)$:

$$\sigma(N, m) = \log g(N, 0) - \frac{2m^2}{N} . \tag{30}$$

---

[9] It would be desirable to treat the ideal gas as our first physical example, but the ideal gas is not a trivial problem. History is littered with incomplete expressions for the entropy of an ideal gas. A correct and relatively simple derivation is the subject of Chapter 11.

This is an approximation valid for $|m|/N \ll 1$. The energy U is given by

$$U = -2m\mu_0 H ,\tag{31}$$

where μ_0 is the magnetic moment of an elementary magnet.

We wish to write the entropy as $\sigma(N, U)$ because our definition of T was expressed with σ as a function of N and U. On squaring (31) we have

$$U^2 = 4\mu_0{}^2 H^2 m^2 ; \qquad m^2 = \frac{U^2}{4\mu_0{}^2 H^2} ,\tag{32}$$

so that (30) becomes

$$\sigma(N, U) = \sigma(N, 0) - \frac{U^2}{2\mu_0{}^2 H^2 N} ,\tag{33}$$

using $\sigma(N, 0) = \log g(N, 0)$.

By definition of the temperature T we find the result

$$\frac{1}{T} = \left(\frac{\partial \sigma}{\partial U}\right)_N = -\frac{U}{\mu_0{}^2 H^2 N} ,\tag{34}$$

so that in thermal equilibrium[10] at temperature T the thermal average energy and the entropy of the system are given by

$$U(T) = -\frac{N\mu_0{}^2 H^2}{T} ; \qquad \sigma = \sigma_0 - \frac{N\mu_0{}^2 H^2}{2T^2} .\tag{35}$$

The energy and the entropy increase as the temperature increases. The entropy decreases as the square of the magnetic field: the magnetic field decreases the randomness of the system (increases the order). The result (35) is an approximation [by virtue of (30)] valid only for $\mu_0 H \ll T$.

The thermal average or ensemble average of a quantity A is denoted by $\langle A \rangle$. By convention we do not usually bother to write the thermal average of the energy as $\langle U(T) \rangle$, but simply write $U(T)$. We take the thermal average of (31) to obtain

$$U(T) = -2\langle m \rangle \mu_0 H .\tag{36}$$

On comparison with (35),

$$-2\langle m \rangle \mu_0 H = -\frac{N\mu_0{}^2 H^2}{T} .\tag{37}$$

Thus the fractional magnetization is

---

[10] Nothing in our treatment has excluded the possibility that the temperature may be negative. In fact, if the energy U in (34) is positive, the temperature will be negative. In Chapter 6 we study this system in both positive and negative temperature regions.

$$\frac{\mathfrak{m}}{N\mu_0} = \frac{2\langle m \rangle}{N} = \frac{\mu_0 H}{\mathcal{T}} \; . \tag{38}$$

We see that the magnetic field tends to align the spins in competition with the temperature which tends to randomize the spin directions.

The total magnetic moment of the system in thermal equilibrium at temperature $\mathcal{T}$ is given in this approximation by

$$\mathfrak{m}(\mathcal{T}) = \frac{N\mu_0^2 H}{\mathcal{T}} \; . \tag{39}$$

The total magnetic moment is directly proportional to the number of particles and to the magnetic field, but inversely proportional to the temperature.

The **magnetic susceptibility** is defined as $\chi = d\mathfrak{m}/dH$. Here χ is the Greek letter chi. From (39) we have

$$\chi = \frac{N\mu_0^2}{\mathcal{T}} \; . \tag{40}$$

This is the result for the susceptibility[11] for systems of spin $\frac{1}{2}$ in the limit $\mu_0 H/\mathcal{T} \ll 1$. Why only in this limit? The derivation of (30) assumed that $|m|/N \ll 1$. We shall see in Problem 6.2 how to lift this restriction most conveniently and thereby to obtain the exact expression for the susceptibility. Often the susceptibility is understood to refer to unit volume; we then take N as the number of particles per unit volume.

EXAMPLE. *Magnetic cooling.* We may cool a system of magnetic moments by switching off a magnetic field in which the moments were bathed. Consider a magnetic system of N moments μ_0 in thermal equilibrium at the initial temperature $\mathcal{T}_i$ in a magnetic field H_i. In the approximation $\mu_0 H \ll \mathcal{T}$ the entropy is

$$\sigma(N, \mathcal{T}_i, H_i) = \sigma_0 - \frac{N\mu_0^2 H_i^2}{2\mathcal{T}_i^2} \; , \tag{40a}$$

according to (35). If we can arrange[12] for the entropy to remain constant while we reduce the magnetic field to the final value H_f, then the temperature must change to a final value $\mathcal{T}_f$ which satisfies the requirement of constant entropy:

$$\sigma_0 - \frac{N\mu_0^2 H_f^2}{2\mathcal{T}_f^2} = \sigma_0 - \frac{N\mu_0^2 H_i^2}{2\mathcal{T}_i^2} \; . \tag{40b}$$

[11] See ISSP, pp. 434–435. The result (40) is a form of the **Curie law** for the magnetic susceptibility, $\chi = C/\mathcal{T}$, where C is called the Curie constant.

[12] We shall see in Chapter 7 that in a constant entropy process no flow of heat is permitted into or out of the specimen; hence the magnetic specimen must be thermally insulated if the entropy is to be constant.

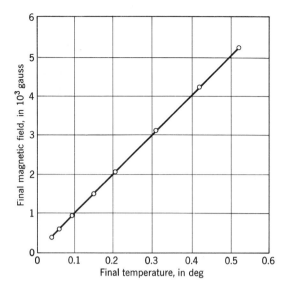

Figure 6 Final magnetic field H_f versus final temperature T_f for magnetic cooling of cerous magnesium nitrate. In these experiments the magnetic field was not removed entirely, but only to the indicated values. The initial fields and temperatures were identical in all runs. [After unpublished results of J. S. Hill and J. H. Milner, as cited by N. Kurti, Nuovo Cimento (Supplemento) **6**, 1109 (1957).]

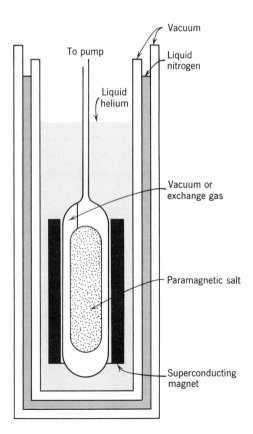

Figure 7 Apparatus for magnetic cooling. From *Heat and thermodynamics*, by M. W. Zemansky, 5th ed. (Copyright © 1968 by McGraw-Hill, Inc. Used with permission of McGraw-Hill Book Company.)

That is,

$$\boxed{\frac{T_f}{T_i} = \frac{H_f}{H_i}}, \tag{40c}$$

so that the temperature is reduced in the ratio by which the magnetic field is reduced. An experimental verification of this relation is shown in Fig. 6; representative apparatus is shown in Fig. 7. This is the magnetic cooling method for the attainment of very low temperatures. (The result (40c) is actually valid irrespective of the ratio of $\mu_0 H$ to $\mathcal{T}$.)

We cannot cool to absolute zero by letting the final field H_f be zero because there are always present local magnetic fields from the magnetic interaction of the moments with each other. These local fields act to give a nonzero limit to H_f. If $H_i = 10^4$ gauss, $H_f = 100$ gauss, and $T_i = 1$ K, then the final temperature will be $T_f = 0.01$ K. In a representative experiment[13] the paramagnetic salt cerous magnesium nitrate was cooled to 0.002 K, starting from $T_i = 1$ K and $H_i = 2 \times 10^4$ gauss. The cerous ion Ce^{3+} is paramagnetic, with six spin orientations possible in a magnetic field.

Problem 2. *Entropy versus magnetic field.* Make a rough plot of σ versus T for the model spin system for $H = 10^3$ and for $H = 10^4$ gauss. Use $N = 10^{22}$ and $\mu_0 = 10^{-20}$ erg gauss^{-1}. Cover the range between 1 and 4 K. Indicate the application of this plot to the magnetic cooling process.

ADDITIVITY OF THE ENTROPY

We showed in (19) that the entropy σ of a combined system with parts not in contact is the sum of the entropies $\sigma_1 + \sigma_2$ of the individual systems. Can we establish a similar result for two systems in thermal contact? We know that the exact expression for the entropy of the combined system is

$$\sigma = \log g(N, U) = \log \left(\sum_{U_1} g_1(N_1, U_1) g_2(N_2, U - U_1) \right), \tag{41}$$

from (15). How can we obtain the additive form $\sigma = \sigma_1 + \sigma_2$?

We should say right out that for typical systems we may replace the value of the summation Σ in (41) by a quantity a billion times smaller without a significant effect on the value of $\log \Sigma$. For a system of N particles the value of $g(N, U)$ is typically of the order of magnitude of 2^N or more. For $N = 10^{22}$ we have

$$\log 2^{(10^{22})} = 10^{22} \log 2 \approx 0.69 \times 10^{22} ,$$

---

[13] R. B. Frankel, D. A. Shirley, and N. J. Stone, *Physical Review* **140A**, 1020 (1965).

whereas the value of the logarithm of an argument that is a billion times smaller is

$$\log\left[10^{-9} \times 2^{(10^{22})}\right] = \log 10^{-9} + \log 2^{(10^{22})} \simeq -20.7 + 0.69 \times 10^{22} .$$

We may always neglect 20.7 in comparison with 0.69×10^{22}.

There is a moral here: some numbers in physics are **very large,** such as 10^9, but other numbers in physics are really **very very large,** such as $2^{(10^{22})}$. For when we take the logarithm of $2^{(10^{22})}$ we obtain 0.69×10^{22}, which is much much larger than the miserable number 20.7 that we obtain when we take the logarithm of 10^9. Much of thermal physics is the study of changes in very very large numbers.

We have considered the number of states in the most probable configuration of two systems in thermal contact, as in (16). But other states are allowed under the condition of constant total energy besides those states that are enumerated in the most probable configuration. The argument of (6) through (14) suggests that for any large system in contact with a reservoir **the average physical properties are very close to the average properties of the states included in the most probable configuration alone.** We assume that this is always true.

One physical property is the entropy. Can we calculate the entropy as $\log (g_1g_2)_{max}$? Are there enough other states (that is, states not in the most probable configuration) to affect the value we calculate for the entropy of the combined systems? Can we replace $\log g(N, m)$ by $\log (g_1g_2)_{max}$, so that

$$\sigma = \log (g_1g_2)_{max} = \log \left[g_1(N_1, \hat{U}_1)g_2(N_2, \hat{U}_2)\right]$$
$$= \log g_1(N_1, \hat{U}_1) + \log g_2(N_2, U_2) \; ? \tag{42a}$$

The answer is yes, as we show below. Notice that only by virtue of the indicated replacement has the total entropy the additive property

$$\sigma(\text{final}) = \sigma_1(\text{final}) + \sigma_2(\text{final}) , \tag{42b}$$

where *final* refers to the condition after thermal contact and equilibrium have been established.

We examine the accuracy of (42a) for two model spin systems in thermal contact. We use the distribution found in (6) for the product $g_1(N_1, m_1)$ $g_2(N_2, m_2)$. It is convenient to set $N_1 = N_2 = \frac{1}{2}N$. By (13) and (14) we have

$$g(N, m) = \sum_{\delta} g_1(N_1, \hat{m}_1 + \delta)g_2(N_2, \hat{m}_2 - \delta) = (g_1g_2)_{max} \int_{-\infty}^{\infty} d\delta \, e^{-8\delta^2/N} , \tag{43}$$

where the sum over δ has been replaced by an integral. We have made an unimportant approximation in taking the limits of integration on δ as $\pm\infty$.

The value of the definite integral is

$$\int_{-\infty}^{\infty} d\delta \, e^{-8\delta^2/N} = \left(\frac{N}{8}\right)^{\frac{1}{2}} \int_{-\infty}^{\infty} dx \, e^{-x^2} = \left(\frac{\pi N}{8}\right)^{\frac{1}{2}} , \tag{44}$$

whence

$$\log g(N, m) = \log (g_1 g_2)_{\max} + \tfrac{1}{2} \log (\pi N / 8) \ . \tag{45}$$

This differs from $\log (g_1 g_2)_{\max}$ by a term of the order of $\log N$. By (2.42) we know that the value of $\log (g_1 g_2)_{\max}$ is of the order of N, because $(g_1 g_2)_{\max}$ is of the order of 2^N. For $N \gg 1$ we can neglect $\log N$ in comparison with N. The former is a small number; the latter is a large number. If $N = 10^{22}$, we have $\log 10^{22} = 22 \log 10 = 22(2.30) \approx 50$, which is negligible in comparison with 10^{22}.

This example supports our assumption that we may take the entropy of a combined system as equal to the sum of the entropies of the component systems when in the most probable configuration. The approximation is accurate so long as one of the systems is large. We do not limit the further theory to two macroscopic systems, but we can treat the thermal properties of even a single particle in weak contact with a macroscopic body (large system) which serves as a heat reservoir. (The definition of macroscopic is "visible to the eye." A body of 10^{20} atoms is macroscopic; a body of 10^4 atoms is not macroscopic.)

Problem 3. *Large and small systems in contact.* Estimate the fractional error in using $\log (g_1 g_2)_{\max}$ in place of $\log g(N, m)$ for the entropy of a combined spin system with $N_1 = 10^{22}$, $N_2 = 10^1$, and $m = 0$. Use (45) generalized by use of (14). *Answer.* 2×10^{-22}, approximately.

NUMBER OF ACCESSIBLE STATES FOR CONTINUOUS DISTRIBUTION OF ENERGY LEVELS

We have defined the entropy for a system with discrete energy levels such that $g(U)$ is well-defined. How do we calculate the entropy if the actual system, because of weak unspecified interactions,[14] has a quasicontinuous distribution of energy levels? What does the quantity $\log g(U)$ or $\log g(N, m)$ mean if the distribution of quantum states is smeared out in energy?

Let the number of quantum states in the quasicontinuous distribution be described by a function

$$\mathfrak{D}(U) \equiv \text{number of states per unit energy range} \ . \tag{46}$$

The number of states in the energy range δU is $\mathfrak{D}(U) \, \delta U$, and this product plays the role of the degeneracy $g(U)$.

---

[14] Such as magnetic dipole-dipole interactions between the magnetic moments.

Let us suppose that the energy of the system is known to lie in a range δU which is small in comparison with U, the total energy of the system. The entropy is

$$\sigma = \log g(U) = \log [\mathfrak{D}(U)\, \delta U] = \log \mathfrak{D}(U) + \log \delta U \ . \qquad (47)$$

The entropy is relatively insensitive to the accuracy δU with which the energy is known. For example, suppose that we perform two entropy experiments, in one of which the energy uncertainty is a million times larger than in the other. The change in $\log \delta U$ between the two experiments is to add $\log 10^6 \approx 14$ to the entropy of one system. However, the entropy is typically of the order of the number of particles,[15] say 10^{22}, so that we may neglect the additive term 14. Thus both experiments will give essentially the identical value of the entropy.

LAW OF INCREASING ENTROPY FOR A CLOSED SYSTEM: THE SECOND LAW

In a closed system the total energy U and the total number of particles N are independent of time.[16] If the mechanical specifications of the system, such as the volume and the nature of the contact between the parts of the system, are also independent of time, the total number of states accessible to the system is independent of time. The entropy of such a closed system is rigorously constant. What could we mean, then, by an increase in the entropy of the system?

It is convenient to discuss this question for a closed system composed of two parts in thermal contact. The entropy of the system is given by

$$\sigma(U) = \log g(U) = \log \sum_{U_1} g_1(U_1) g_2(U - U_1) \ , \qquad (48)$$

or, to an excellent approximation, by

$$\sigma(U) = \log (g_1 g_2)_{\text{max}} \ , \qquad (49)$$

which is smaller than (48), but only by a negligible amount, as we saw in (45). Both of these expressions are independent of time. But there is a sense in which it may be said that the entropy of a closed system tends to increase with time or to remain constant.

We need a definition of a generalized entropy that will apply to any particular partition or division of the total energy among the two parts of the system,

---

[15] We have not proved that the entropy is typically of the order of the number of particles, and this is not always true, but 14 is always a negligible number in comparison with the entropy of a macroscopic system.

[16] For the present we neglect the possibility of chemical or nuclear reactions among the particles. We treat reactions in Chapter 21.

U_1 with part 1 and $U - U_1$ with part 2. The total number of states consistent with this particular partition of the energy is

$$g_1(U_1)g_2(U - U_1) \ .$$

We can define the entropy as the logarithm of this quantity, but such an entropy would not refer to the equilibrium configuration of the system (unless the value of U_1 happened to correspond to the most probable partition of the energy). We reserve the use of the term entropy for the most probable configuration, as in (49). Thus we are led to define the **generalized entropy** σ_G for an arbitrary partition of the energy of the system as

$$\sigma_G(U_1, U - U_1) = \log g_1(U_1)g_2(U - U_1) \ . \tag{50}$$

For a macroscopic system there will never occur spontaneously large differences or even significant differences between the value of the entropy and the value of the generalized entropy. We showed this for the model system in the argument following (14), with "never" used in the sense of not once in the entire age of the universe, 10^{18} sec. In practice this result means that we can find a significant difference between the entropy and the generalized entropy of a macroscopic system only if we have prepared the system at the initial instant of time in some special way, as by lining up all the spins parallel to one another or with a gas by collecting all the molecules in the air of the room into a small volume in one corner of the room. Such extreme situations never arise in a system that has been left undisturbed for some time, but it is still of interest to discuss them.

The generalized entropy is a maximum when the system is in the equilibrium configuration. The maximum value is very close to the value of the exact entropy (48), and we treat the two values as equal. For other configurations the value of σ_G may be very much smaller than $\sigma(U)$.

It is written that **the entropy of a closed system tends to remain constant or to increase.** In greater detail the statement reads: If two systems (of which one is a large reservoir) at some instant of time are in a configuration other than the most probable configuration, then the most probable consequence will be that the configuration will change in such a way that the generalized entropy will increase monotonically in successive instants of time. The statement is one formulation of the **second law of thermodynamics.**[17]

This statement must be interpreted with delicacy, as it leads to both meaningful and meaningless conclusions. Our fundamental assumption that all accessible states are equally likely implies that a closed system[18] in the course of

---

[17] The first law of thermodynamics is an expression of the principle of conservation of energy. It is introduced and used in Chapter 7, at the end of which the laws of thermodynamics are summarized.
[18] The closed system here is spoken of as composed of a body and a reservoir, with thermal contact between the two. The reservoir always contains a macroscopic number of particles.

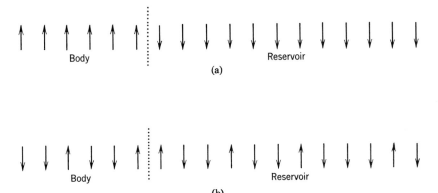

Figure 8 (a) The system is prepared initially in the special state shown. The spin excess $2m = n(\uparrow) - n(\downarrow)$ is equal to -6. There is only one state for a configuration with $m_1 = 3$ and $m_2 = -6$, so that the generalized entropy is zero. (b) The system is in one of the many states that belong to the most probable configuration for a spin excess of -6, just as in (a). For the most probable configuration $\hat{m}_1 = -1$ and $\hat{m}_2 = -2$.

time will pass through all the accessible states and, consequently, through all the partitions of the energy U between the body and the reservoir. The generalized entropy of the system changes with time because of random fluctuations in the exchange of energy across the surface of thermal contact. What can we say about the occurrence of perceptible fluctuations in σ_G?

For two spin systems in thermal contact we know that $\log g_1g_2$ is extremely sharply peaked with respect to energy exchange between the reservoir and the system. The probability of a significant fluctuation in the value of σ_G during the age of the universe is negligible. Apart from the unobservable fluctuations, the value of σ_G is constant and for all practical purposes is equal to the value of the entropy.

We are at liberty, however, to prepare the system initially in a special configuration that is very different from the most probable configuration. For example, with the model spin system in a magnetic field we can align all the spins in the body in one direction and all the spins in the reservoir in the opposite direction, as in Fig. 8a. In the course of time the interactions between the two sets of spins will redistribute the spin directions as in Fig. 8b. The state in (b) belongs to the most probable configuration for the identical value of the spin excess as in (a); here "most probable" refers to the distribution of the spin excess between the body and the reservoir.

As another example, we consider the gas in a room: the gas in one half of the room might be prepared initially (by cooling) with a low value of the average energy per molecule, while the gas in the other half of the room might be prepared initially with a ten times higher value of the average energy per molecule. If now the two halves are allowed to interact, assuming equal num-

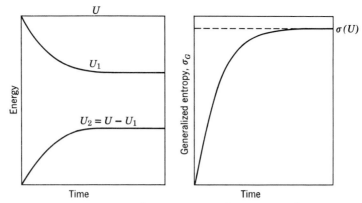

Figure 9   A system with two parts, 1 and 2, is prepared at zero time with $U_2 = 0$ and $U_1 = U$. Exchange of energy takes place between two parts and presently the system will be found in or close to the most probable configuration with respect to energy exchange. The generalized entropy increases as the system attains configurations of increasing probability. The generalized entropy eventually reaches the entropy $\sigma(U)$ of the most probable configuration.

bers of molecules initially in each half, the room will come very quickly to a most probable configuration in which the molecules in both halves have an average energy equal to $\frac{1}{2}(1 + 10)$ times the initial average energy of a molecule in the first half.

In both examples the value of the generalized entropy increases by a large amount after the initial constraints are removed and the combined system is allowed to explore all the states normally accessible to it. The result is that after a certain lapse of time the system will arrive at or very close to the most probable configuration. The increase in the value of the generalized entropy (Fig. 9) in the process is what is meant by the tendency of the entropy to increase. In these examples it is overwhelmingly probable that the generalized entropy will increase with time following removal of the initial constraints. Nothing else will ever be observed to happen, and in particular we will never observe the system to leave the most probable configuration and appear later in what we have called the initial configuration.

The equations of motion of physics are believed to be reversible in time and do not distinguish past and future. This reversibility does not contradict the result of everyday observation: if we find or prepare a closed system in a configuration with a relatively low value of the generalized entropy, the system will evolve in time in such a way that the generalized entropy increases toward a maximum value, that of the most probable configuration. The reversibility suggests that if we wait long enough, a given system will inevitably appear in any accessible configuration, however improbable. But this is wrong, because "long enough" is so long as to be never. In practice, in everyday observation, we may well feel that an event is possible only if it appears in the

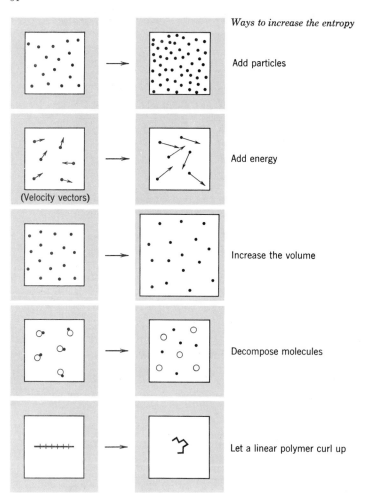

Figure 10 Factors that tend to increase the entropy of a system.

lifetime of a man ($\sim 2 \times 10^9$ sec), or in the lifetimes of all men now alive ($\sim 8 \times 10^{18}$ sec), or in the age of the universe ($\sim 10^{18}$ sec). An event that is expected to occur only once in 10^{160} sec, for example, can be said to be impossible. Thus when we release the helium atoms from a balloon in the middle of the room, the atoms will diffuse through the room and attain configurations of higher values of some generalized form of the entropy. The laws of mechanics allow the atoms to reverse their motions and crowd back into the balloon, with a resultant decrease in the value of the generalized entropy. But in any practical operational sense it is impossible that this could happen, for no one would ever see it happen in the age of the universe. The tendency of the generalized entropy to increase has a real meaning only for systems that are prepared by us initially in a nonequilibrium configuration and from which a restraint is then lifted, thereby allowing the system to move to the most probable configuration.

The entropy is constant in a closed system, that is, in a system of constant energy and constant number of particles. The sun, for example, is not a closed system: it loses energy by radiation and is cooling down. The entropy of the sun is decreasing. It is not clear, according to geophysicists, whether the total entropy of the earth is decreasing or increasing at this moment. The earth receives, generates, and radiates entropy.

Fifty years ago it was common to say that the entropy of the universe is increasing, which may very well be true. This is a cosmological question. The "Big Bang" models of the expansion of the universe imply an increase in the entropy at the present epoch. If the universe contracts at a later epoch, the entropy will probably decrease.

We do know that in laboratory scale experiments on closed systems the generalized entropy increases. We also know that the entropy increases when two systems at different temperatures are brought into thermal contact. Processes that tend to increase the entropy of a system are shown in Fig. 10; the arguments involved will be developed in the chapters which follow.

Problem 4. *The meaning of never.* It has been said[19] that "six monkeys, set to strum unintelligently on typewriters for millions of millions of years, would be bound in time to write all the books in the British Museum." This statement is misleading nonsense, for it gives a misleading conclusion about very very large numbers. Could all the monkeys in the world have typed out a single specified book in the age of the universe?[20]

Suppose that 10^{10} monkeys have been seated at typewriters throughout the age of the universe, 10^{18} sec. This number of monkeys is about three times greater than the present human population[21] of the earth. We suppose that a monkey can hit 10 typewriter keys per second. A typewriter may have 44 keys; we accept lowercase letters in place of capital letters. Assuming that Shakespeare's *Hamlet* has 10^5 characters, will the monkeys hit upon *Hamlet?*

---

[19] J. Jeans, *Mysterious universe*, Cambridge University Press, 1930, p. 4. The statement is attributed to Huxley.

[20] For a related mathematico-literary study, see "The Library of Babel," by the fascinating Argentine writer Jorge Luis Borges, in *Ficciones*, Grove Press, Evergreen paperback, 1962, pp. 79–88.

[21] For every person now alive, some thirty persons have once lived. This figure is quoted by R. C. Clarke in "2001." I am grateful to the Population Reference Bureau and to Dr. Roger Revelle for explanations of the evidence. The cumulative number of man-seconds is 2×10^{20}, if we take the average lifetime as 2×10^9 sec and the number of lives as 1×10^{11}. The cumulative number of man-seconds is much less than the number of monkey-seconds (10^{28}) taken in the problem.

(a) Show that the probability that any given sequence of 10^5 characters typed at random will come out in the correct sequence (the sequence of *Hamlet*) is

$$\left(\tfrac{1}{44}\right)^{100\,000} = 10^{-164\,345} \; ,$$

where we have used $\log_{10} 44 = 1.64345$.

(b) Show that the probability that a *monkey-Hamlet* will be typed in the age of the universe is approximately $10^{-164\,316}$. The probability of *Hamlet* is therefore zero in any operational sense of an event, so that the original statement at the beginning of this problem is nonsense: one book, much less a library, will never occur in the total literary production of the monkeys.

(c) What happens to the result (b) if we do not specify the title of the book, but agree to accept any known book? There may be about 30×10^6 distinct titles of books: the largest library, the Library of Congress, contains about 15×10^6 books and pamphlets. Note that the total production of the monkeys is equivalent to 10^{24} short volumes of 10^5 characters each, but you will find that none of these duplicate any existing book.

REFERENCES

A helpful reference is Chapter 4 of J. E. Mayer and M. G. Mayer, *Statistical mechanics,* Wiley, 1940. A short discussion of magnetic cooling is given in Chapter 14 of ISSP.

Summary of Arguments on Entropy and Temperature

(a) The degeneracy of two systems in thermal contact is given by

$$g(N, U) = \sum_{U_1} g_1(N_1, U_1)g_2(N_2, U - U_1) \ .$$

(b) The maximum product in the sum defines the most probable configuration:

$$(g_1 g_2)_{\max} \equiv g_1(N_1, \hat{U}_1)g_2(N_2, U - \hat{U}_1) \ ;$$

for this configuration

$$\left(\frac{\partial \log g_1}{\partial U_1}\right)_{N_1} = \left(\frac{\partial \log g_2}{\partial U_2}\right)_{N_2} \ .$$

(c) The average physical properties of a large system may be calculated as averages over the most probable configuration. In particular, for the entropy we have to high accuracy that

$$\sigma(N, U) \equiv \log g(N, U) \cong \log (g_1 g_2)_{\max}$$
$$= \log g_1(N_1, \hat{U}_1) + \log g_2(N_2, \hat{U}_2) = \sigma_1 + \sigma_2 \ .$$

The most probable configuration is called the equilibrium configuration.

(d) For systems in thermal contact:

$$\frac{1}{T_1} \equiv \left(\frac{\partial \sigma_1}{\partial U_1}\right)_{N_1} = \left(\frac{\partial \sigma_2}{\partial U_2}\right)_{N_2} \equiv \frac{1}{T_2} \ .$$

(e) The generalized entropy of a configuration is

$$\sigma_G \equiv \log g_1(N_1, U_1)g_2(N_2, U - U_1) \ ;$$

it is a maximum for the most probable configuration. Fluctuations of σ_G from $\sigma = \log (g_1 g_2)_{\max}$ are extremely small. If the system is prepared initially in a configuration quite far from the most probable configuration, then σ_G will increase with time until it becomes equal to the entropy.

Two Systems in Diffusive Contact:
The Chemical Potential

CHAPTER 5   TWO SYSTEMS IN DIFFUSIVE CONTACT:
THE CHEMICAL POTENTIAL

In Chapter 4 we considered the behavior of two systems in thermal contact, and we were led to a natural definition of the temperature of a system. The definition of temperature has the important consequence that the number of accessible states of the combined systems is a maximum when the systems are at the same temperature.

Let us now consider two systems in thermal contact and also in diffusive contact. **Diffusive contact** means that atoms or molecules can move from one system to the other by diffusion through a permeable boundary or membrane. Systems in thermal and diffusive contact may exchange both particles and energy. We shall not treat systems that are only in diffusive contact and not in thermal contact.[1] Our considerations will lead to a natural definition of the chemical potential of a system. The chemical potential is just as important a quantity as the temperature. We shall find (particularly in Chapter 11) that the chemical potential usually is made up of two kinds of contributions, one that is the potential energy of a particle and one that involves the concentration of particles.

We have seen that two systems which can exchange energy will be in equilibrium when they are at the same temperature. What can we say about the equilibrium condition if the systems can exchange particles? We shall find a new equilibrium condition that leads to the introduction of the chemical potential. This allows us to discuss the concentration gradients of particles in external electric, magnetic, and gravitational fields (Chapter 11), and it is the basis for the discussion of the equilibrium conditions in chemical reactions (Chapter 21).

Under the constraints

$$U = U_1 + U_2 = \text{constant} \; ; \qquad N = N_1 + N_2 = \text{constant} \; ,$$

the most probable configuration[2] of the combined systems is that for which the number of accessible states is a maximum. This is a maximum when the product

$$g_1(N_1, U_1)g_2(N - N_1; U - U_1) \tag{1}$$

of the number of accessible states of the separate systems is a maximum with

---

[1] We can imagine an exchange of particles without an exchange of energy. To accomplish such an exchange we might take a particle of zero energy from one system and carry it through the interface membrane to the second system, where it is released at zero energy.

[2] **Configuration** is used here to denote a particular allocation of the total energy U among the two systems and a particular allocation of the total number of particles N among the two systems.

respect to the independent variation of N_1 and U_1. This conclusion follows from the fundamental assumption, Chapter 3.

The condition that (1) be an extremum is that

$$d(g_1 g_2) = \left(\frac{\partial g_1}{\partial N_1} dN_1 + \frac{\partial g_1}{\partial U_1} dU_1 \right) g_2 + g_1 \left(\frac{\partial g_2}{\partial N_2} dN_2 + \frac{\partial g_2}{\partial U_2} dU_2 \right) = 0 \ . \tag{2}$$

The partial derivative $\partial g_1 / \partial N_1$ is understood to denote $(\partial g_1 / \partial N_1)_{U_1}$, the partial derivative taken with respect to the number of particles, but at constant energy. We indicated the possibility of such a process in footnote 1. Because $N = N_1 + N_2 = $ constant and $U = U_1 + U_2 = $ constant, we have

$$dN_2 = d(N - N_1) = -dN_1 \ ; \qquad dU_2 = d(U - U_1) = -dU_1 \ , \tag{2a}$$

so that

$$\frac{\partial g_2}{\partial N_1} = -\frac{\partial g_2}{\partial N_2} \ ; \qquad \frac{\partial g_2}{\partial U_1} = -\frac{\partial g_2}{\partial U_2} \ . \tag{3}$$

If we divide (2) by $g_1 g_2$ and use (2a), we obtain

$$\left(\frac{1}{g_1} \frac{\partial g_1}{\partial N_1} - \frac{1}{g_2} \frac{\partial g_2}{\partial N_2} \right) dN_1 + \left(\frac{1}{g_1} \frac{\partial g_1}{\partial U_1} - \frac{1}{g_2} \frac{\partial g_2}{\partial U_2} \right) dU_1 = 0 \tag{4}$$

as the condition for the two systems to be in equilibrium.

We define the entropies σ_1, σ_2 of the two systems as in Chapter 4:

$$\sigma_1(N_1, U_1) = \log g_1(N_1, U_1) \ ; \qquad \sigma_2(N_2, U_2) = \log g_2(N_2, U_2) \ . \tag{5}$$

We may write (4) as

$$d\sigma = \left[\left(\frac{\partial \sigma_1}{\partial N_1} \right) - \left(\frac{\partial \sigma_2}{\partial N_2} \right) \right] dN_1 + \left[\left(\frac{\partial \sigma_1}{\partial U_1} \right) - \left(\frac{\partial \sigma_2}{\partial U_2} \right) \right] dU_1 = 0 \ , \tag{6}$$

where $\sigma = \sigma_1 + \sigma_2$. The entropy is a maximum in equilibrium.

We see from (6) that the condition for thermal and diffusive equilibrium of the two systems is that the terms in the two brackets $[\cdots]$ must vanish:

$$\left(\frac{\partial \sigma_1}{\partial U_1} \right)_{N_1} = \left(\frac{\partial \sigma_2}{\partial U_2} \right)_{N_2} \ ; \qquad \left(\frac{\partial \sigma_1}{\partial N_1} \right)_{U_1} = \left(\frac{\partial \sigma_2}{\partial N_2} \right)_{U_2} \ , \tag{7}$$

where now we have indicated explicitly the variables that are held constant in the differentiation. If no diffusion is allowed, $dN_1 = 0$, and the discussion in Chapter 4 is sufficient. The equation on the left tells us that the two temperatures must be equal. But if dN_1 is not restricted to zero, then to obtain equilibrium not only must the two systems have the same temperature, but they must also satisfy the new condition given by the equation on the right in (7).

The condition on the left in (7) is familiar to us from Chapter 4, for it reads

$$\frac{1}{T_1} = \frac{1}{T_2} \; , \tag{8}$$

or $T_1 = T_2$. The other condition is new. To discuss it we need to introduce a symbol and a name related to the quantity $(\partial\sigma/\partial N)_U$.

DEFINITION OF THE CHEMICAL POTENTIAL

The **chemical potential** μ of a system[3] is defined by

$$\boxed{-\frac{\mu}{T} \equiv \frac{1}{g}\left(\frac{\partial g}{\partial N}\right)_U \equiv \left(\frac{\partial\sigma}{\partial N}\right)_U .} \tag{9}$$

The chemical potential is also and quite properly called the **electrochemical potential**. Particularly in books on transistor electronics it is necessary to be careful to find out whether the author uses chemical potential in the present or in some different sense,[4] usually one in which the electrostatic potential energy of a particle has been subtracted from the quantity defined by (9).

The chemical potential is related to the fractional change of the number of accessible states with a change in the number of particles. For two systems at the same temperature T the new equilibrium condition (7) now reads

$$-\frac{\mu_1}{T} = -\frac{\mu_2}{T} \; , \tag{10}$$

or

$$\mu_1 = \mu_2 \; . \tag{11}$$

Two systems that can exchange energy and particles are in equilibrium when the temperatures and the chemical potentials are equal.

What is the direction of flow of particles if initially $\mu_2 > \mu_1$? Consider two systems at temperature T: the entropy change when δN particles are taken from system 2 and are added to system 1 is

$$\delta\sigma = \delta(\sigma_1 + \sigma_2) = \left(\frac{\partial\sigma_1}{\partial N_1}\right)_{U_1}\delta N - \left(\frac{\partial\sigma_2}{\partial N_2}\right)_{U_2}\delta N = \left(-\frac{\mu_1}{T} + \frac{\mu_2}{T}\right)\delta N \; ,$$

[3] Do not confuse μ for the chemical potential with μ for the magnetic moment. Henceforth we shall usually denote the magnetic moment by μ_0.

[4] For details, see the discussion by T. C. Harman and J. M. Honig, *Thermoelectric and thermomagnetic effects and applications*, McGraw-Hill, 1967, pp. 9, 15, 122, 130, and 140.

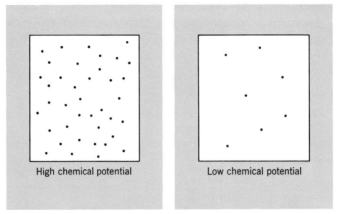

Figure 1 At the same temperature the system with a high concentration of particles will have a higher chemical potential than the system with a low concentration of particles.

using (6). Now $\mu_1 = \mu_2$ in equilibrium, and $\delta\sigma = 0$. If initially $\mu_2 > \mu_1$, then $\delta\sigma$ will be positive under the transfer δN. The total entropy will increase as particles flow from 2 to 1. We shall assume that the temperature is always positive $(\mathcal{T} > 0)$ except where specifically stated otherwise.

As the combined systems approach equilibrium, there is a net flow of particles from the system of high chemical potential to the system of low chemical potential. The minus sign in the definition of the chemical potential arranges that particles shall flow from a high chemical potential to a low chemical potential. We shall see later that a system of high particle concentration has a higher value of the chemical potential than does a system of low particle concentration (Fig. 1). Particles tend therefore to diffuse from the region of high concentration to the region of low concentration.

Notice that μ will be negative if σ increases as N increases, with U constant. This is a common situation. The temperature $\mathcal{T}$ was included in the definition (9) to give the chemical potential the dimensions of energy: $\mathcal{T}$ is an energy, and both σ and N are dimensionless.

A similar argument can be carried through separately for each distinct chemical species in a mixture of diverse atoms and molecules, for example, H_2, O_2, and H_2O. We then define the chemical potential of chemical component r of the mixture by

$$-\frac{\mu_r}{\mathcal{T}} \equiv \left(\frac{\partial\sigma}{\partial N_r}\right)_{U,\,N_s} , \tag{13}$$

where the derivative is taken with the numbers N_s of all other species in the system held constant.

The chemical potential is a very important quantity. We will get a physical feeling for it as we go along, particularly in Chapters 11 and 21. For the present, think of the chemical potential as the quantity defined by (9); it measures the dependence of the number of accessible states on the number of particles in the system. The chemical potential is somewhat less accessible than the temperature to experimental measurement, but for charged particles it can be determined with a voltmeter and for neutral particles it can be determined by osmotic pressure measurements. Such experiments are quite routine in physical chemistry laboratories.

Problem 1. Chemical potential. Suppose that $g = BV^N$, where B is a constant and V is the volume. Show that $\mu = -T \log V$.

EXAMPLE. Chemical potential of a spin system in a magnetic field. Find the chemical potential of the model spin system.

From (4.33) for the entropy of the model spin system we have

$$-\frac{\mu}{T} = \left(\frac{\partial \sigma(N, U)}{\partial N}\right)_U = \left(\frac{\partial \sigma(N, 0)}{\partial N}\right)_U + \frac{U^2}{2\mu_0^2 H^2 N^2} = -\frac{\mu(0)}{T} + \frac{U^2}{2\mu_0^2 H^2 N^2} , \tag{14}$$

where $\mu(0)$ is the chemical potential evaluated at zero magnetic energy U. The magnetic moment of each spin is μ_0. We use (4.35) for $U(T)$ to obtain the chemical potential as a function of the temperature and the magnetic field:

$$-\frac{\mu}{T} = -\frac{\mu(0)}{T} + \frac{\mu_0^2 H^2}{2T^2} , \tag{15}$$

or

$$\mu(T, H) = \mu(0) - \frac{\mu_0^2 H^2}{2T} . \tag{16}$$

The chemical potential decreases as the magnetic field intensity increases.

Suppose that each spin is mounted on an atom of a gas. Let there be N atoms in a volume V. Will the local concentration c of atoms be affected by the magnetic field? We shall see in Chapter 11 that the translational motion of the atoms of an ideal gas gives a contribution to the chemical potential of the form $T \log c$ plus terms that are independent of the concentration c. If we anticipate this result and add a term $T \log c$ to (16), we have for the chemical potential of the magnetic gas

$$\mu(T, H) = T \log c - \frac{\mu_0^2 H^2}{2T} + \text{constant} . \tag{17}$$

The dependence of the chemical potential on concentration and on magnetic

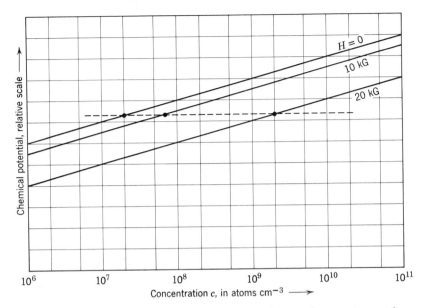

Figure 2 Dependence of the chemical potential of a gas of magnetic particles on the concentration, at several values of the magnetic field intensity. If $c = 2 \times 10^7$ cm^{-3} for $H = 0$, then at a point where $H = 20$ kilogauss the concentration will be 2×10^9 cm^{-3}.

field is illustrated in Fig. 2. We divide by T and raise e to powers of both sides. We find

$$e^{\mu/T} \propto c e^{-\mu_0{}^2 H^2/2T^2} \ . \tag{18}$$

Let the system be immersed in an inhomogeneous magnetic field. The atoms diffuse freely. In equilibrium the chemical potential and temperature must be constant throughout the volume. Thus the left-hand side of (18) must be constant; it follows that the local concentration c must depend on the magnetic field H as

$$c \propto e^{\mu_0{}^2 H^2/2T^2} \ . \tag{19}$$

At first sight the puzzle is not why the particles tend to concentrate in regions of high field, for that is readily explained as being energetically favorable: it lowers the energy of the system. The puzzle is, why are any left at all in the low field regions? The answer is that the entropy is higher when the particles are distributed in space than when they are concentrated. The same reasoning applies to cream stirred into coffee: why doesn't the cream, being less dense than the coffee, float to the top? We have assumed that the local concentration can be defined to sufficient accuracy; we shall see later that this depends on keeping the square root of the number of particles in the element of volume

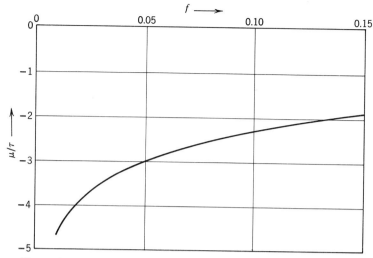

Figure 3   Dependence of the chemical potential μ on fraction of occupied sites $f = N/N_0$ for a lattice gas in the approximation $f \ll 1$ for which $\mu \cong T \log f$.

large in comparison with unity, while keeping H essentially constant over the same element of volume. We see from (19) that magnetic particles tend to congregate in regions of high magnetic field at the expense of regions of low magnetic field.

Problem 2. *Chemical potential of the lattice gas.* The ideal lattice gas is discussed in Problem 2.3 and in Appendix B. Use the result (B.4) for the entropy to show that in the limit $N \ll N_0$ the chemical potential is given by

$$\mu \cong T \log f , \tag{20}$$

where $f = N/N_0$ is the fraction of lattice sites occupied by atoms. In this approximation the chemical potential is plotted in Fig. 3. Notice that μ is negative in this problem. A generalization of this result is given in (6.82).

Problem 3. *Magnetic concentration.* Estimate the value of the magnetic moment needed to produce the effect of magnetic field on particle concentration shown in Fig. 2. The temperature is 300 K. Express the magnetic moment in units of the Bohr magneton, 0.927×10^{-20} erg gauss^{-1}. (The result is appropriate to very fine ferromagnetic particles in suspension, as used in studies of the flux structure of superconductors and the domain structure of ferromagnetic materials.)

CHAPTER 6

Gibbs and Boltzmann Factors

SYSTEMS AND RESERVOIRS

We consider a very large body which has a constant energy U_0 and a constant number of particles N_0. Let us imagine that the body is constructed of two parts (Fig. 1). The part of primary interest to our investigation is called the **system.** The other part, very much larger, is called the **reservoir.**

The system and the reservoir are in thermal and diffusive contact with each other. They may exchange particles and energy. The contact assures that the temperature of the system is equal to that of the reservoir and the chemical potential of the system is equal to that of the reservoir. When the system has N particles, the reservoir has $N_0 - N$ particles. When the system has energy ϵ, the reservoir has energy $U_0 - \epsilon$.

We wish to obtain statistical properties of the system. To carry out our program we make observations as discussed in Chapter 3 on the members of an ensemble which consists of identical copies of the **system + reservoir,** one copy for each accessible quantum state of the combination. The most useful question to ask is "What is the probability in a given observation that the system will be found to contain N particles and to be in a state l of energy ϵ_l?"

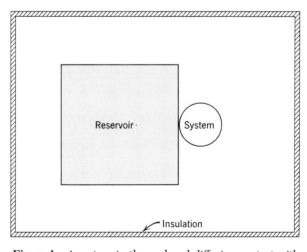

Figure 1   A system in thermal and diffusive contact with a large reservoir of energy and of particles. The complex is insulated from the external world, so that the total energy and total number of particles are constant. The temperature of the system is equal to the temperature of the reservoir, and the chemical potential of the system is equal to the chemical potential of the reservoir.

The probability $P(N, \epsilon_l)$ that the system has N particles and is in the particular state l of energy ϵ_l is proportional to the number of accessible states of the reservoir. For when we specify the state of the system, the number of accessible states of the complex is just the number of accessible states of the reservoir:

$$g(\text{complex}) = g(\text{reservoir}) . \tag{1}$$

These states of the reservoir have $N_0 - N$ particles and have energy $U_0 - \epsilon_l$.

It follows that the probability $P(N, \epsilon_l)$ is proportional to the number of accessible states of the reservoir:

$$P(N, \epsilon_l) \propto g(N_0 - N, U_0 - \epsilon_l) , \tag{2}$$

where g now refers to the reservoir alone. We have displayed explicitly in (2) the dependence of $g(\text{reservoir})$ on the number of particles in the reservoir and on the energy of the reservoir. Notice that questions about the system appear to depend on the constitution of the reservoir, but only (as we shall see) on the temperature and chemical potential of the reservoir.

Because we have not determined the constant of proportionality in (2), we express the result as a ratio of two probabilities, one that the system is in a state 1 and the other that the system is in a state 2:

$$\frac{P(N_1, \epsilon_1)}{P(N_2, \epsilon_2)} = \frac{g(N_0 - N_1, U_0 - \epsilon_1)}{g(N_0 - N_2, U_0 - \epsilon_2)} . \tag{3}$$

The two situations are shown in Fig. 2.

The g's are very large numbers; to avoid the inconvenience of large numbers we work instead with $\log g$, the entropy of the reservoir. By definition

$$g(N_0, U_0) \equiv e^{\sigma(N_0, U_0)} , \tag{4}$$

so that the probability ratio in (3) may be written as

$$\frac{P(N_1, \epsilon_1)}{P(N_2, \epsilon_2)} = \frac{e^{\sigma(N_0 - N_1, U_0 - \epsilon_1)}}{e^{\sigma(N_0 - N_2, U_0 - \epsilon_2)}} , \tag{5}$$

or

$$\frac{P(N_1, \epsilon_1)}{P(N_2, \epsilon_2)} = e^{\sigma(N_0 - N_1, U_0 - \epsilon_1) - \sigma(N_0 - N_2, U_0 - \epsilon_2)} = e^{\Delta\sigma} , \tag{6}$$

where $\Delta\sigma$ is defined as the entropy difference:

$$\Delta\sigma \equiv \sigma(N_0 - N_1, U_0 - \epsilon_1) - \sigma(N_0 - N_2, U_0 - \epsilon_2) . \tag{7}$$

The reservoir is assumed to be very large in comparison with the system, and thus we may approximate $\Delta\sigma$ quite accurately by the first-order terms in

Reservoir System

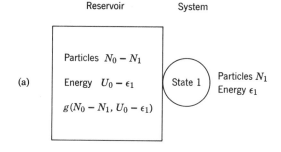

(a)

Particles $N_0 - N_1$

Energy $U_0 - \epsilon_1$

$g(N_0 - N_1, U_0 - \epsilon_1)$

State 1

Particles N_1
Energy ϵ_1

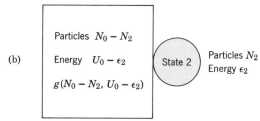

(b)

Particles $N_0 - N_2$

Energy $U_0 - \epsilon_2$

$g(N_0 - N_2, U_0 - \epsilon_2)$

State 2

Particles N_2
Energy ϵ_2

Figure 2 The reservoir is in thermal and diffusive contact with the system. In (a) the system is in quantum state 1, and the reservoir has $g(N_0 - N_1, U_0 - \epsilon_1)$ states accessible to it. In (b) the system is in the quantum state 2, and the reservoir has $g(N_0 - N_2, U_0 - \epsilon_2)$ states accessible to it. If, as here, we specify the exact state of the system, then the total number of states accessible to the complex of system + reservoir is just the number of states accessible to the reservoir.

a power series. The Taylor series expansion of $f(x + a)$ about $f(x)$ is

$$f(x + a) = f(x) + a\frac{df}{dx} + \frac{1}{2!}a^2\frac{d^2f}{dx^2} + \frac{1}{3!}a^3\frac{d^3f}{dx^3} + \cdots . \qquad (8)$$

Thus for the reservoir

$$\sigma(N_0 - N, U_0 - \epsilon) = \sigma(N_0, U_0) - N\left(\frac{\partial\sigma}{\partial N_0}\right)_{U_0} - \epsilon\left(\frac{\partial\sigma}{\partial U_0}\right)_{N_0}$$
$$+ \text{ terms of higher order.} \qquad (9)$$

For $\Delta\sigma$ as defined by (7) we have, to first order in $N_1 - N_2$ and $\epsilon_1 - \epsilon_2$,

$$\Delta\sigma = [(N_0 - N_1) - (N_0 - N_2)]\left(\frac{\partial\sigma}{\partial N_0}\right)_{U_0} + [(U_0 - \epsilon_1) - (U_0 - \epsilon_2)]\left(\frac{\partial\sigma}{\partial U_0}\right)_{N_0}$$
$$= -(N_1 - N_2)\left(\frac{\partial\sigma}{\partial N_0}\right)_{U_0} - (\epsilon_1 - \epsilon_2)\left(\frac{\partial\sigma}{\partial U_0}\right)_{N_0} . \qquad (10)$$

We use the definition of the temperature and chemical potential,

$$\frac{1}{T} \equiv \left(\frac{\partial \sigma}{\partial U}\right)_N ; \qquad -\frac{\mu}{T} \equiv \left(\frac{\partial \sigma}{\partial N}\right)_U , \qquad (11)$$

to write the entropy difference (10) as

$$\Delta\sigma = \frac{(N_1 - N_2)\mu}{T} - \frac{(\epsilon_1 - \epsilon_2)}{T} . \qquad (12)$$

Here $\Delta\sigma$ refers to the reservoir, but N_1, N_2, ϵ_1, ϵ_2 refer to the system.

GIBBS FACTOR

The single most useful result of statistical mechanics is found on combining (6) and (12):

$$\boxed{\frac{P(N_1, \epsilon_1)}{P(N_2, \epsilon_2)} = \frac{e^{(N_1\mu - \epsilon_1)/T}}{e^{(N_2\mu - \epsilon_2)/T}} .} \qquad (13)$$

The probability is the ratio of two exponential factors, each of the form

$$e^{(N\mu - \epsilon)/T} .$$

A term of this form will be called a **Gibbs factor.**[1] The Gibbs factor is proportional to the probability that the system is in a state l of energy ϵ_l and number of particles N. Several aspects of the Gibbs factor should be noted:

Boltzmann factor. If the number of particles in the system is fixed, then $N_1 = N_2$ and the Gibbs factor reduces to

$$\frac{P(\epsilon_1)}{P(\epsilon_2)} = \frac{e^{-\epsilon_1/T}}{e^{-\epsilon_2/T}} . \qquad (14)$$

This gives the ratio of the probability that a system will be in a state of energy ϵ_1 to the probability the system will be in a state of energy ϵ_2, the number of particles always being N. A term of the form $e^{-\epsilon/T}$ is called a **Boltzmann factor.**[2] In actual applications the ratio (14) is nearly as useful as (13).

Degeneracy. The probabilities in (13) refer to single quantum states, **each ennumerated separately.** We can give an alternate form of (13) that applies to degenerate energy levels: if the system has ρ_a states of energy ϵ_a and ρ_b states of energy ϵ_b, then the ratio of the probability of finding the system

---

[1] The result (13) was first given by J. W. Gibbs and was referred to by him as the **grand canonical distribution.**

[2] Gibbs refers to this as a **canonical distribution.**

with energy ϵ_a to that of finding the system with energy ϵ_b is

$$\frac{W(N_a, \epsilon_a)}{W(N_b, \epsilon_b)} = \frac{\rho_a e^{(N_a\mu - \epsilon_a)/T}}{\rho_b e^{(N_b\mu - \epsilon_b)/T}} . \tag{15}$$

We have changed notation for probability from P to W to emphasize the change in the meaning of the ratio.

Quantum statistics. The energy ϵ_l that enters the Gibbs factor is the energy of the quantum state l of an N particle system. All questions of which "statistics," such as Bose-Einstein or Fermi-Dirac, happen to be satisfied by the particles, are irrelevant when using (13), for these questions are involved only in the original determination of the allowed quantum states of the many-particle system. We shall explore these matters in later chapters, beginning with the Fermi-Dirac and Bose-Einstein problems in Chapter 9.

EXAMPLE. *Accuracy of the expansion of $\Delta\sigma$.* For $N = 0$ the series expansion (9) for the entropy of the reservoir may be written as

$$\sigma(N_0, U_0 - \epsilon) = \sigma(N_0, U_0) - \epsilon \left(\frac{\partial\sigma}{\partial U_0} \right)_{N_0} + \tfrac{1}{2}\epsilon^2 \left(\frac{\partial^2\sigma}{\partial U_0^2} \right)_{N_0} + \cdots , \tag{16}$$

where now we have retained the term of order ϵ^2. We may write this result as

$$\sigma(N_0, U_0 - \epsilon) = \sigma(N_0, U_0) - \frac{\epsilon}{T} + \tfrac{1}{2}\epsilon^2 \frac{\partial}{\partial U} \left(\frac{1}{T} \right) + \cdots . \tag{17}$$

We remark that

$$\left[\frac{\partial}{\partial U_0} \left(\frac{1}{T} \right) \right]_{N_0} = -\frac{1}{T^2} \left(\frac{\partial T}{\partial U_0} \right)_{N_0} . \tag{18}$$

The ratio of the third term to the second term on the right-hand side of (17) is

$$\frac{\epsilon}{2T} \left(\frac{\partial T}{\partial U_0} \right)_{N_0} . \tag{19}$$

The entropy that we expanded in (16) and (17) is the entropy of the reservoir, hence the derivative $\partial T/\partial U_0$ also refers to the reservoir. The reciprocal is $\partial U_0/\partial T$, and this increases in value without limit as the size of the reservoir is increased without limit. Thus $\partial T/\partial U_0 \to 0$ as the size of the reservoir is increased, and the quantity in (19) also $\to 0$.

We see that the term in ϵ in the expansion (16) becomes the dominant term for a sufficiently large reservoir (Fig. 3). This argument justifies the neglect of the higher-order terms in (9) and (10).

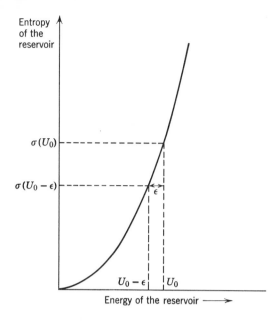

Entropy of the reservoir

$\sigma(U_0)$

$\sigma(U_0 - \epsilon)$

ϵ

$U_0 - \epsilon$ U_0

Energy of the reservoir ⟶

Figure 3   The variation of σ with U_0 gives the change of entropy when the reservoir transfers energy ϵ to the system. The fractional effect of the transfer on the reservoir is small when the reservoir is large: a large reservoir tends to have a high entropy.

GRAND SUM

The sum of Gibbs factors taken over all states of the system for all numbers of particles is a most useful function to have in calculations. We observe first that the sum is the normalizing factor that converts relative probabilities to absolute probabilities:

$$\mathcal{Z}(\mu, T) = \sum_{N=0}^{\infty} \sum_{l} e^{[N\mu - \epsilon_l(N)]/T} \quad . \tag{20}$$

This is called[3] the **grand sum**. Here $\mathcal{Z}$ is the script capital letter z. The sums are to be carried out over all states of the system for all numbers of particles. We have written ϵ_l as $\epsilon_l(N)$ to emphasize the dependence of the state on the number of particles N. *Beware:* if two states are degenerate in energy, both will contribute separate identical terms $e^{[N\mu - \epsilon_l(N)]/T}$ which must be included in the grand sum.

The absolute probability that we will find the system in a state N_1, ϵ_1 is given by the Gibbs factor divided by the grand sum:

$$P(N_1, \epsilon_1) = \frac{e^{[N_1\mu - \epsilon_1(N_1)]/T}}{\mathcal{Z}} \quad . \tag{21}$$

[3] Also called the **grand partition function**; perhaps a better name would be Gibbs sum.

This applies to a system that is at temperature T and chemical potential μ. To prove (21), first observe that the ratio of any two P's is consistent with our central result (13) for the Gibbs factors:

$$\frac{P(N_1, \epsilon_1)}{P(N_2, \epsilon_2)} = \frac{e^{(N_1\mu - \epsilon_1)/T}}{e^{(N_2\mu - \epsilon_2)/T}} \ . \tag{22}$$

Thus (21) gives the correct relative probabilities for the states N_1, ϵ_1 and N_2, ϵ_2. Second, observe that the sum of the probabilities of all states of the system is unity:

$$\sum_N \sum_l P(N, \epsilon_l) = \frac{\sum_N \sum_l e^{(N\mu - \epsilon_l)/T}}{\mathcal{Z}} = \frac{\mathcal{Z}}{\mathcal{Z}} = 1 \ , \tag{23}$$

by the definition of $\mathcal{Z}$. Thus (21) gives the correct absolute probability.

Average values over the systems in the ensemble are easily defined, as we saw in Chapter 3. We use the notation $\langle A \rangle$ to denote the average of a physical quantity A taken over the ensemble of systems. We call $\langle A \rangle$ the **thermal average** or **ensemble average** of A. If $A(N, l)$ is the value of A when the system has N particles and is in the quantum state l, then the ensemble average of A is

$$\langle A \rangle = \sum_N \sum_l A(N, l) P(N, \epsilon_l) = \frac{\sum_N \sum_l A(N, l) e^{(N\mu - \epsilon_l)/T}}{\mathcal{Z}} \ . \tag{24}$$

We shall use this result to calculate $\langle A \rangle$. Several important applications of (24) follow.

Number of particles. The number of particles in the system can vary because the system is in diffusive contact with a reservoir. The ensemble average of the number of particles in the system is

$$\langle N \rangle = \frac{\sum_N \sum_l N e^{(N\mu - \epsilon_l)/T}}{\mathcal{Z}} \ , \tag{25}$$

according to (24). To obtain the numerator, each Gibbs factor in the grand sum has been multiplied by N.

We can write (25) for $\langle N \rangle$ in forms more convenient for calculation. We note from the definition of $\mathcal{Z}$ that

$$\frac{\partial \mathcal{Z}}{\partial \mu} = \frac{1}{T} \sum_N \sum_l N e^{(N\mu - \epsilon_l)/T} \ , \tag{26}$$

whence we have the general relation

$$\langle N \rangle = T \frac{\partial \mathcal{Z}/\partial \mu}{\mathcal{Z}} = T \frac{\partial \log \mathcal{Z}}{\partial \mu} \ . \tag{27}$$

The thermal average number of particles is easily found from the grand sum $\mathfrak{Z}$ by direct use of (27). When no confusion arises, we shall write N for the thermal average $\langle N \rangle$.

Chemists often employ the handy notation

$$\lambda \equiv e^{\mu/T} , \tag{28}$$

where λ is called the **absolute activity.**[4] Here λ is the Greek letter lambda. The grand sum is written as

$$\mathfrak{Z} = \sum_N \sum_l \lambda^N e^{-\epsilon_l/T} , \tag{29}$$

and the ensemble average number of particles is

$$\boxed{\langle N \rangle = \lambda \frac{\partial}{\partial \lambda} \log \mathfrak{Z} .} \tag{30}$$

This relation is useful. With N written for $\langle N \rangle$, it will appear as $N = \lambda \frac{\partial}{\partial \lambda} \log \mathfrak{Z}$.

In many actual problems we determine λ by finding the value that makes $\langle N \rangle$ come out equal to the given number of particles; see, for example, Chapter 11.

Energy. The thermodynamic average of the energy of the system is

$$\langle \epsilon \rangle = \frac{\displaystyle\sum_N \sum_l \epsilon_l e^{\beta(N\mu - \epsilon_l)}}{\mathfrak{Z}} , \tag{31}$$

where we have temporarily introduced the notation $\beta \equiv 1/T$. We shall usually write U for $\langle \epsilon \rangle$: $U \equiv \langle \epsilon \rangle$. Now observe that

$$\langle N\mu - \epsilon \rangle = \langle N \rangle \mu - U = \frac{1}{\mathfrak{Z}} \frac{\partial \mathfrak{Z}}{\partial \beta} = \frac{\partial}{\partial \beta} \log \mathfrak{Z} , \tag{32}$$

so that (27) and (31) may be combined to give

$$U = \left(\mu T \frac{\partial}{\partial \mu} - \frac{\partial}{\partial \beta} \right) \log \mathfrak{Z} . \tag{33}$$

A simpler expression widely used in calculations is obtained in (36) below.

---

[4] We shall find in Chapter 11 that λ for an ideal gas is directly proportional to the concentration; it is also proportional to $1/T^{\frac{3}{2}}$. Thus λ is high when the concentration is high and the temperature is low. We notice that for a state of zero energy the Gibbs factor is $P(N, 0) = \lambda^N/\mathfrak{Z}$.

PARTITION FUNCTION

The expression (33) for the average energy is not particularly tidy, although its use is entirely practical. When the number of particles in the system is fixed, it is of advantage to consider the function

$$Z(N, T) = \sum_l e^{-\epsilon_l/T} .$$

(34)

This is called the **partition function.** The summation is over the Boltzmann factor for all states l for which the number of particles in the system is constant and equal to N. The ratio of successive terms in the sum is consistent with (14) which introduced the Boltzmann factor.

Just as the grand sum is the proportionality factor between $P(N, \epsilon_l)$ and the Gibbs factor, so the partition function is the proportionality factor between the probability $P(\epsilon_l)$ and the Boltzmann factor $\exp{(-\epsilon_l/T)}$:

$$P(\epsilon_l) = \frac{e^{-\epsilon_l/T}}{Z} .$$

(35)

The average energy for a fixed number of particles is

$$U \equiv \langle \epsilon \rangle = \frac{\sum_l \epsilon_l e^{-\epsilon_l/T}}{Z} = \frac{T^2}{Z} \frac{\partial Z}{\partial T} = T^2 \frac{\partial}{\partial T} \log Z ,$$

(36)

where Z is the partition function. The average energy in (36) is for an ensemble of systems which can exchange energy but not particles with the reservoir. We use the same notation $\langle \cdots \rangle$ for the average value over the ensemble with thermal contact as for the average value over the ensemble with both thermal and diffusive contact.

EXAMPLE. *Energy, heat capacity, and entropy of a two state system.* We treat a system of N independent particles. Each particle has two states, one of energy 0 and one of energy ϵ. We want to find the energy and the heat capacity of the system as a function of the temperature T.

We consider a single particle in thermal contact with a reservoir at temperature $\mathcal{T}$. The partition function for the two states of the particle is

$$Z = 1 + e^{-\epsilon/\mathcal{T}} . \tag{37}$$

The average energy of the single particle is

$$\langle \epsilon \rangle = \frac{0 \cdot 1 + \epsilon \cdot e^{-\epsilon/\mathcal{T}}}{Z} = \epsilon \frac{e^{-\epsilon/\mathcal{T}}}{1 + e^{-\epsilon/\mathcal{T}}} , \tag{38}$$

which may be calculated independently by (36).

The thermal average energy of the system of N independent particles is just N times $\langle \epsilon \rangle$ for a single particle:

$$U = N\langle \epsilon \rangle = N\epsilon \frac{e^{-\epsilon/\mathcal{T}}}{1 + e^{-\epsilon/\mathcal{T}}} = \frac{N\epsilon}{e^{\epsilon/\mathcal{T}} + 1} . \tag{39}$$

This function is plotted in Fig. 4.

The **heat capacity** C_V of the system at constant volume is defined as

$$C_V \equiv \left(\frac{\partial U}{\partial T} \right)_{N, V} . \tag{40}$$

From (39) and (40) we have

$$C_V = Nk_B\epsilon \frac{\partial}{\partial \mathcal{T}} \frac{1}{e^{\epsilon/\mathcal{T}} + 1} = Nk_B \left(\frac{\epsilon}{\mathcal{T}} \right)^2 \frac{e^{\epsilon/\mathcal{T}}}{(e^{\epsilon/\mathcal{T}} + 1)^2} . \tag{41}$$

For practical reasons we have used the conventional definition of C_V as $(\partial U/\partial T)_V$, where $T = \mathcal{T}/k_B$, with k_B as the Boltzmann constant. It would have been tidier, but less useful, to have defined C_V as $(\partial U/\partial \mathcal{T})_V$. We defer until Chapter 8 the discussion of the exact definition of the conventional absolute temperature T.

The result for U and for C_V is plotted in Fig. 4. We notice a hump in the plot of heat capacity versus temperature. Such a hump is called a **Schottky anomaly**; it is often of help in the determination of energy levels in solids.

In the high temperature limit the temperature is large in comparison with the energy level spacing ϵ. For $\mathcal{T} \gg \epsilon$ the heat capacity becomes

$$C_V \cong \tfrac{1}{4}Nk_B \left(\frac{\epsilon}{\mathcal{T}} \right)^2 . \tag{42}$$

Notice that $C_V \propto \mathcal{T}^{-2}$ in this high temperature limit. In the low temperature limit the temperature is small in comparison with the energy level spacing. For $\mathcal{T} \ll \epsilon$ we have

$$C_V \cong Nk_B(\epsilon/\mathcal{T})^2 e^{-\epsilon/\mathcal{T}} . \tag{43}$$

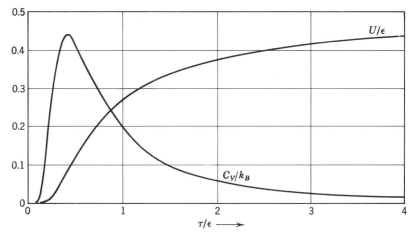

Figure 4   Energy and heat capacity of a two state system as functions of the temperature T in units of the energy splitting ϵ. The energy is plotted in units of ϵ, and the heat capacity is plotted in units of k_B.

The presence of the exponential factor $e^{-\epsilon/T}$ reduces C_V rapidly as T decreases, for $e^{-1/x} \to 0$ as $x \to 0$.

The entropy[5] is obtained from the definition

$$\frac{1}{T} = \left(\frac{\partial \sigma}{\partial U}\right)_N ; \qquad \sigma(U) = \int_0^U \frac{dU}{T} . \qquad (44)$$

To evaluate the integral we first solve (39) for $1/T$ in terms of the energy U:

$$\frac{\epsilon}{T} = \log\left(\frac{N\epsilon}{U} - 1\right) = \log(N\epsilon - U) - \log U . \qquad (45)$$

We perform the integration in (44) to find

$$\int_0^U \frac{dU}{T} = \frac{1}{\epsilon}[-(N\epsilon - U)\log(N\epsilon - U) + (N\epsilon - U) - U\log U + U]_0^U , \qquad (46)$$

or

$$\sigma(U) = \frac{1}{\epsilon}[N\epsilon \log N\epsilon - U\log U - (N\epsilon - U)\log(N\epsilon - U)] . \qquad (47)$$

In Fig. 5 we plot this function for the special case $N = 1$ and $\epsilon = 1$:

$$\sigma(U) = [-U\log U - (1 - U)\log(1 - U) . \qquad (48)$$

---

[5] We most often use the entropy as a property of the reservoir. When we speak of the entropy of a system it is often in the sense of the system when used as a reservoir and composed of a large number of particles. If the system consists of one or a few particles, we imagine a collection of a large number of similar systems in mutual contact and employed as a reservoir.

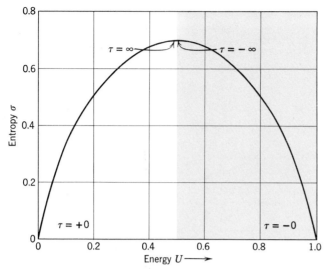

Figure 5 Entropy as function of energy for a two state system. The separation of the states is $\epsilon = 1$ in this example. In the left-hand side of the figure $\partial\sigma/\partial U$ is positive, so that T is positive. On the right-hand side $\partial\sigma/\partial U$ is negative and T is negative.

We usually want to express the entropy as a function of T, rather than as a function of U, because $\sigma(T)$ is more accessible experimentally. We form the differential

$$dU = \frac{dU}{dT}\, dT \ , \tag{49}$$

whence (44) becomes

$$\sigma(T) = \int_0^T dT \,\frac{1}{T}\frac{dU}{dT} \ . \tag{50}$$

In terms of the conventional entropy $S \equiv k_B\sigma$ we have

$$S(T) = \int_0^T dT \,\frac{C_V}{T} \ , \tag{51}$$

where C_V is the heat capacity $(\partial U/\partial T)_{N,\,V}$.

We now evaluate $\sigma(T)$ for the two state system. It is convenient to integrate (50) by parts to obtain

$$\sigma(T) = \left[\frac{U}{T}\right]_0^T + \int_0^T dT \,\frac{U}{T^2} \ . \tag{52}$$

The integral on the right-hand side is, by (39) with $x \equiv \epsilon/T$ and $dx = -(\epsilon/T^2)\, dT$,

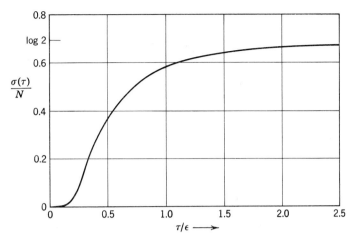

Figure 6   Entropy of a two state system as a function of T/ϵ. Only positive temperatures are plotted. Notice that $\sigma(T) \to N \log 2$ as $T \to \infty$.

$$N \int_0^T dT \; \frac{\epsilon}{T^2} \frac{1}{e^{\epsilon/T} + 1} = -N \int_\infty^{\epsilon/T} dx \; \frac{1}{e^x + 1}$$

$$= N [\log (1 + e^{-x})]_\infty^{\epsilon/T} = N \log (1 + e^{-\epsilon/T}) \; . \quad (53)$$

Thus

$$\sigma(T) = N \left[\frac{\epsilon/T}{e^{\epsilon/T} + 1} + \log (1 + e^{-\epsilon/T}) \right], \quad (54)$$

where we have used the result that $U/T \to 0$ as $T \to 0$. The function $\sigma(T)$ is plotted in Fig. 6. As $T \to \infty$ we have $\sigma \to N \log 2 = \log 2^N$; in this limit all 2^N states are accessible.

In Chapter 18 we shall find another and more convenient method for finding the entropy directly as a function of T, starting from the partition function [see (18.13) and (18.18)]. We shall find $\sigma = -\partial F/\partial T$, where the free energy $F = -T \log Z$ as in (7.53).

EXAMPLE. *Grand sum for two independent systems.* If $\mathcal{Z}_i$ is the grand sum for a system i and $\mathcal{Z}_j$ is the grand sum for an independent system j, where the two systems i, j are in thermal and diffusive contact with a reservoir at temperature T and chemical potential μ, then

$$\mathcal{Z} = \mathcal{Z}_i \mathcal{Z}_j \quad (55)$$

is the grand sum of the combined systems i and j.

Proof. The probability that system i has N_i particles and is in a state

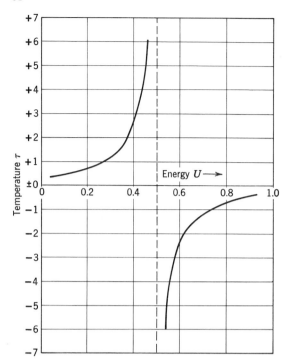

Figure 7 Temperature versus energy for the two state system of Fig. 5. Here

$$T = \frac{1}{\left(\dfrac{\partial \sigma}{\partial U}\right)_N} = \frac{1}{\log \dfrac{1-U}{U}} \, .$$

Notice that the energy is not a maximum at $T = +\infty$, but is a maximum at $T = -0$.

$\epsilon_l(N_i)$, while at the same time system j has N_j particles and is in a state $\epsilon_m(N_j)$, is the product of the separate probabilities:

$$P(N_i, N_j; \epsilon_l, \epsilon_m) = P(N_i, \epsilon_l)P(N_j, \epsilon_m) = \frac{e^{(N_i\mu - \epsilon_l)/T}}{\mathfrak{Z}_i} \times \frac{e^{(N_j\mu - \epsilon_m)/T}}{\mathfrak{Z}_j}$$

$$= \frac{e^{[(N_i + N_j)\mu - (\epsilon_l + \epsilon_m)]/T}}{\mathfrak{Z}_i \mathfrak{Z}_j} \, . \tag{56}$$

Thus the product $\mathfrak{Z} = \mathfrak{Z}_i \mathfrak{Z}_j$ is the grand sum for the Gibbs factor of the combined systems i and j.

NEGATIVE TEMPERATURE[6]

Equation (48) as plotted in Fig. 5 has a region in which $(\partial \sigma/\partial U)_N$ is negative. In this region T must be negative, if we take the definition of T literally. The negative temperature region is shown in Fig. 7. In terms of the Boltzmann factors for the two states we have the population ratio

$$\frac{P(\text{upper})}{P(\text{lower})} = e^{-\epsilon/T} \, . \tag{57}$$

---

[6] This section may be saved for a second reading of the text.

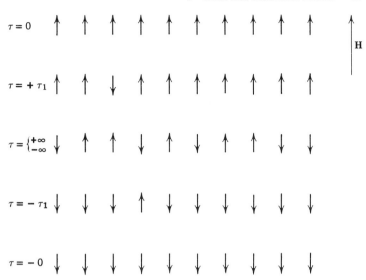

Figure 8 Possible spin distributions for various positive and negative temperatures. The magnetic field is directed upward. The negative spin temperatures cannot last indefinitely because of weak coupling between spins and the lattice. The lattice can only be at a positive temperature because its energy level spectrum is unbounded on top. The downward-directed spins, as at $\tau = -\tau_1$, turn over one by one, thereby releasing energy to the lattice and approaching equilibrium with the lattice at a common positive temperature. A nuclear spin system at negative temperature may relax quite slowly, over a time of minutes or hours; during this time experiments at negative temperatures may be carried out.

Negative τ means that the population of the upper state is greater than the population of the lower state. When this condition obtains we say that the population is inverted, as illustrated in Fig. 8.

The concept of negative temperature is physically meaningful for a system that satisfies the following restrictions: (a) There must be a finite upper limit to the spectrum of energy states, for otherwise a system at a negative temperature would have an infinite energy. A freely moving particle or a harmonic oscillator cannot have negative temperatures, for there is no upper bound on their energies. Thus only certain degrees of freedom of a particle can be at a negative temperature: the nuclear spin orientation in a magnetic field is the degree of freedom most commonly considered in experiments at negative temperatures. (b) The system must be in internal thermal equilibrium. This means the states must have occupancies in accord with the Boltzmann factor taken for the appropriate negative temperature. (c) The states that are at a negative temperature must be isolated and inaccessible to those states of the body that are at a positive temperature.

We shall see in later chapters that the ordinary translational and vibrational degrees of freedom of a body have an entropy that increases without limit as the energy increases, in contrast to the two state or spin system of

Fig. 5. If σ increases without limit, then τ is always positive. The exchange of energy between a system at a negative temperature and a system that can only have a positive temperature (because of an unbounded spectrum) will lead always to an equilibrium configuration in which both systems are at a positive temperature.

Negative temperatures correspond to higher energies than positive temperatures. When a system at a negative temperature is brought into contact with a system at a positive temperature, energy will be transferred from the negative temperature to the positive temperature. Negative temperatures are *hotter* than positive temperatures.

The temperature scale from cold to hot runs $+0$ K, ..., $+300$ K, ..., $+\infty$ K, $-\infty$ K, ..., -300 K, ..., -0 K. Note that if a system at -300 K is brought into thermal contact with an identical system at 300 K, the final equilibrium temperature is not 0 K, but is $\pm\infty$ K.

Nuclear and electron spin systems can be promoted to negative temperatures by suitable radio frequency techniques. If a spin resonance experiment is carried out on a spin system at negative temperature, resonant emission of energy is obtained instead of resonant absorption.[7] A negative temperature system is useful as an rf amplifier in radio astronomy where weak signals must be amplified.

Abragam and Proctor[8] have carried out an elegant series of experiments on calorimetry with systems at negative temperatures. Working with a LiF crystal, they established one temperature in the system of Li nuclear spins and another temperature in the system of F nuclear spins. In a strong static magnetic field the two thermal systems are essentially isolated, but in the earth's magnetic field the energy levels overlap and the two systems rapidly approach equilibrium among themselves (mixing). It is possible to determine the temperature of the systems before and after the systems are allowed to mix. Abragam and Proctor found that if both systems were initially at positive temperatures they attained a common positive temperature on being brought into thermal contact (mixing). If both systems were prepared initially at negative temperatures, they attained a common negative temperature on being brought into thermal contact. If prepared one at a positive temperature and the other at a negative temperature, then an intermediate temperature was attained on mixing, warmer than the initial positive temperature and cooler than the initial negative temperature.

Further References on Negative Temperature

N. F. Ramsey, "Thermodynamics and statistical mechanics at negative absolute temperature," Physical Review **103**, 20 (1956).

M. J. Klein, "Negative absolute temperature," Physical Review **104**, 589 (1956).

[7] E. M. Purcell and R. V. Pound, Physical Review **81**, 279 (1951).
[8] A. Abragam and W. G. Proctor, Physical Review **106**, 160 (1957); **109**, 1441 (1958).

Problem 1. *Concentration fluctuations.* The number of particles is not constant in a system in diffusive contact with a reservoir. We have seen that

$$\langle N \rangle = \frac{T}{\mathfrak{Z}} \frac{\partial \mathfrak{Z}}{\partial \mu} , \tag{58}$$

from (27). (a) Show that

$$\langle N^2 \rangle = \frac{T^2}{\mathfrak{Z}} \frac{\partial^2 \mathfrak{Z}}{\partial \mu^2} . \tag{59}$$

The mean square deviation $\langle (\Delta N)^2 \rangle$ of N from $\langle N \rangle$ is defined by

$$\langle (\Delta N)^2 \rangle = \langle (N - \langle N \rangle)^2 \rangle = \langle N^2 \rangle - 2\langle N \rangle\langle N \rangle + \langle N \rangle^2 = \langle N^2 \rangle - \langle N \rangle^2 , \tag{60}$$

or, by (58) and (59),

$$\langle (\Delta N)^2 \rangle = T^2 \left[\frac{1}{\mathfrak{Z}} \frac{\partial^2 \mathfrak{Z}}{\partial \mu^2} - \frac{1}{\mathfrak{Z}^2} \left(\frac{\partial \mathfrak{Z}}{\partial \mu} \right)^2 \right] . \tag{61}$$

(b) Show that (61) may be written as

$$\langle (\Delta N)^2 \rangle = T \frac{\partial \langle N \rangle}{\partial \mu} . \tag{62}$$

In Chapter 11 we shall apply this result to the ideal gas to find that

$$\frac{\langle (\Delta N)^2 \rangle}{\langle N \rangle^2} = \frac{1}{\langle N \rangle}$$

is the mean square fractional fluctuation in the population of an ideal gas in diffusive contact with a reservoir. Now $\langle N \rangle$ may well be of the order of 10^{20} atoms, so that the fractional fluctuation is exceedingly small. We conclude that in such a system the number of particles is well-defined even though it is not held rigorously constant.

Problem 2. *Magnetic susceptibility.* Use the partition function to find an exact expression for the magnetization and the susceptibility as a function of temperature and magnetic field for the model system of magnetic moments in a magnetic field. We treated this problem in Chapter 4 in the approximation that the fractional magnetization was small. The exact result for the magnetization is

$$\mathfrak{M} = N\mu_0 \tanh (\mu_0 H/T)$$

and is plotted in Fig. 9. The susceptibility may be defined as $\chi = d\mathfrak{M}/dH$, where χ is the Greek letter chi.

Hint. We can consider a system of a single moment having energy states $\pm\mu_0 H$. The single spin is in thermal contact with a reservoir at temperature

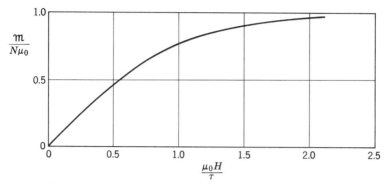

Figure 9 Plot of the total magnetic moment as a function of $\mu_0 H/T$. Notice that at low H/T the moment is a linear function of H/T, but at high H/T the moment tends to saturate.

T. The fractional magnetization of the single moment will be found to be $\tanh(\mu_0 H/T)$. This is also the fractional magnetization of the set of N magnetic moments, because on our model they do not interact with each other.

Problem 3. *Energy fluctuations and energy as function of temperature.* Give an alternate proof to that of Chapter 4 that the thermal average energy $U \equiv \langle\epsilon\rangle$ of any system increases as the temperature increases. It is convenient to consider a system of fixed volume in thermal, but not diffusive, contact with a reservoir. (a) Show first that

$$\langle(\epsilon - \langle\epsilon\rangle)^2\rangle = T^2\left(\frac{\partial U}{\partial T}\right)_V . \tag{63}$$

Hint. Use the partition function Z to relate $\partial U/\partial T$ to the mean square fluctuation. (b) Complete the proof.

Note. We shall find it convenient to treat the temperature T of a system as a quantity that does not fluctuate in value when the system is in thermal contact with a reservoir. This attitude is consistent with our definition of

$$\frac{1}{T} = \left(\frac{\partial \sigma}{\partial U}\right)_N , \tag{64}$$

where $(\partial\sigma/\partial U)_N$ is evaluated in the most probable or equilibrium configuration of the system. The energy of the system may fluctuate, but the temperature does not. Some workers do not give as strict a definition of temperature. Thus Landau and Lifshitz[9] give the result

$$\langle(\Delta T)^2\rangle = \frac{k_B T^2}{C_V} , \tag{65}$$

[9] L. D. Landau and E. M. Lifshitz, *Statistical physics*, Pergamon, 1958, Eq. (111.6).

but this may be viewed as just another form of (63). For $\Delta U = C_V \Delta T$, whence (64) becomes

$$\langle (\Delta U)^2 \rangle = k_B T^2 C_V , \tag{66}$$

which is our result (63).

Problem 4. *Grand sum for a two level system.* (a) Consider a system that may be unoccupied with energy zero or occupied by one particle in either of two states, one of energy zero and one of energy ϵ. Show that the grand sum for this system is

$$\mathfrak{Z} = 1 + \lambda + \lambda e^{-\epsilon/T} .$$

Our assumption excludes the possibility of one particle in each state at the same time.

(b) Show that the thermal average occupancy of the system is

$$\langle N \rangle = \frac{\lambda + \lambda e^{-\epsilon/T}}{\mathfrak{Z}} . \tag{67}$$

(c) Show that the thermal average occupancy of the state at energy ϵ is

$$\langle N(\epsilon) \rangle = \frac{\lambda e^{-\epsilon/T}}{\mathfrak{Z}} . \tag{68}$$

(d) Find an expression for the thermal average energy of the system.

(e) If we now allow the possibility that each state may be occupied by one particle at the same time, show that

$$\mathfrak{Z} = 1 + \lambda + \lambda e^{-\epsilon/T} + \lambda^2 e^{-\epsilon/T} = (1 + \lambda)(1 + \lambda e^{-\epsilon/T}) . \tag{69}$$

Note that $\mathfrak{Z}$ can be factored as shown, so that now the two states are in effect independent of each other.

Problem 5. *Harmonic oscillator.* A one-dimensional harmonic oscillator in quantum mechanics has an infinite series of equally spaced energy states, with

$$\epsilon_s = s\hbar\omega , \tag{70}$$

where s is a positive integer or zero and ω is the classical frequency of the oscillator. We have chosen the zero of energy at the state $s = 0$, but for some purposes it is more useful to write $\epsilon_s = (s + \tfrac{1}{2})\hbar\omega$, where $\tfrac{1}{2}\hbar\omega$ is called the zero-point energy.

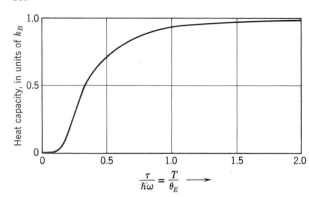

Figure 10 Heat capacity versus temperature for harmonic oscillator of frequency ω. The horizontal scale is in units of $T/\hbar\omega$, which is identical with T/θ_E, where θ_E is called the **Einstein temperature**. In the high temperature limit $C_V \to k_B$. This value is known as the classical value. At low temperatures C_V decreases exponentially.

(a) Find the partition function and show that it may be summed to give

$$Z = \frac{1}{1 - e^{-\hbar\omega/T}} . \tag{71}$$

(b) Show that the ensemble-average energy is

$$U = \frac{\hbar\omega}{e^{\hbar\omega/T} - 1} . \tag{72}$$

Show that $U \to T$ when $T \gg \hbar\omega$. To include the zero-point energy, we simply add $\frac{1}{2}\hbar\omega$ to the result (72). With this inclusion, show that at high temperatures the energy of the oscillator is more accurately equal to the temperature T.

(c) Show that the heat capacity of a single oscillator is

$$C_V = \left(\frac{\partial U}{\partial T}\right)_V = k_B \left(\frac{\hbar\omega}{T}\right)^2 \frac{e^{\epsilon/T}}{(e^{\epsilon/T} - 1)^2} . \tag{73}$$

This function is plotted in Fig. 10. The zero-point energy makes no contribution to the heat capacity.

(d) The vibrational frequency of the hydrogen molecule H_2 is given in Table 1 as $\nu = 4395$ cm^{-1}. In the vibration the two nuclei move in and out along the line between them. Evaluate the energy in ergs per mole of H_2 at a temperature $T = \theta_E$ such that $k_B \theta_E \equiv \hbar\omega$. *Note.* A "frequency" given in cm^{-1} or wavenumbers is usually understood to be equal to $1/\lambda$, where λ is the wavelength. To convert to radians per second, multiply by $2\pi c$, where c is the speed of light.

The entropy of the harmonic oscillator is discussed in Chapter 15 and again in Chapter 18.

Table 1   **Vibrational Frequencies of Diatomic Molecules**

Molecule	ν, in cm^{-1}	ω, in 10^{14} sec^{-1}	θ_E, in deg
H_2	4395	8.279	6300
He_2	1811	3.411	2600
O_2	1580	2.976	2260
F_2	892	1.680	1280
Cl_2	565	1.064	810
Br_2	323	0.608	460
HF	4139	7.797	5930
HCl	2990	5.632	4280
HBr	2650	4.992	3800
CO	2170	4.088	3110
CN	2069	3.897	2960

Problem 6. *Overhauser effect.* Suppose that by a suitable external mechanical or electrical arrangement one can increase the energy of the heat reservoir by $\alpha\epsilon$ whenever the reservoir passed to the system the quantum of energy ϵ. Here α is some numerical factor, positive or negative. Show that the effective Boltzmann factor for this abnormal system is given by

$$P(\epsilon) \propto e^{-(1-\alpha)\epsilon/T} . \tag{74}$$

This reasoning is the statistical basis of the Overhauser effect whereby the nuclear polarization in a magnetic field can be enhanced above the thermal equilibrium polarization. For further details see A. W. Overhauser, Physical Review **92**, 411 (1953); C. Kittel, Physical Review **95**, 589 (1954); T. Carver and C. P. Slichter, Physical Review **92**, 212 (1953).

Problem 7. *States of positive and negative ionization.* Consider a lattice of fixed hydrogen atoms; suppose that each atom can exist in four states:

State	Number of electrons	Energy
Ground	1	$-\frac{1}{2}\Delta$
Positive ion	0	$-\frac{1}{2}\delta$
Negative ion	2	$\frac{1}{2}\delta$
Excited	1	$\frac{1}{2}\Delta$

Find the quite simple condition that the average number of electrons per atom be unity.

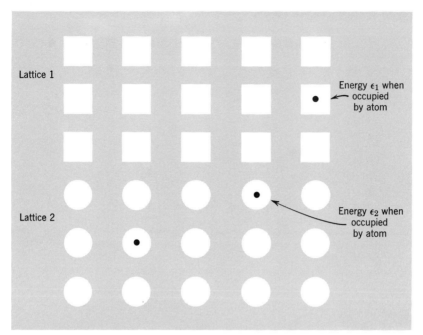

Lattice 1

Energy ϵ_1 when
occupied
by atom

Lattice 2

Energy ϵ_2 when
occupied
by atom

Figure 11 Two lattices in thermal and diffusive contact. Each lattice has N_0 sites. Each site may adsorb one atom. The total number of atoms present is N, and these are distributed between the two different lattices in such a way that the chemical potential of lattice 1 is equal to the chemical potential of lattice 2. The temperatures are also equal.

EXAMPLE. *Two different lattice gases in contact.* Consider as in Fig. 11 two lattices, 1 and 2, each with N_0 independent sites. The energy of an atom when on a site on lattice 1 is ϵ_1; when the atom is on a site on lattice 2 the energy is ϵ_2. The atoms themselves on the two lattices are identical. We want first to find the ratio of the fraction f_1 of occupied sites on lattice 1 to the fraction f_2 of occupied sites on lattice 2. The two lattices are in thermal and diffusive contact.

The grand sum for a single site on lattice 1 is

$$\mathfrak{Z}_1 = 1 + \lambda e^{-\epsilon_1/\tau} , \tag{75}$$

whence

$$f_1 = \frac{\lambda e^{-\epsilon_1/\tau}}{1 + \lambda e^{-\epsilon_1/\tau}} = \frac{1}{\lambda^{-1} e^{\epsilon_1/\tau} + 1} . \tag{76}$$

Similarly, for lattice 2,

$$\mathfrak{Z}_2 = 1 + \lambda e^{-\epsilon_2/\tau} ; \qquad f_2 = \frac{1}{\lambda^{-1} e^{\epsilon_2/\tau} + 1} . \tag{77}$$

The two λ's are identical because the lattices are in diffusive contact. Thus

$$\frac{f_1}{f_2} = \frac{e^{\epsilon_2/T} + \lambda}{e^{\epsilon_1/T} + \lambda} \ . \tag{78}$$

Given the total number of atoms N, we can find λ from

$$N = (f_1 + f_2)N_0 \ . \tag{79}$$

Now suppose that both lattices are sparsely occupied, so that $f_1, f_2 \ll 1$. In this limit

$$f_1 \cong \lambda e^{-\epsilon_1/T} \ ; \qquad f_2 \cong \lambda e^{-\epsilon_2/T} \ , \tag{80}$$

and

$$N \cong \lambda(e^{-\epsilon_1/T} + e^{-\epsilon_2/T})N_0 \ , \tag{81}$$

which may be solved for λ.

It is of interest to solve (80) for λ and the chemical potential $\mu \equiv T \log \lambda$. We have

$$\mu_1 \cong T \log f_1 + \epsilon_1 \ ; \qquad \mu_2 \cong T \log f_2 + \epsilon_2 \ , \tag{82}$$

and the two μ's are of course equal when the lattices are in contact. This form of the chemical potential is characteristic of problems in which the fractional occupancy is low: we find a term of the form $T \log$ (fractional concentration); this term is always negative. There is also a term in the potential energy: a positive energy increases the chemical potential, whereas a negative energy (which corresponds to binding of an atom to the site) lowers the chemical potential.

If after mature reflection and after reviewing the earlier sections you do not understand statements 1 to 3 below, you should seek help. The derivation of the Gibbs factor is the most important derivation in the book.

Consider the model system of N independent spins in a magnetic field:

1. If *nothing* is known about the energy,

(a) the probability of finding the system in *any particular state* is

$$\frac{1}{2^N} \; ; \tag{83}$$

(b) the probability of finding the system with quantum number m is

$$\frac{g(N, m)}{2^N} \; . \tag{84}$$

2. If the energy of the system is known to be $U(m) = -2m\mu_0 H$, the probability of finding the system in *any particular state* is

$$\frac{1}{g(N, m)} \; . \tag{85}$$

3. If the system is in thermal contact with a reservoir at temperature $\mathcal{T}$, the probability of finding the system with quantum number m is

$$\frac{g(N, m)e^{2m\mu_0 H/\mathcal{T}}}{\sum_m g(N, m)e^{2m\mu_0 H/\mathcal{T}}} \; . \tag{86}$$

4. Derive the Gibbs factor

$$P(N, \epsilon_l) = \frac{e^{(N\mu - \epsilon_l)/\mathcal{T}}}{\mathcal{Z}} \; ,$$

starting from the definitions of σ, μ, and $\mathcal{T}$.

CHAPTER 7

Pressure and the Thermodynamic Identity

<hr>

"We are able to distinguish in mechanical terms the thermal action of one system on another from that which we call mechanical in the narrower sense ... so as to specify cases of thermal action and cases of mechanical action." (Gibbs)

"The laws of thermodynamics may easily be obtained from the principles of statistical mechanics, of which they are the incomplete expression." (Gibbs)

PRESSURE AND ENTROPY

We first derive an important relation between the pressure exerted on a system and the change of entropy with volume. The system we consider may be any liquid or gas. A solid is more complicated because the state of elastic strain of a solid is not completely specified by the volume alone. First we prove that the pressure is related to the volume derivative of the energy by

$$p = -\left(\frac{\partial U}{\partial V}\right)_{\sigma, N},$$

where the derivative is taken at constant entropy. We then show that this relation leads to

$$\frac{p}{T} = \left(\frac{\partial \sigma}{\partial V}\right)_{U, N}.$$

This equation resembles the definitions of $1/T$ and $-\mu/T$, but it is **not a definition** of the pressure p. We are not free to define the pressure, for the pressure (or the force) is a mechanical quantity that satisfies the acceleration equation of Newton or the equation for the conservation of energy. Thus the status of p in the theory is quite different from the status of μ or T.

We consider an ensemble of identical closed systems, each of energy U, volume V, and a constant number N of particles. We change the volume of each system from V to $V + \Delta V$, where ΔV is the same for each system. We change the volume slowly enough so that each system remains in its *initial quantum state*. Such a slow change is an extreme example of a reversible process at constant entropy.

A process will be **reversible**[1] if

(a) it is performed quasistatically (see below);
(b) it is not accompanied by dissipative effects, such as turbulence, friction, or electrical resistance.

Examples of **irreversible** processes include turbulent stirring of a viscous liquid; passage of electric charge through a resistor; free expansion of a gas into a vacuum; mixing of two dissimilar gases. A process such as an expansion is **quasistatic** if it is carried out so slowly that at every instant the body can be considered to be in the equilibrium configuration which corresponds to the

---

[1] Another statement is that a reversible process is performed in such a way that if we reverse the direction of the process both the system and the local surroundings may be restored to their initial configurations without any changes in the rest of the universe.

external conditions at that moment. That is, the process is carried out so slowly that the system remains at all times arbitrarily close to equilibrium. In any step of the process the system has time to adjust itself to the new equilibrium configuration.

We consider a process in which a system in a state l at volume V remains in the same state[2] when the volume is increased quasistatically to $V + \Delta V$; the energy of the state l will change from $\epsilon_l(V)$ to $\epsilon_l(V + \Delta V)$. In such a special process the systems in the ensemble remain in their initial states. This is a special case of a quasistatic process at constant entropy—the entropy is constant because the number of states in the ensemble is not changed. If all systems in the ensemble have the same final energy, as for the special example of Fig. 1, then the ensemble that was accurately representative of the system at volume V is also accurately representative at volume $V + \Delta V$.

We expand the energy of a system in the state l:

$$\epsilon_l(V + \Delta V) = \epsilon_l(V) + \frac{\partial \epsilon_l(V)}{\partial V} \Delta V + \cdots \ , \tag{1}$$

to first order in ΔV. By elementary mechanics (Fig. 2) the pressure p_l on a system in the state ϵ_l at volume V is

$$p_l = -\frac{\partial \epsilon_l}{\partial V} \ . \tag{2}$$

If we think of the force F_l exerted on a piston, then

$$F_l = -\frac{\partial \epsilon_l}{\partial x} \ , \tag{3}$$

where x is the displacement of the piston. The partial derivatives are taken at constant number of particles. The result (2) is an expression of the principle of conservation of energy, for $-p_l \, dV$ is the work done on the system in the volume change dV, and $d\epsilon_l$ is the change of energy of the system.

We then write (1) as

$$\epsilon_l(V + \Delta V) \cong \epsilon_l(V) - p_l \, \Delta V \ , \tag{4}$$

so that a volume increase ΔV lowers the energy of the state l by the product of the pressure p_l times the volume change. We average the ϵ_l and p_l in (4) over all the systems in the ensemble to obtain

$$U(V + \Delta V) \cong U(V) - p \, \Delta V = U(V) + \left(\frac{\partial U}{\partial V}\right)_{\sigma, N} \Delta V \ , \tag{5}$$

to first order in ΔV. This establishes the result

$$p = -\left(\frac{\partial U}{\partial V}\right)_{\sigma, N} \ . \tag{6}$$

[2] Thus a free particle (Fig. 1 in an orbital labeled by $n_x = 1$, $n_y = 2$, $n_z = 3$ will remain in the same orbital throughout the volume change.

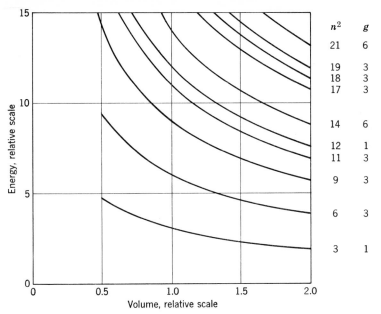

Figure 1 Dependence of energy on volume, for the energy levels of a free particle confined to a cube. The curves are labeled by $n^2 = n_x^2 + n_y^2 + n_z^2$, as in Fig. 1.2. The degeneracies are also given. A volume change here is isotropic: a cube remains a cube.

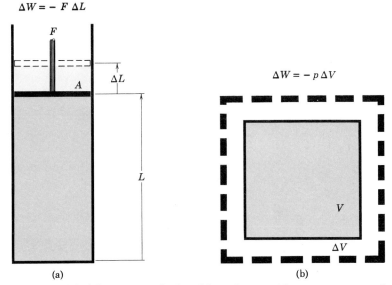

Figure 2 (a) If the piston is displaced by a distance ΔL against an external force $|F|$, the gas does work $|F|\,\Delta L$ on the external world. The work ΔW done *on the gas* is the negative of the work done by the gas, whence

$$\Delta W = -|F|\,\Delta L \ .$$

The volume change is $\Delta V = A\,\Delta L$, where A is the area of the piston. Thus

$$\Delta W = -(|F|/A)\,\Delta V = -p\,\Delta V \ ,$$

where $p \equiv |F|/A$ is the pressure. (b) If the expansion ΔV occurs against a uniform hydrostatic pressure p, the work done on the gas is $\Delta W = -p\,\Delta V$, as in (a).

The fluctuations in the pressure exerted by a macroscopic system are not significant. The fluctuations are discussed in Chapter 11. They are detectable only by especially sensitive experiments, usually with small systems.

At constant number of particles the entropy is a function of the energy and the volume:

$$\sigma = \sigma(U, V) , \tag{7}$$

and the differential[3] of the entropy is

$$d\sigma = \left(\frac{\partial \sigma}{\partial U}\right)_V dU + \left(\frac{\partial \sigma}{\partial V}\right)_U dV . \tag{8}$$

In a process at constant entropy

$$d\sigma = 0 ,$$

and when we divide (8) by dV we obtain

$$\left(\frac{\partial \sigma}{\partial V}\right)_\sigma = 0 = \left(\frac{\partial \sigma}{\partial U}\right)_V \left(\frac{\partial U}{\partial V}\right)_\sigma + \left(\frac{\partial \sigma}{\partial V}\right)_U . \tag{9}$$

By use of the relation (6) and the definition of T we have

$$0 = \frac{(-p)}{T} + \left(\frac{\partial \sigma}{\partial V}\right)_U , \tag{10}$$

whence

$$\boxed{\frac{p}{T} = \left(\frac{\partial \sigma}{\partial V}\right)_{N, U} .} \tag{11}$$

This relates the pressure to the volume dependence of the entropy. It is a consequence of the principle of conservation of energy.

Problem 1. *Equilibrium in mechanical contact.* Two systems are defined to be in mechanical contact when they are free to adjust their volumes V_1, V_2 subject to $V_1 + V_2 = V$, a constant. Show that in equilibrium $p_1 = p_2$ for two systems in thermal and mechanical contact.

---

[3] This expansion is a generalization to a function of two variables $f(x, y)$ of the series expansion of $f(x)$.

THERMODYNAMIC IDENTITY

We now consider the differential of the entropy:

$$d\sigma = \left(\frac{\partial \sigma}{\partial U}\right)_{V,\,N} dU + \left(\frac{\partial \sigma}{\partial N}\right)_{U,\,V} dN + \left(\frac{\partial \sigma}{\partial V}\right)_{U,\,N} dV \ . \tag{12}$$

We use the definitions of μ and T and the result (11) for p to write this expression as

$$d\sigma = \frac{1}{T} dU - \frac{\mu}{T} dN + \frac{p}{T} dV \ . \tag{13}$$

We multiply both sides by T to obtain the **thermodynamic identity**[4]:

$$\boxed{T\, d\sigma = dU - \mu\, dN + p\, dV \ ,} \tag{14}$$

or, in conventional notation,

$$T\, dS = dU - \mu\, dN + p\, dV \ . \tag{15}$$

The equation is valid only for reversible changes because (11) for p/T is valid only for reversible changes.

We have written the thermodynamic identity in terms of the pressure and volume: that is, $-p\, dV$ is the work done on the system. There are many other ways by which an external agency can do work on a system. For example, the stress on a crystal[5] is specified in terms of six independent stress components, and the elastic strain is specified by six independent strain components. The most general expression for the work done on an elastic crystal contains twenty-one terms, in place of the single term $-p\, dV$ for the problem of the gas or liquid. Of special importance to solid state physics and low temperature physics are processes in which work is done on the system by external electric or magnetic fields. We treat these processes in Chapters 22 and 23. For the present it is simpler to develop the subject in terms of $-p\, dV$ as the work done on the system. Let us look, however, at the more general expression of (14).

---

[4] The usage is that of Landau and Lifshitz. The equation is often referred to as the second law of thermodynamics, but in our present approach it is better to reserve this title for the law of increase of entropy (Chapter 4).

[5] The general thermodynamics of elastic crystals is treated by H. B. Callen, *Thermodynamics*, Wiley, 1960, Chapter 13.

If X_ν denotes a generalized force with the property that

$$-\frac{X_\nu}{T} = \left(\frac{\partial\sigma}{\partial x_\nu}\right)_{U,\, x\mu,\, N} \quad , \tag{16}$$

where x_ν is a generalized coordinate, then the thermodynamic identity becomes

$$T\, d\sigma = dU - \mu\, dN - \sum_\nu X_\nu\, dx_\nu \, . \tag{17}$$

For the pressure and volume, the generalized force[6] and coordinate are

$$X \equiv -p \; ; \qquad x \equiv V \, . \tag{18}$$

For a stretched linear polymer, the polymeric chain of Appendix A, we have

$$X = f \; ; \qquad dx = dl \, , \tag{19}$$

where f is the force we apply to one end of the line and dl is the extension of the line. In a reversible process we have

$$T\, d\sigma = dU - \mu\, dN - f\, dl \, . \tag{20}$$

EXAMPLE. *Maxwell relation.* We may write the thermodynamic identity as

$$dU = T\, d\sigma + \mu\, dN - p\, dV \, . \tag{21a}$$

This is equivalent to

$$dU = \left(\frac{\partial U}{\partial\sigma}\right)_{N,\, V} d\sigma + \left(\frac{\partial U}{\partial N}\right)_{\sigma,\, V} dN + \left(\frac{\partial U}{\partial V}\right)_{\sigma,\, N} dV \, , \tag{21b}$$

so that on comparison with (21a) we have

$$\left(\frac{\partial U}{\partial\sigma}\right)_{N,\, V} = T \; ; \qquad \left(\frac{\partial U}{\partial N}\right)_{\sigma,\, V} = \mu \; ; \qquad \left(\frac{\partial U}{\partial V}\right)_{\sigma,\, N} = -p \, . \tag{22}$$

Now $\partial^2 U/\partial V\, \partial\sigma = \partial^2 U/\partial\sigma\, \partial V$, so that we can derive at once from (22) relations such as

$$\left(\frac{\partial T}{\partial V}\right)_{\sigma,\, N} = -\left(\frac{\partial p}{\partial\sigma}\right)_{V,\, N} \, . \tag{23}$$

This relation between thermodynamic quantities is one of a group known as Maxwell relations. Other Maxwell relations are given in Chapters 18 and 19.

---

[6] The minus sign occurs with the pressure in this equivalence because there is a special convention about the sign of the pressure, particularly and most simply in a gas. The pressure is always taken to be positive when the boundary *pushes* inward on the body or medium. The convention is different when we speak of a force, for a positive force *pulls* on the body: an external force applied on the body is positive when directed outward across the boundary of the body. Thus the external force applied to a unit area of the boundary of a gas is opposite in sign to the pressure of the gas.

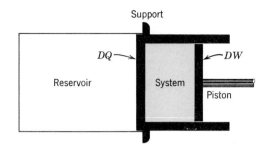

Figure 3 Energy added to the system as a result of thermal contact with the reservoir is called **heat;** energy added by the displacement of the piston is called **work.**

HEAT

What is heat and what is work? We consider a system in thermal contact with a reservoir. We can make changes in the system such that the work done on the system by external agencies is not equal to the increase in the energy of the system. With the number of particles held constant, the difference between the energy change dU and the external work DW done on the system is the **quantity of heat** DQ added to the system in an infinitesimal change:

$$dU - DW = DQ \ , \tag{24}$$

or

$$\boxed{dU = DQ + DW \ .} \tag{25}$$

The first law of thermodynamics states that energy is conserved. Equation (25) is just a statement of the first law. (A summary of the laws of thermodynamics is given at the end of this chapter.)

Heat is added to the system as the result of the transfer of energy by thermal contact from the reservoir to the system (Fig. 3). That is, **energy added to the system by thermal contact with a reservoir is called heat. Energy added by other agencies is called work.** If the system is not in thermal contact with a reservoir, no heat is added. The energy of an isolated system changes only if external work is done, as by the displacement of a piston.

The symbols DQ and DW have the following definitions:

$$DQ \equiv \text{quantity of heat added to the system} \ ; \tag{26}$$
$$DW \equiv \text{amount of external work done on the system} \ . \tag{27}$$

We use the capital D to denote an infinitesimal change as in (25) when the change is not in a well-defined physical property of the system. A change in the energy U, entropy σ, or volume V is a change in a well-defined physical property of the system, and we write dU, $d\sigma$, or dV for the differential. But there is no such property as the *work content* W or *heat content* Q of a system, therefore the notations DW and DQ are used in place of dW and dQ.

Here "well-defined" has a very strong, specific meaning: Any quantity A is well-defined (and will be called a **state variable**) if and only if with two systems physically alike in all macroscopic properties the same value of A is always associated. But W and Q fail this test.

Thus far we have kept the number of particles in the system constant. If now dN particles are added to the system from the reservoir at chemical potential μ, the energy of the system is increased by $\mu\, dN$, for each particle contributes its chemical potential μ to the energy of the system. We call $\mu\, dN$ the amount of **chemical work** DW_c done on the system.

We may combine (14) and (25) to obtain

$$dU = T\, d\sigma + \mu\, dN - p\, dV = DQ + DW_c + DW \ , \tag{28}$$

in a reversible change. The change in the internal energy of the system is the sum of three contributions:

$T\, d\sigma$ from energy transfer by thermal contact with the reservoir ;

$\mu\, dN$ from particle transfer from the reservoir ;

$-p\, dV$ from work done on the system by a piston .

We make the following identifications for reversible changes:

Heat added to system:	$DQ = T\, d\sigma$
Chemical work done on system:	$DW_c = \mu\, dN$
Mechanical work done on system:	$DW = -p\, dV$.

$$\tag{29}$$

For work done by electrical and magnetic fields, see (22.10), (22.15), (23.1), and (23.2).

As an example of the addition of heat and the performance of work, we consider in Fig. 4 the expansion of a gas at constant temperature. In this process heat is added to the system from the reservoir, and work is done by the system on the piston. The curves were calculated with the help of results which are derived in Chapters 11 and 12.

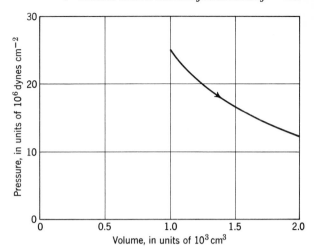

Figure 4a Pressure versus volume for the isothermal expansion of 1 mole (6.02×10^{23} molecules) of an ideal gas at 300 K. The volume is increased from 1 to 2 liters.

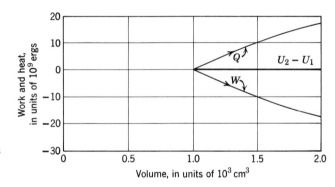

Figure 4b Heat supplied to the system and work done on the system in an isothermal reversible expansion from 1 to 2 liters, for 1 mole of an ideal gas at 300 K. The change in internal energy of the ideal gas is zero. The calculations were carried out using the results of Chapters 11 and 12:

$$Q = \mathcal{T}\,\Delta\sigma = Nk_BT \log\,(V_2/V_1) \ ;$$
$$W = -\textstyle\int p\,dV = -Nk_BT \log\,(V_2/V_1) \ ;$$
$$U_2 - U_1 = Q + W = 0 \ .$$

IRREVERSIBLE PROCESSES

Which of the relations above hold if a change in the system is accomplished by an irreversible process and not by a reversible process? Suppose, for example, that a gas expands into a vacuum (Fig. 5). This is irreversible, in contrast to a reversible expansion in which the gas does work on a piston. Both processes are analyzed quantitatively in Chapter 12.

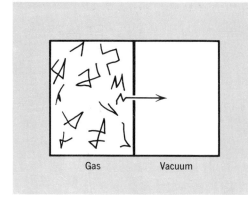

Figure 5 Expansion of a gas into a vacuum. This process is not reversible: if we replace the plug in the opening between the gas and the vacuum, the molecules that have already entered the vacuum will not be restored to their initial configuration in the gas container. This expansion is not quasistatic, for after the initial removal of the plug the molecules are not in the equilibrium configuration of the new larger system, but are in a nonequilibrium configuration with nearly all the molecules on the left-hand side of the new larger system.

We can argue quite generally that the work done on the system to accomplish a given change is always greater for an irreversible process than for a reversible process:

$$W(\text{irreversible}) > W(\text{reversible}) \ . \tag{30}$$

The inequality follows for either of two reasons:

(a) The process is irreversible because of dissipative effects, for then extra work must be provided to overcome the dissipation.

(b) The process is irreversible because it is carried out in an apparatus which does not permit work to be performed by the system. This is illustrated by an example.

Consider the reversible expansion of a gas from volume V_1 to volume V_2 at constant temperature. The work done on the system is

$$W(\text{reversible}) = -\int_{V_1}^{V_2} p \, dV \ , \tag{31}$$

and this quantity will be negative because V_2 is larger than V_1. The expansion of the gas into a vacuum is irreversible. The work done is zero, because the apparatus (Fig. 5) does not permit work to be done by the system (there is no piston):

$$W(\text{irreversible}) = 0 \ . \tag{32}$$

Equations (31) and (32) satisfy the inequality (30).

For either a reversible process or an irreversible process, we always have, provided the number of particles is constant,

$$dU = DQ + DW \ , \tag{33}$$

by virtue of the conservation of energy. But because of the inequality (30), which for a gas or liquid may be written as

$$DW(\text{irreversible}) > -p\,dV\ ,\tag{34}$$

it then follows from (33) that

$$DQ(\text{irreversible}) < T\,d\sigma\ .\tag{35}$$

Both the left- and right-hand sides of (34) and (35) must add up to dU, the difference of energy between the initial and the final configurations of the system.

We have from (35) that

$$
\boxed{
\begin{array}{ll}
T\,d\sigma > DQ & \text{irreversible process}\ , \\[2ex]
T\,d\sigma = DQ & \text{reversible process}\ .
\end{array}
}
\tag{36}
$$

HEAT AND WORK AT FIXED NUMBER OF PARTICLES

It is helpful to consider the difference between heat and work in another way. We treat a system with a constant number of particles in thermal contact with a reservoir at temperature T. The equilibrium value of the energy of the system is

$$U = \sum_l \epsilon_l P_l\ ,\tag{37}$$

where P_l is the probability that the system is in the state l. The value of P_l is given by the Boltzmann factor.

In an infinitesimal quasistatic change in an external parameter (such as the volume) which describes the system we have

$$dU = \underbrace{\sum_l \epsilon_l\,dP_l}_{T\,d\sigma} + \underbrace{\sum_l P_l\,d\epsilon_l}_{-p\,dV}\ .\tag{38}$$

One contribution to dU is from the change in the probabilities P_l and the other is from the change in the energies ϵ_l. The first is heat; the second is work. The term $\Sigma P_l\,d\epsilon_l$ agrees with our result (5) for the mechanical work done on the system. The term $\Sigma \epsilon_l\,dP_l$ must then be identified as $T\,d\sigma$ in (28).

EXAMPLE. *Boltzmann definition of the entropy.* We take the logarithm of the Boltzmann factor

$$P_l = \frac{e^{-\epsilon_l/\tau}}{Z} \tag{39}$$

to obtain

$$\log P_l = -\frac{\epsilon_l}{\tau} - \log Z , \tag{40}$$

$$\epsilon_l = -\tau(\log P_l + \log Z) . \tag{41}$$

This relation between ϵ_l and P_l applies only if the system is in equilibrium. If we are to use (41) to describe changes in the system, the changes must be reversible.

From (38) and (41) we find

$$\tau \, d\sigma = \sum_l \epsilon_l \, dP_l = -\tau \sum_l (\log P_l) \, dP_l - \tau(\log Z) \sum_l dP_l . \tag{42}$$

But the probabilities are normalized with $\Sigma P_l \equiv 1$, so that $\Sigma \, dP_l = 0$. We note further that

$$d \sum_l (P_l \log P_l) = \sum_l (\log P_l) \, dP_l + \sum_l dP_l = \sum_l (\log P_l) \, dP_l . \tag{43}$$

Thus (42) may be written as

$$\tau \, d\sigma = \sum_l \epsilon_l \, dP_l = \tau \, d\left(-\sum_l P_l \log P_l\right) . \tag{44}$$

We may identify the change of entropy (42) as

$$d\sigma = d\left(-\sum_l P_l \log P_l\right) . \tag{45}$$

We may go one step further and identify the entropy as

$$\boxed{\sigma = -\sum_l P_l \log P_l .} \tag{46}$$

If the system is known to be in a nondegenerate ground state, then $P_0 = 1$ and all other P's are zero. Thus $\sigma = -1 \log 1 = 0$. This shows that no additive constant appears on going from (45) to (46).

This relation is known as the **Boltzmann definition of the entropy.**[7] We have derived it from a line of reasoning that goes back to the definition $\sigma = \log g$ of Chapter 4. The form (46) is often useful.

We can apply (46) to a closed system (U constant, N constant): If we have a closed system with g accessible states, then

---

[7] It is sometimes also called the Gibbs entropy.

$$P_l = \frac{1}{g} \tag{47}$$

for each of the g states. Thus

$$- P_l \log P_l = -\frac{1}{g} \log \frac{1}{g} = -\frac{1}{g} (\log 1 - \log g) = \frac{1}{g} \log g , \tag{48}$$

because $\log 1 = 0$. There are g terms equal to this in value, whence (46) becomes

$$\sigma = g \cdot \frac{1}{g} \log g = \log g , \tag{49}$$

which is our original definition.

For a system in thermal and diffusive contact with a reservoir we may obtain the result

$$\sigma = -\sum_N \sum_l P(N, l) \log P(N, l) , \tag{50}$$

by retracing the argument of (38) through (46), with appropriate modifications.

Problem 2. Entropy at constant T and N. Consider a system with a constant number of particles in thermal contact with a reservoir at temperature T. (a) From the result (46) and (6.36) show that

$$\sigma = \frac{U}{T} + \log Z = T \frac{\partial}{\partial T} (T \log Z) . \tag{51}$$

The last is a very useful form for the entropy because only T and not U is involved.

The expression (51) for the entropy may be solved for the partition function:

$$Z = e^{-(U - T\sigma)/T} = e^{-F/T} , \tag{52}$$

where we define

$$F \equiv U - T\sigma . \tag{53}$$

This is the **free energy** as discussed in Chapter 18. (b) Show from the thermodynamic identity that

$$p = -\left(\frac{\partial U}{\partial V}\right)_T + T\left(\frac{\partial \sigma}{\partial V}\right)_T . \tag{53a}$$

It follows from $F \equiv U - T\sigma$ that

$$p = -\left(\frac{\partial F}{\partial V}\right)_T , \tag{53b}$$

so that the free energy F acts as an effective energy in an isothermal change of volume. These results for the pressure are strikingly different from (6) where the derivative was taken at constant entropy. The two terms on the right-hand side of (53a) represent the energy and the entropy contributions to the pressure. The energy contribution $-(\partial U/\partial V)_T$ is dominant in crystals, and the entropy contribution $T(\partial\sigma/\partial V)_T$ is dominant in gases and in elastic polymers (rubber). The presence of the entropy contribution is testimony of the importance of the entropy—the naive feeling that $\partial U/\partial V$ must tell everything about the pressure is seriously incomplete for a process at constant temperature.

Problem 3.[8] *Extremal property of the entropy.* We know that the entropy of a closed system is a maximum when the P_l of all accessible states are equal.

(a) For a system in thermal contact with a reservoir, find the form of P_l such that the entropy according to the Boltzmann definition is an extremum. The extremum is subject to the constraints

$$\Sigma P_l = 1 \ ; \qquad \Sigma\epsilon_l P_l = U$$

The result is

$$P_l = \frac{e^{-\beta\epsilon_l}}{\Sigma e^{-\beta\epsilon_m}} \ .$$

(b) Use the result for P_l to solve for σ and show that $\beta = (\partial\sigma/\partial U)_N$, so that β may be identified with $1/T$.

GENERATION OF THERMODYNAMIC RELATIONS

Throughout the book we shall make use of relations among partial derivatives of thermodynamic quantities. The generation of these relations is a pleasant game, and many relations are helpful in obtaining the quantities of interest from the quantities that are easiest to measure. Suppose that we know the volume as a function of the pressure and the temperature:

$$V = V(p, T) \ . \tag{54}$$

Then

$$dV = \left(\frac{\partial V}{\partial p}\right)_T dp + \left(\frac{\partial V}{\partial T}\right)_p dT \ . \tag{55}$$

---

[8] This problem requires a knowledge of the mathematical method of Lagrangian multipliers.

(a) Consider a change that takes place at constant volume. Then $dV = 0$, and (55) may be written as

$$\left(\frac{\partial V}{\partial p}\right)_T = -\left(\frac{\partial V}{\partial T}\right)_p\left(\frac{\partial T}{\partial p}\right)_V , \qquad (56)$$

or as

$$\left(\frac{\partial V}{\partial T}\right)_p = -\left(\frac{\partial V}{\partial p}\right)_T\left(\frac{\partial p}{\partial T}\right)_V . \qquad (57)$$

(b) We may divide (56) by (57) to find

$$\left(\frac{\partial p}{\partial T}\right)_V = \frac{1}{(\partial T/\partial p)_V} . \qquad (58)$$

(c) Suppose that we want $(\partial V/\partial p)_\sigma$. From (55) we form

$$\left(\frac{\partial V}{\partial p}\right)_\sigma = \left(\frac{\partial V}{\partial p}\right)_T + \left(\frac{\partial V}{\partial T}\right)_p\left(\frac{\partial T}{\partial p}\right)_\sigma . \qquad (59)$$

We defer until Chapter 19 the derivation of a beautiful relationship between the heat capacities at constant pressure and constant volume.

Problem 4. *Dependence of the entropy on temperature.* Use

$$d\sigma = \left(\frac{\partial \sigma}{\partial U}\right)_N dU + \left(\frac{\partial \sigma}{\partial N}\right)_U dN , \qquad (60)$$

and the result of (4.28) to show that

$$\left(\frac{\partial \sigma}{\partial T}\right)_N > 0 \qquad \text{for } \mathcal{T} \text{ positive ;} \qquad (61)$$

$$\left(\frac{\partial \sigma}{\partial T}\right)_N < 0 \qquad \text{for } \mathcal{T} \text{ negative .} \qquad (62)$$

Problem 5. *Experimental determination of the entropy.* How would you determine the entropy of a solid at room temperature? (a) What quantities would you measure experimentally? (b) What apparatus would you need? (c) How would you determine the energy of the solid at room temperature, referred to $U = 0$ at the absolute zero of temperature?

Problem 6. *Elasticity of polymers.* The thermodynamic identity for a one-dimensional system was given in (20):

$$\mathcal{T} \, d\sigma = dU - \mu \, dN - f \, dl \tag{63}$$

when f is the external force exerted on the line and dl is the extension of the line. By analogy with (11) we form the derivative to find

$$-\frac{f}{\mathcal{T}} = \left(\frac{\partial \sigma}{\partial l}\right)_{U,N} . \tag{64}$$

We consider a polymeric chain of N links each of length ρ, with each link equally likely to be directed to the right and to the left. The number of arrangements that give an overall (straight) head-to-tail length of $l = 2|m|\rho$ is found from the generating function (A.3) to be

$$g(N, m) = \frac{2N!}{(\frac{1}{2}N + m)! \, (\frac{1}{2}N - m)!} , \tag{65}$$

essentially as for (2.12), but with m positive.

(a) For $|m| \ll N$ show that

$$\sigma(l) = \log g(N, 0) - \frac{l^2}{2N\rho^2} . \tag{66}$$

(b) Show that the force at extension l is

$$f = \frac{l\mathcal{T}}{N\rho^2} . \tag{67}$$

Notice that the force is proportional to the temperature. The force arises because the polymer wants to curl up: the entropy is higher in a random coil than in an uncoiled configuration. Warming a rubber band (as with a match) makes it contract; warming a steel wire makes it expand.

The theory of rubber elasticity is discussed by H. M. James and E. Guth, Journal of Chemical Physics **11**, 455 (1943); Journal of Polymer Science **4**, 153 (1949); see also L. R. G. Treloar, *Physics of rubber elasticity*, Oxford, 1958.

Summary of the Laws of Thermodynamics

Thermodynamics is often studied as a deductive subject with no reference to the statistical definition of entropy, Eq. (4.18). When thermodynamics is studied in this way a number of postulates are introduced, which are called the laws of thermodynamics. These are:

Zeroth law. If two systems are in thermal equilibrium with a third system, then they must be in thermal equilibrium with each other.

First law. Energy is conserved; heat is a particular form of energy.

Second law. If a closed system is at some instant of time in a macroscopic configuration that is not the equilibrium configuration, then the most probable consequence will be that the entropy of the system will increase monotonically in successive instants of time.

There are many forms[9] of the second law. We have given the form that has the closest connection with the present approach. An important form, the Kelvin postulate, emphasizes the limits on the conversion of heat into work: it is impossible to devise an engine that, working in a cycle, shall produce no effect other than the extraction of heat from a reservoir and the performance of an equal amount of mechanical work.

Third law. The entropy of a system has the property that $\sigma \to \sigma_0$ as $\mathcal{T} \to 0$, where σ_0 is a constant independent of the external parameters which act on the system.

These laws play the following roles in the statistical formulation of thermal physics:

(a) The zeroth law is the basis of the definition of temperature.

(b) The first law is a statement of the conservation of energy.

(c) The second law remains a plausible postulate, although a postulate that corresponds to reality and is confirmed by all our everyday observations.

(d) The third law has no special significance; it is an assumption about the invariance of the degeneracy of the ground level or of the accessible ground level.

The statistical formulation of thermal physics rests on several assumptions:

(a) All accessible states of a closed system are equally likely to occur.

(b) Averages of physical quantities taken over an ensemble are equal to the equilibrium values of the quantities.

(c) In a sufficiently large system the average value of a quantity may be replaced by its value in the most probable configuration of the system. This assumption is convenient, but not essential: with small systems we always calculate the averages themselves. The reservoir is always taken as macroscopic.

(d) The law of increase of entropy (the second law).

[9] Good discussions are given by H. B. Callen, *Thermodynamics*, Wiley, 1960, and A. B. Pippard, *Elements of classical thermodynamics*, Cambridge, 1957, Chapter 4.

CHAPTER 8

The Thermodynamic Temperature

In this chapter we discuss the definition of the conventional absolute temperature scale, called Kelvin thermodynamic temperature scale. A temperature determined on this scale is given in degrees Kelvin or in kelvin, which may be abbreviated as °K or as K. We shall often write deg for degrees Kelvin. We now discuss the relation of a temperature T on the Kelvin scale to the fundamental temperature τ defined in Chapter 4 as

$$\frac{1}{\tau} \equiv \left(\frac{\partial \sigma}{\partial U}\right)_{V, N} .$$

CARNOT CYCLE AND THERMODYNAMIC TEMPERATURE

On first exposure the definition given below of the Kelvin scale will strike the reader as quite peculiar. The definition is in terms of an imaginary experiment. Temperatures are never actually determined in this way, but they are determined in ways that are equivalent in a thermodynamic sense.

We take a fixed quantity of any liquid or gas around the cycle of reversible operations shown in Fig. 1. This describes the **Carnot cycle** which is used to define the Kelvin thermodynamic temperature scale. The special feature of the Carnot cycle which makes it convenient for this purpose is that exchanges of energy occur at two temperatures only. By analogy with the relation $DQ = \tau \, d\sigma$ we *assume* that the Kelvin temperature T has the property $DQ = k_B T \, d\sigma$, for a reversible process; here k_B is a constant to be determined and σ is the entropy. We *define* the conventional entropy S as $S \equiv k_B \sigma$, whence $DQ = T \, dS$.

The steps in the cycle are:

$1 \rightarrow 2$ Isothermal expansion at temperature T_h. Heat Q_h flows into the system from the reservoir, with entropy change $\int_1^2 dS = Q_h/T_h$, with Q_h positive.

$2 \rightarrow 3$ Expansion at constant entropy.

$3 \rightarrow 4$ Isothermal compression at temperature T_c. Heat flows from the system into a reservoir. The entropy change of the system is $\int_3^4 dS = Q_c/T_c$; here Q_c is negative.

$4 \rightarrow 1$ Compression at constant entropy to the original condition 1.

This completes one Carnot cycle.

The net entropy change when the system is taken once around the cycle

128

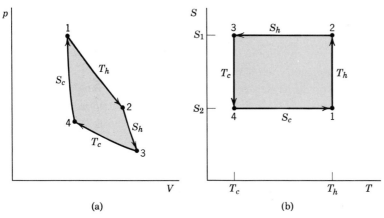

(a)   (b)

Figure 1 A Carnot cycle, illustrated in (a) as a plot of pressure versus volume and in (b) as a plot of entropy versus temperature. The Carnot cycle defined by the vertical and horizontal steps in (b) is applicable to any working substance. Observe that the entropy decreases on the leg $3 \rightarrow 4$. There is no law against the entropy decreasing if the system is not closed. We have to do work on the system to achieve the decrease in entropy on the leg $3 \rightarrow 4$. The sense in which the cycle is traversed determines whether we have a heat engine or a refrigerator. The sense as shown corresponds to a **heat engine:** a quantity of heat is taken in at the temperature T_h and there is output of mechanical work. A smaller quantity of heat is given out at the lower temperature T_c. If the cycle is reversed, we have a **refrigerator:** the system extracts heat from the low temperature reservoir and gives out a larger quantity of heat to the high temperature reservoir. Mechanical work must be done on the refrigerator in order to provide the excess heat given out at T_h.

is zero, because the system ends up in the same configuration 1 in which it began. The temperature, pressure, and volume are restored to their initial values. That is,

$$\oint dS = \int_1^2 dS + \int_3^4 dS = 0 \ . \tag{1}$$

We note that there is no change of entropy along each constant entropy leg $2 \rightarrow 3$ or $4 \rightarrow 1$. If the cycle is carried out reversibly, (1) is equivalent to

$$\frac{Q_h}{T_h} + \frac{Q_c}{T_c} = 0 \ , \tag{2}$$

for

$$Q_h = T_h(S_2 - S_1) = T_h(S_h - S_c) \ ; \qquad Q_c = T_c(S_4 - S_3) = T_c(S_c - S_h) \ , \tag{3}$$

where Q_h is a positive quantity and Q_c is negative. The entropies S_h and S_c are defined by Fig. 1.

The operation of a Carnot cycle as a heat engine and a refrigerator are

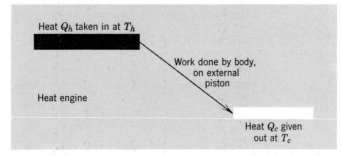

Figure 2   In a heat engine the combustion of fuel leads to heat input to the working substance which expands at the upper temperature T_h. The working substance then does more work in an expansion at constant entropy, and as it does work, it cools. At the lower temperature T_c the heat Q_c is given up to a reservoir.

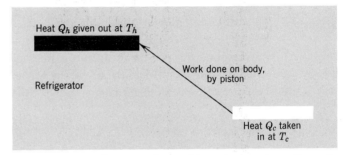

Figure 3   In a refrigerator the working substance in cooling coils take up heat Q_c from the reservoir at the temperature T_c. An external mechanism does mechanical work on the working substance, and the heat Q_h is given up in an isothermal expansion at the higher temperature T_h.

contrasted in Figs. 2 and 3. The only difference is in the sense in which the cycle is carried out.

$$\frac{T_h}{T_c} = \frac{|Q_h|}{|Q_c|} \ . \tag{4}$$

This relation is used to define a **thermodynamic temperature scale** independent of the properties of the particular working substance that is taken around the Carnot cycle. The quantities of heat Q_h, Q_c are accessible to direct experimental measurement. The Carnot cycle is defined by the four steps (two with $\Delta T = 0$; two with $\Delta S = 0$), and because the definition does not depend on the particular liquid or gas used in the cycle, the Carnot cycle makes an excellent definition of the ratio of any two thermodynamic temperatures. To obtain an absolute temperature T_c we only have to assign a value of T_h to some easily established fixed point.

Problem 1. *Efficiency of a Carnot cycle.* We know that $\oint p \, dV$ taken in a clockwise direction is the work performed by a heat engine. (a) Show that the work performed by a heat engine that operates in a Carnot cycle is

$$|W| = \oint p \, dV = Q_h \left(1 - \frac{T_c}{T_h} \right) . \tag{5}$$

The fraction of the heat withdrawn from the reservoir that can be delivered as work is called the **thermodynamic engine efficiency.** For the Carnot heat engine the thermodynamic efficiency is

$$\frac{|W|}{Q_h} = 1 - \frac{T_c}{T_h} . \tag{6}$$

It can be shown[1] that the Carnot efficiency is the maximum efficiency of a cyclic process with which heat can be converted into external work, on the assumption that we throw away the heat Q_c extracted from the system at the temperature T_c. (b) Show that for the Carnot refrigerator the ratio of the heat extracted from the surroundings to the work done on the refrigerator is

$$\frac{Q_c}{W} = \frac{T_c}{T_h - T_c} . \tag{7}$$

The ratio Q_c/W is called the **coefficient of refrigerator performance.** (c) Notice that for the Carnot refrigerator the value of this coefficient may be greater than unity. Why does this conclusion not violate the principle of conservation of energy?

KELVIN THERMODYNAMIC TEMPERATURE SCALE

The **absolute thermodynamic temperature scale** or **Kelvin scale** was defined in 1954 with the triple point[2] of water (Fig. 4) as the fundamental fixed point, by assigning to it the temperature 273.16 K, exactly. All temperatures in this book when given simply as deg or as K are understood to be in degrees Kelvin.

---

[1] Consider a plot in the *T-S* plane of any other heat engine cycle. Such a plot may be represented in terms of a series of Carnot cycle rectangles, as in Fig. 1b. Any rectangle that takes in heat at a temperature below T_h or gives it out above T_c will have a lower efficiency than (6). The maximum efficiency is that of a single cycle with only two temperatures involved; that is, a Carnot cycle.
[2] The triple point of water is the unique temperature and unique pressure at which water, ice, and water vapor coexist. This is discussed in Chapter 20.

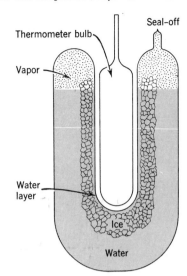

Figure 4 A triple-point cell maintains the thermometer at the temperature at which ice, water, and water vapor coexist in equilibrium. This temperature is defined as 273.16 K. (From *Heat and thermodynamics* by M. W. Zemansky, 5th ed. Copyright © 1968 by McGraw-Hill, Inc. Used by permission of the McGraw-Hill Book Company.)

The zero of the **thermodynamic centigrade scale** is defined as 0.01 K below the triple point of pure water. Thus absolute zero on the centigrade scale, is −273.15 exactly, by definition. The centigrade zero is intended to coincide closely with the equilibrium temperature of ice and air-saturated water, at 1 atm pressure. The name Celsius is recommended by international authorities as the name of the degree, in place of centigrade.

The **international practical temperature scale** is not a thermodynamic scale. It is intended to represent the thermodynamic scale in a practical manner with sufficient accuracy to serve the everyday needs of scientific and industrial laboratories. It is based on six fixed and reproducible equilibrium temperatures, *all under a pressure of one standard atmosphere:*

0 C	Ice point of water (ice and air-saturated water)
100 C	Liquid water and its vapor
−182.970 C	Boiling point of oxygen
444.600 C	Boiling point of sulfur
960.8 C	Freezing point of silver
1063.0 C	Freezing point of gold.

CONNECTION OF T AND $\mathcal{T}$

How can we establish the connection between a known value of the Kelvin temperature T and the corresponding value of the fundamental temperature $\mathcal{T}$? The Kelvin temperature is determined by a Carnot engine that operates between T and the triple point of water. The fundamental temperature is defined by

$$\frac{1}{\mathcal{T}} = \left(\frac{\partial \sigma}{\partial U}\right)_{N, V} , \tag{8}$$

where for a closed system $\sigma = \log g$.

We first recall that the two scales are directly proportional to each other, for $\mathcal{T}$ has been shown in Chapter 7 to have the property $DQ = \mathcal{T} \, d\sigma$ in a reversible process and T is assumed to have the property $DQ = k_B T \, d\sigma$ where k_B is a constant to be determined. Thus

$$\boxed{\mathcal{T} = k_B T .} \tag{9}$$

We call k_B the **Boltzmann constant.**

The form assumed for the connection of DQ and T is equivalent to, and was used in, the definition of T in terms of the Carnot cycle. The Carnot cycle cleverly enables one to determine T without the explicit determination of $d\sigma$.

The experimental determination of the connection between T and $\mathcal{T}$ can be done by means of any system for which $\mathcal{T}$ can be measured at a known temperature T. Among possible systems are:

(a) For an ideal gas of N atoms we know (Chapter 11) that

$$pV = N\mathcal{T} , \tag{10}$$

where p, V, and N may be measured. For real (imperfect) gases there enter small corrections to the ideal gas law (10), and these corrections may be determined by independent experiments.

(b) The ratio of the occupancy probabilities of two states 1 and 2 of a system in thermal contact with a reservoir is

$$\frac{P_1}{P_2} = e^{-(\epsilon_1 - \epsilon_2)/\mathcal{T}} . \tag{11}$$

With paramagnetic systems we can determine P_1 and P_2 by measurement of the magnetic moment of the system, and $\epsilon_1 - \epsilon_2$ may be measured by radio-frequency or microwave spectroscopy. From (11) we calculate $\mathcal{T}$. A number of variations of the method are possible.

Method (a) above is used in practice to determine the Boltzmann constant. The present experimental value of the Boltzmann constant is[3]

$$k_B = 1.38054(\pm 6) \times 10^{-16} \text{ erg deg}^{-1} \text{ .} \qquad (12)$$

The conventional entropy S is defined so that

$$DQ = T\,dS \qquad (13)$$

in a quasistatic process with N constant. But

$$DQ = \mathcal{T}\,d\sigma = k_B T\,d\sigma \text{ ,} \qquad (14)$$

by the definition of T, whence

$$S = k_B\,\sigma \text{ .} \qquad (15)$$

This relates the conventional entropy S to the fundamental entropy σ.

REFERENCES

H. F. Stimson, "Heat units and temperature scales for calorimetry," American Journal of Physics **23**, 614 (1955).

Temperature—Its measurement and control in science and industry, Vol. III, Ed.-in-Chief C. M. Herzfeld; Part I, *Basic concepts, standards and methods*, ed. F. G. Brickwedde, Reinhold, 1962. See particularly the article "The thermodynamic temperature scale, its definition and realization," by C. M. Herzfeld, pp. 41–50.

[3] See E. R. Cohen and J. W. M. DuMond, "Our knowledge of the fundamental constants of physics and chemistry in 1965," Reviews of Modern Physics **37**, 537 (1965). A more recent survey is given by B. N. Taylor, W. H. Parker, and D. N. Langenberg, Reviews of Modern Physics **41**, July 1969; they propose $k_B = 1.380622(\pm 59) \times 10^{-16}$ erg deg^{-1}.

Summary of the Current Practice
in Temperature Measurements

The methods used in the laboratory for the practical measurement of temperature and the calibration of the devices are discussed in standard textbooks and reference books. In the table below we list some of the methods most commonly employed. Accuracy is used here as the precision to which one can assign an absolute temperature to a measurement made on a thermometer that has been calibrated against some primary standard.

Thermometer	Useful temperature range, in deg	Accuracy, in deg
Magnetic susceptibility of cerous magnesium nitrate°	0.002 to 1	
Constant volume gas	He^3: 0.2 to 3.3	± 0.002
	He^4: 0.8 to 5.4	± 0.002
Vapor pressure manometer	He^3: 0.3 to 3.2	± 0.01
	He^4: 1 to 4.2	± 0.01
Carbon resistance	0.1 to 10	± 0.01
Germanium resistance	0.3 to 20 (can be used to 77 K)	± 0.0001
Thermocouples		± 0.002 at 300 K
(a) Copper-constantan	20 to above 300	decreasing to ± 0.1 at 77 K.
(b) Gold alloys	4 up	
Platinum resistance	10 to 903	Varies with temperature; commonly ± 0.01 at 300 K.
GaAs *pn* junction diodes	10 to 300	± 1
Optical pyrometer	1000 up	

° See R. B. Frankel, D. A. Shirley, and N. J. Stone, Physical Review **140A**, 1020 (1965). Courtesy of Gene Rochlin.

CHAPTER 9

Fermions and Bosons: Distribution Functions

"In an atom there can never be two or more equivalent electrons for which all quantum numbers coincide. If an electron is in a state with definite values of all quantum numbers, then this state is occupied."

(W. Pauli, Jr., 1925)[1]

[1] This is a free translation of the original statement of the exclusion principle given in Pauli's famous paper "Über den Zusammenhang des Abschlusses der Elektronengruppen im Atom mit der Komplexstruktur der Spektren," Zeitschrift für Physik **31**, 765–783 (1925).

FERMIONS AND THE PAULI PRINCIPLE

The original statement of the **Pauli exclusion principle** is the most elementary statement, although not the most general. In our present language the statement is that at most one electron can occupy any orbital l. But what is an orbital?

The term **orbital** is very useful here and in atomic and molecular theory. It denotes a state of the Schrödinger equation or wave equation for a **system of only one particle.** In our actual applications the orbitals will be those of a free particle confined within a volume, usually a cube. These orbitals are simple in form, as we shall see in Chapter 10. The term[2] allows us to distinguish between an exact quantum state of the Schrödinger equation of a system of N electrons and an approximate quantum state that we construct by assigning the N electrons to N different orbitals, where each orbital is a solution of a one-particle Schrödinger equation. There are usually an infinite number of orbitals available for occupancy; N electrons will occupy N of them.

The orbital model is an exact solution of a problem only if there are no interactions between particles. There are more general formulations of the Pauli principle which enable us to discuss the restrictions it imposes on the allowed forms of the exact N-particle wavefunctions. Here we use only the elementary statement of the principle, that the allowed occupancies of an orbital are 0 or 1.

The Pauli principle is responsible for the shell structure of atoms; that is, for the regularities of the periodic table of elements. As an example of the application of the principle, an atom of lithium with three electrons has in the ground state, one electron in the $1s\uparrow$ orbital, one electron in the $1s\downarrow$ orbital, and one electron in a $2s\uparrow$ or $2s\downarrow$ orbital. Here the number 1 or 2 is the principal quantum number of the orbital; the letter s by convention denotes a state of zero orbital angular momentum; and the arrow $\uparrow$ or $\downarrow$ denotes the direction of the electron spin assigned to the orbital. The total energy of the lithium atom (Fig. 1.1) would be lower if the electron in the $2s$ orbital could be put into either of the two $1s$ orbitals, but this would require double occupancy of an orbital and is forbidden by the Pauli principle.

The exclusion principle has nothing to do with the fact that two electrons repel each other by the electrostatic repulsion e^2/r. The exclusion principle is a

---

[2] The term orbital as used here has no relation with the orbital angular momentum of an atom.

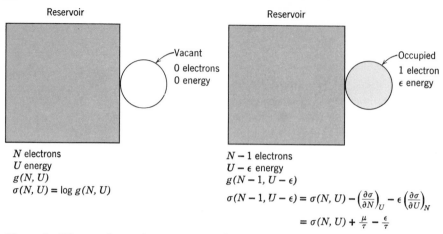

Figure 1 We consider as the system a single orbital that may be occupied at most by one electron. The system is in weak thermal and diffusive contact with the reservoir at temperature T. The energy ϵ of the occupied orbital might be the kinetic energy of a free electron of a definite spin orientation and confined to a fixed volume. Other allowed quantum states may be considered as forming the reservoir. The reservoir will contain N particles if the system is unoccupied and $N - 1$ particles if the system is occupied by one electron.

principle of physics that must be taken into account, together with the interactions between particles, in any correct description of the behavior of a system of many electrons. The exclusion principle would be true even if the charge on the electron were zero. It is true, for example, for neutrons in the nucleus.

The exclusion principle is known to be satisfied by every particle, elementary or composite, whose spin or intrinsic angular momentum is a half-integral multiple of $\hbar$. Neutrons and neutrinos, as well as electrons, positrons, and protons, have spin $\frac{1}{2}\hbar$ and obey the exclusion principle. An example of a composite particle that obeys the exclusion principle is an atom of He^3, with two electrons, two protons, and one neutron: the total spin is $\frac{1}{2}\hbar$. The principle does not apply to particles whose spin is zero or an integral multiple of $\hbar$. Particles that obey the exclusion principle are called **fermions**.

There are two classes of particles known in physics: fermions and bosons.[3] Their properties are summarized in Table 1. There is no limit to the number of bosons that may occupy the same orbital. We treat bosons later in this chapter. An atom of He^4 is a boson: the atom has two electrons, two protons, and two neutrons. The total spin is zero.

[3] The connection between the spin of a particle and the "statistics" (whether it is a fermion or a boson) is a consequence of the requirement of relativistic invariance. The proof of the connection unfortunately involves a knowledge of quantum field theory. A proof was first given by W. Pauli, Physical Review **58**, 716 (1940); see also F. Mandl, *Introduction to quantum field theory*, Interscience, 1959, p. 53.

Table 1   Distribution Laws for Fermions and Bosons

Particle type	Spin in units of $\hbar$	Examples	Occupancy of an orbital	Distribution function
Fermion	$\frac{1}{2}, \frac{3}{2}, \frac{5}{2}, \ldots$	Electron, proton, neutron, He^3 atom	0, 1	Fermi-Dirac $$f(\epsilon) = \frac{1}{e^{(\epsilon - \mu)/T} + 1}$$
Boson	0, 1, 2, \ldots	Photon, deuteron, He^4 atom	0, 1, 2, 3, \ldots	Bose-Einstein $$n(\epsilon) = \frac{1}{e^{(\epsilon - \mu)/T} - 1}$$

The Pauli principle refers to occupancy of an orbital by identical fermions: neutrons are not identical with protons, so that one neutron and one proton can occupy the same orbital in a nucleus, but two neutrons cannot occupy the same orbital. Electrons are not identical with positrons, although both are fermions. One electron and one positron can occupy the same orbital. Many experiments[4] carried out on positrons (positively charged electrons) in metals show conclusively that a positron in a metal can occupy a conduction electron orbital that is also occupied by an electron.

FERMI-DIRAC DISTRIBUTION FUNCTION

We consider a system composed of a single orbital[5] that may be occupied by an electron. The system is placed in thermal and diffusive contact with a reservoir, as in Figs. 1 and 2. A real system may consist of a large number N of electrons, but it is very helpful to focus on one orbital and call it the system. All other orbitals of the real system, occupied by a total of N or $N - 1$ electrons, are thought of as the reservoir. Our problem is to find the thermal average occupancy of the orbital thus singled out.

Electrons are fermions, so that an orbital can be occupied by zero or by one electron. No other occupancy is allowed by the Pauli exclusion principle.

---

[4] The subject of positrons in metals is reviewed by A. T. Stewart in *Positron annihilation*, eds. A. T. Stewart and L. O. Roellig, Academic Press, 1966. The conduction electron orbitals in metals are usually called Bloch states.

[5] We reemphasize that the orbital is a state of a one particle problem and not a state of an entire interacting system of N particles. An orbital is a solution of the Schrödinger equation for one particle. If interactions between particles are important, then it will not be a good approximation to construct an N particle state by superposing N orbitals, one for each particle. But for many problems the approximation is excellent. The more general statements of the Pauli principle that apply to the exact many-body wavefunctions are discussed in textbooks on quantum mechanics and are touched on in Chapter 10.

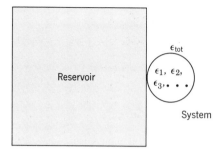

Figure 2a The obvious method of viewing a system of noninteracting particles is shown in (a). The energy levels each refer to an orbital that is a solution of a single particle Schrödinger equation. The total energy of the system is

$$\epsilon_{tot} = \sum_i n_i \epsilon_i,$$

where n_i is the number of particles in the orbital i of energy ϵ_i. For fermions $n_i = 0$ or 1.

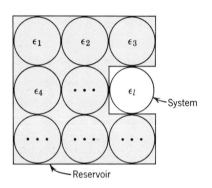

Figure 2b It is much simpler than (a), and equally valid, to treat a single orbital as the system. The system in this scheme may be the orbital l of energy ϵ_l. All other orbitals are viewed as the reservoir. The total energy of this one-orbital system is $\epsilon = n_l \epsilon_l$, where n_l is the number of particles in the orbital. This device of using one orbital as the system works because the particles are supposed to interact only weakly with each other. If we think of the fermion system associated with the orbital l, there are two possibilities: either the system has 0 particles and energy 0 or the system has 1 particle and energy ϵ_l. Thus the grand sum consists of only two terms:

$$\mathcal{Z} = 1 + \lambda e^{-\epsilon_l/\tau}.$$

The first term arises from the orbital occupancy $n_l = 0$, and the second term arises from $n_l = 1$.

If the orbital is unoccupied, the energy of the system will be taken to be zero; if the orbital is occupied by one electron, the energy is ϵ.

The grand sum is simple: from the definition of the grand sum in (6.20) we have

$$\mathcal{Z} = 1 + \lambda e^{-\epsilon/\tau} . \tag{1}$$

Here the term 1 comes from the configuration with occupancy $n = 0$ and energy $\epsilon = 0$. The term $\lambda e^{-\epsilon/\tau}$ comes when the orbital is occupied by one particle, so that $n = 1$ and the energy is ϵ.

The thermal average value of the occupancy of the orbital is the ratio of the term in the grand sum with $n = 1$ to the sum of the terms with $n = 0$ and $n = 1$:

$$\langle n(\epsilon) \rangle = \frac{\lambda e^{-\epsilon/\tau}}{1 + \lambda e^{-\epsilon/\tau}} = \frac{1}{\lambda^{-1} e^{\epsilon/\tau} + 1} . \tag{2}$$

We introduce for the average fermion occupancy the conventional symbol $f(\epsilon)$ defined by

$$f(\epsilon) \equiv \langle n(\epsilon) \rangle . \tag{3}$$

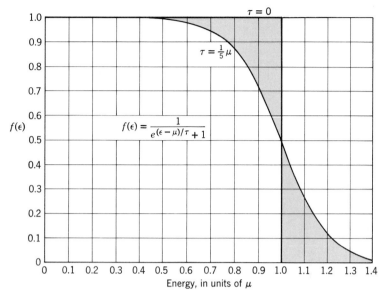

Figure 3   Plot of the Fermi-Dirac distribution function $f(\epsilon)$ versus ϵ/μ, for zero temperature and for a temperature $\mathcal{T} = \frac{1}{5}\mu$. The value of $f(\epsilon)$ gives the fraction of orbitals at a given energy which are occupied when the system is in thermal equilibrium. When the system is heated from absolute zero, electrons are transferred from the shaded region at $\epsilon/\mu < 1$ to the shaded region at $\epsilon/\mu > 1$. For a metal μ might correspond to 50,000 K.

Recall that $\lambda \equiv e^{\mu/\mathcal{T}}$, where μ is the chemical potential. We may write (2) in the standard form

$$f(\epsilon) = \frac{1}{e^{(\epsilon - \mu)/\mathcal{T}} + 1} \, . \tag{4}$$

This result is known as the Fermi-Dirac distribution function.[6] Equation (4) gives the average number of fermions in a single orbital of energy ϵ. The value of f is always between zero and one. The FD distribution function is plotted in Fig. 3 for particular values of the parameters μ and $\mathcal{T}$.

In the field of solid state physics the chemical potential μ is often called the **Fermi level.** The chemical potential usually depends on the temperature.

---

[6] This distribution function was discovered independently by E. Fermi, Zeitschrift für Physik **36**, 902 (1926), and P. A. M. Dirac, Proceedings of the Royal Society of London **A112**, 661 (1926). Both workers drew on Pauli's paper of the preceding year in which the exclusion principle was discovered. The paper by Dirac is concerned with the new quantum mechanics and contains a general statement of the form assumed by the Pauli principle on this theory.

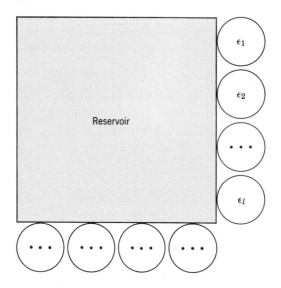

Figure 4 A convenient pictorial way to think of a system composed of independent orbitals that do not interact with each other, but interact with a common reservoir.

The value of μ at zero temperature is often written as ϵ_F; that is,

$$\mu(T = 0) \equiv \mu(0) \equiv \epsilon_F \ . \tag{5}$$

We call ϵ_F the **Fermi energy**.

Consider a system of many independent orbitals, as in Fig. 4. At the temperature $T = 0$ all orbitals of energy below the Fermi energy are occupied by exactly one electron each, and all orbitals of higher energy are unoccupied. At nonzero temperatures the value of the chemical potential μ departs from the Fermi energy, as we may see in detail in Chapter 14 and in Appendix C.

At any temperature an orbital of energy equal to the chemical potential $(\epsilon = \mu)$ is exactly half-filled:

$$f(\epsilon = \mu) = \frac{1}{1 + 1} = \tfrac{1}{2} \ . \tag{6}$$

Orbitals of lower energy are more than half-filled and orbitals of higher energy are less than half-filled.

We shall discuss the physical consequences of the FD distribution in Chapter 14. Right now we go on to discuss the statistical mechanics of non-interacting bosons, and then in Chapter 11 we establish the ideal gas law for both fermions and bosons in the appropriate limit.

Problem 1. *Derivative of Fermi-Dirac function.* (a) Show that $-\partial f/\partial \epsilon$ evaluated at the Fermi level $\epsilon = \mu$ has the value $(4k_B T)^{-1}$. Thus the lower the temperature, the steeper the slope of the Fermi-Dirac function. (b) Make a careful plot of $-\partial f/\partial \epsilon$ versus ϵ/k_B for the case $\mu/k_B = 5 \times 10^4$ K and $T = 5 \times 10^2$ K, or $\mathcal{T} = 0.01\,\mu$. This plot exhibits the limited region of energy in which filled and vacant states coexist for values of μ characteristic of a metal.

Problem 2. Symmetry of filled and vacant orbitals. Let $\epsilon = \mu + \delta$; show that

$$f(\delta) = 1 - f(-\delta) \ . \tag{7}$$

Thus the probability that an orbital δ above the Fermi level is occupied is equal to the probability an orbital δ below the Fermi level is vacant. A vacant orbital is sometimes known as a **hole**.

Problem 3. Distribution function in event of strong electron-electron repulsion. Consider a system which has two orbitals, both of the same energy. When both orbitals are unoccupied the energy of the system is zero; when one orbital or the other is occupied by one electron, the energy is ϵ. We suppose that the energy of the system is much higher, say infinitely high, when both orbitals are occupied. (a) Show that the grand sum is

$$\mathfrak{Z} = 1 + 2\lambda e^{-\epsilon/\mathcal{T}} \ . \tag{8}$$

(b) Show that the ensemble average number of electrons in the level is

$$\langle n \rangle = \frac{1}{\frac{1}{2} e^{(\epsilon - \mu)/\mathcal{T}} + 1} \ . \tag{9}$$

This situation may be encountered in the ionization of impurity atoms in semiconductors.[7]

Problem 4. Distribution function for double occupancy statistics. Let us imagine a new mechanics in which the allowed occupancies of an orbital are 0, 1, and 2. The values of the energy associated with these occupancies are assumed to be 0, ϵ, and 2ϵ, respectively.

(a) Derive an expression for the ensemble average occupancy $\langle n \rangle$, when the system composed of this orbital is in thermal and diffusive contact with a reservoir at temperature $\mathcal{T}$ and chemical potential μ.

(b) Return now to the usual quantum mechanics, and derive an expression for the ensemble-average occupancy of an energy level which is doubly degenerate; that is, two orbitals have the identical energy ϵ. If both orbitals are occupied the total energy is 2ϵ. The result is different from the result of (a) and also from the result of Problem 3.

---

[7] See ISSP, Chapter 10.

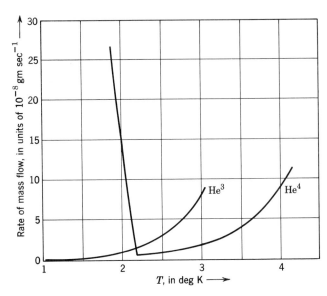

Figure 5 Comparison of rates of flow of liquid He³ and liquid He⁴ under gravity through a fine hole. Notice the sudden onset of high fluidity or superfluidity in He⁴. The absence of superfluidity in He³ has been confirmed down to 0.1 K and indicates the importance of the distinction between bosons and fermions. [After D. W. Osborne, B. Weinstock, and B. M. Abraham, Physical Review **75**, 988 (1949).]

BOSONS

We have seen that an orbital can be occupied by at most one fermion of a given species. A fermion is a particle with a half-integral value of the spin, measured in units of $\hbar$. We are now concerned with bosons. A **boson** is a particle with an integral value of the spin. **An orbital can be occupied by any number of bosons.**

Because an orbital can be occupied by only one fermion, we see that bosons have an essentially different quality than fermions. Systems of bosons can have rather different physical properties than systems of fermions. Atoms of He⁴ are bosons; atoms of He³ are fermions. For example, the remarkable superfluid properties of the low temperature ($T < 2.17$ K) phase of liquid helium can be attributed to the properties of a boson gas. There is a sudden increase in the fluidity and in the heat conductivity of liquid He⁴ below this temperature (Fig. 5). But no superfluidity has ever been observed in liquid He³. In experiments by Kapitza the flow viscosity of He⁴ below 2.17 K was found to be less than 10^{-7} of the viscosity of the liquid above 2.17 K.

Photons (the quanta of the electromagnetic field) and phonons (the quanta of elastic waves in solids) are other examples of bosons; their thermal properties are quite unlike those of fermions. There are two classes of bosons to be considered. In this chapter and in Chapter 17 we treat bosons whose number is constant in the combined system and reservoir. An example is a closed vessel containing atoms of He⁴: unless there is a leak, the number of atoms is conserved. In Chapters 15 and 16 we treat bosons such as photons and phonons whose number may vary, even in a container or in an isolated

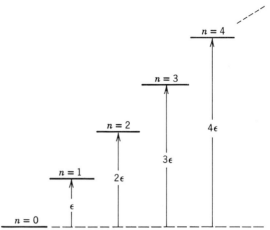

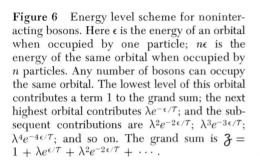

Figure 6 Energy level scheme for noninteracting bosons. Here ϵ is the energy of an orbital when occupied by one particle; $n\epsilon$ is the energy of the same orbital when occupied by n particles. Any number of bosons can occupy the same orbital. The lowest level of this orbital contributes a term 1 to the grand sum; the next highest orbital contributes $\lambda e^{-\epsilon/T}$; and the subsequent contributions are $\lambda^2 e^{-2\epsilon/T}$; $\lambda^3 e^{-3\epsilon/T}$; $\lambda^4 e^{-4\epsilon/T}$; and so on. The grand sum is $\mathfrak{Z} = 1 + \lambda e^{\epsilon/T} + \lambda^2 e^{-2\epsilon/T} + \cdots$.

specimen. For example, the number of photons in a closed cavity increases as the temperature of the cavity is increased.

BOSE-EINSTEIN DISTRIBUTION FUNCTION

We consider the distribution function for a system of noninteracting bosons. The system is in thermal and diffusive contact with a reservoir. In our treatment of bosons we shall let ϵ denote the energy of a single orbital when occupied by one particle. When there are n particles in the orbital, the energy is $n\epsilon$, as in Fig. 6.

We now treat one orbital as the system; we may neglect all other orbitals or view them as part of the reservoir. We may put any arbitrary number of particles into this orbital. The grand sum taken over the one orbital is

$$\mathfrak{Z} = \sum_{n=0}^{\infty} \lambda^n e^{-n\epsilon/T} = \sum_{n=0}^{\infty} (\lambda e^{-\epsilon/T})^n \ . \tag{10}$$

The upper limit on n should be the total number of particles in the combined system and reservoir, but we are free to let the reservoir be very large, so that n may be summed to high accuracy from zero to infinity. The series may be summed in closed form. We write $x \equiv \lambda e^{-\epsilon/T}$, so that the grand sum becomes

$$\mathfrak{Z} = \sum_{n=0}^{\infty} x^n = \frac{1}{1-x} = \frac{1}{1 - \lambda e^{-\epsilon/T}} \ , \tag{11}$$

provided that $\lambda e^{-\epsilon/T} < 1$. In all applications $\lambda e^{-\epsilon/T}$ will satisfy this inequality, for otherwise the number of bosons in the system would not be bounded.

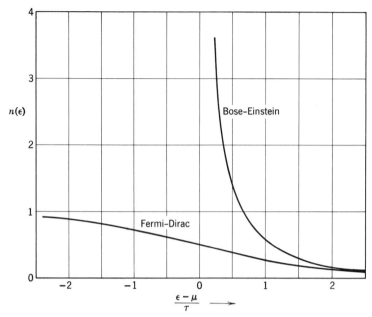

Figure 7 Comparison of Bose-Einstein and Fermi-Dirac distribution functions. The classical regime is attained for $(\epsilon - \mu)/\tau \gg 1$, where the two distributions become nearly identical. We shall see in later chapters that in the degenerate regime at low temperature the chemical potential μ for a FD distribution is positive, and changes to negative at high temperature. The chemical potential of a BE distribution is always negative if the zero of energy is chosen to coincide with the energy of the lowest orbital.

The ensemble average[8] of the number of particles in the orbital is, by the definition of the average value,

$$\langle n(\epsilon)\rangle = \frac{\sum\limits_{n=0}^{\infty} n x^n}{\sum\limits_{n=0}^{\infty} x^n} = \frac{x\dfrac{d}{dx}\sum\limits_{n=0}^{\infty} x^n}{\sum\limits_{n=0}^{\infty} x^n} = \frac{x\dfrac{d}{dx}(1-x)^{-1}}{(1-x)^{-1}} , \tag{12}$$

by virtue of (11). We carry out the differentiation to obtain

$$\langle n(\epsilon)\rangle = \frac{x}{1-x} = \frac{1}{x^{-1}-1} = \frac{1}{\lambda^{-1}e^{\epsilon/T}-1} , \tag{13}$$

or, with $n(\epsilon)$ written for $\langle n(\epsilon)\rangle$,

$$\boxed{n(\epsilon) = \frac{1}{\lambda^{-1}e^{\epsilon/T}-1} = \frac{1}{e^{(\epsilon-\mu)/T}-1} .} \tag{14}$$

This defines the **Bose-Einstein distribution function.** It differs mathematically from the Fermi-Dirac distribution function only by having -1 instead of $+1$ in the denominator. The change may have very significant physical consequences, as we shall see in later chapters. The two distribution functions are compared in Fig. 7. But we go on next to treat the ideal gas, which represents the limit[9] $\epsilon - \mu \gg T$ in which the two distribution functions are approximately equal.

The quantity $n(\epsilon)$ is called the **occupancy** of the orbital. For bosons $n(\epsilon)$ is not the same as the probability that an orbital be occupied. For fermions the occupancy and the probability are the same, because only 0 or 1 particles may occupy an orbital.

---

[8] We may use (6.30) to derive $n(\epsilon)$ in fewer steps:

$$\langle n(\epsilon)\rangle = \lambda\frac{\partial}{\partial\lambda}\log \mathcal{Z} = -x\frac{d}{dx}\log(1-x) = \frac{x}{1-x} = \frac{1}{\lambda^{-1}e^{\epsilon/T}-1} .$$

[9] We bear in mind that the choice of the zero of the energy ϵ is always arbitrary. The particular choice made in any problem will affect the value of the chemical potential μ. The value of the difference $\epsilon - \mu$ is independent of the choice of the zero of ϵ.

Summary: Derivation of Fermi-Dirac Distribution Function

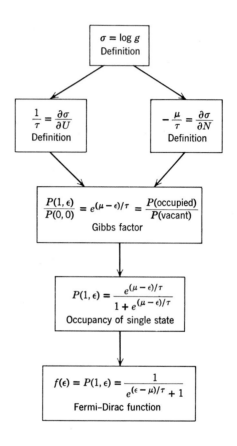

CHAPTER 10

Free Particles: Enumeration of Orbitals

ORBITALS OF FREE PARTICLES IN ONE DIMENSION

We consider the quantum theory of a free particle in a one-dimensional world. A particle of mass M is confined to a length L by means of infinite barriers (Fig. 1). On quantum theory the particle is described by a wavefunction $\varphi_n(x)$, where the index n labels the orbital of the particle. We use φ to denote the wavefunction of a one particle problem, and we reserve ψ for the wavefunction of a system of N particles. We call φ an **orbital** and ψ a **state**. We must carefully distinguish the two classes of problems, that for one particle and that for N particles.

The wavefunction is a solution of the Schrödinger equation

$$\mathcal{H}\varphi_n = \epsilon_n\varphi_n \ , \tag{1}$$

where $\mathcal{H}$ is the hamiltonian or energy operator. We assume that $\mathcal{H}$ is independent of time. For a free particle the classical energy is $\mathcal{H} = p^2/2M$, where p is the momentum.

One postulate of quantum mechanics is that the momentum p is represented in the wave equation (1) by the operator $-i\hbar \, d/dx$, where $\hbar$ is Planck's constant divided by 2π. Thus the kinetic energy operator is

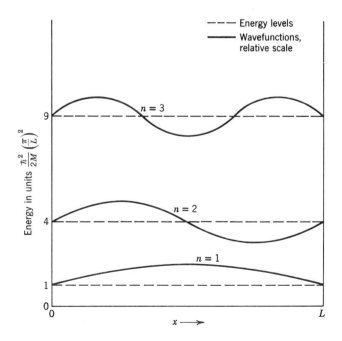

Figure 1   First three orbitals of a free particle of mass M confined to a length L in a one-dimensional world. The orbitals are labeled according to the quantum number n which gives the number of half-wavelengths in the wavefunction. The energy ϵ_n of the level of quantum number n is equal to $(\hbar^2/2M)(\pi n/L)^2$.

$p^2/2M = -(\hbar^2/2M)(d^2/dx^2)$, and the Schrödinger equation for a free particle is

$$-\frac{\hbar^2}{2M}\frac{d^2\varphi_n}{dx^2} = \epsilon_n\varphi_n \; , \tag{2}$$

where ϵ_n is the energy of the particle in the orbital n.

Because of the infinite potential energy barriers at $x = 0$ and $x = L$, the boundary conditions are that the wavefunction be zero[1] at these positions:

$$\varphi_n(0) = 0 \; ; \qquad \varphi_n(L) = 0 \; . \tag{3}$$

The boundary conditions are automatically satisfied if the wavefunction is a sine function with an integral number n of half-wavelengths in the distance L:

$$\varphi_n \propto \sin\left(\frac{2\pi x}{\lambda_n}\right) \; ; \qquad \tfrac{1}{2}\lambda_n \times n = L \; ; \qquad \lambda_n = \frac{2L}{n} \; . \tag{4}$$

The wavelength is λ_n. From (4) we have for the wavefunction

$$\varphi_n = C\sin\left(\frac{n\pi x}{L}\right) , \tag{5}$$

where C is a constant determined by (9) below.

We confirm by substitution in (2) that the wavefunction (5) is a solution of the Schrödinger equation, because

$$\frac{d\varphi_n}{dx} = C\left(\frac{n\pi}{L}\right)\cos\left(\frac{n\pi x}{L}\right) \; ; \qquad \frac{d^2\varphi_n}{dx^2} = -C\left(\frac{n\pi}{L}\right)^2\sin\left(\frac{n\pi x}{L}\right) , \tag{6}$$

whence the energy ϵ_n of the orbital is given by

$$\epsilon_n = \frac{\hbar^2}{2M}\left(\frac{n\pi}{L}\right)^2 \tag{7}$$

The energy is a quadratic function of the quantum number n, where n gives the number of half-wavelengths of the wavefunction, as in Fig. 1.

We choose the constant C in the expression (5) for the wavefunction φ_n so that the probability of finding the particle somewhere in the interval 0 to L is equal to unity. The physical significance of the wavefunction is that

$$\varphi_n^*(x)\,\varphi_n(x)\,dx \tag{8}$$

is the probability that the electron is in the segment dx at x. We require that

---

[1] This boundary condition follows because $\varphi^*(x)\,\varphi(x)\,dx$ is the probability of finding the particle in an element of length dx at x. There is zero probability of finding the particle inside an infinitely high potential barrier.

the probability the electron is somewhere in the interval is unity:

$$\int_0^L dx\, \varphi_n^*(x)\, \varphi_n(x) = 1 \ . \tag{9}$$

The integral of $\sin^2$ over the interval L is $\frac{1}{2}L$, so that $\frac{1}{2}LC^2 = 1$ and the normalized wavefunction of the particle in the quantum state n is

$$\varphi_n(x) = \left(\frac{2}{L}\right)^{\frac{1}{2}} \sin\frac{n\pi x}{L} \ . \tag{10}$$

EXAMPLE. *States and orbitals.* The two lowest-energy orbitals of the one-dimensional problem are

$$\varphi_1(x) = \left(\frac{2}{L}\right)^{\frac{1}{2}} \sin\frac{\pi x}{L} \ ; \qquad \epsilon_1 = \frac{\hbar^2}{2M}\left(\frac{\pi}{L}\right)^2 \ ; \tag{11}$$

$$\varphi_2(x) = \left(\frac{2}{L}\right)^{\frac{1}{2}} \sin\frac{2\pi x}{L} \ ; \qquad \epsilon_2 = \frac{\hbar^2}{2M}\left(\frac{2\pi}{L}\right)^2 \ , \tag{12}$$

from (10). If the particles have zero spin, there will be no need to add spin indices to the wavefunctions. If the particles have spin $\frac{1}{2}$, then the four lowest orbitals are

$$\varphi_{1\uparrow}(x) = \left(\frac{2}{L}\right)^{\frac{1}{2}}\left(\sin\frac{\pi x}{L}\right)\alpha \ ; \qquad \varphi_{1\downarrow}(x) = \left(\frac{2}{L}\right)^{\frac{1}{2}}\left(\sin\frac{\pi x}{L}\right)\beta \ ; \tag{13}$$

$$\varphi_{2\uparrow}(x) = \left(\frac{2}{L}\right)^{\frac{1}{2}}\left(\sin\frac{2\pi x}{L}\right)\alpha \ ; \qquad \varphi_{2\downarrow}(x) = \left(\frac{2}{L}\right)^{\frac{1}{2}}\left(\sin\frac{2\pi x}{L}\right)\beta \ . \tag{14}$$

Here α denotes spin up and β denotes spin down.

These are orbitals and refer to one particle. The ground state ψ_0 of a boson system of two particles of zero spin in the absence of interactions is as follows, for the state with both particles in the orbital $n = 1$:

$$\psi_0(x_a, x_b) = \frac{2}{L}\sin\frac{\pi x_a}{L}\sin\frac{\pi x_b}{L} \ ; \qquad \epsilon_0 = \frac{\hbar^2}{2M}\left(\frac{\pi}{L}\right)^2(1^2 + 1^2) \ , \tag{15}$$

where x_a, x_b are the coordinates of particles a and b. This is the lowest-energy state. It is constructed from two particles in the lowest orbital. Another boson state has one particle with $n = 1$ and one particle with $n = 2$:

$$\psi_1(x_a, x_b) = \frac{1}{\sqrt{2}} \cdot \frac{2}{L}\left[\sin\frac{\pi x_a}{L}\sin\frac{2\pi x_b}{L} + \sin\frac{\pi x_b}{L}\sin\frac{2\pi x_a}{L}\right] . \tag{16}$$

This is the first excited state of the system; the energy is

$$\epsilon_1 = \left(\frac{\hbar^2}{2M}\right)\left(\frac{\pi}{L}\right)^2 (2^2 + 1^2) \; .$$

The form (16) was constructed to be symmetrical[2] with respect to exchange of x_a and x_b: the function does not change sign when x_a and x_b are exchanged. This symmetry is required in quantum mechanics for bosons.

ORBITALS OF FREE PARTICLES IN THREE DIMENSIONS

We consider a free particle confined to a cube of volume $V = L^3$, where L is an edge of the cube. The wave equation of an orbital φ_n is found by the generalization of (3) to be

$$-\frac{\hbar^2}{2M}\left(\frac{\partial^2}{\partial x^2} + \frac{\partial^2}{\partial y^2} + \frac{\partial^2}{\partial z^2}\right)\varphi_n(\mathbf{r}) = \epsilon_n \varphi_n(\mathbf{r}) \; , \tag{17}$$

where $\mathbf{n}$ denotes the triplet of positive integers n_x, n_y, n_z. These integers are the quantum numbers, together with the spin quantum number which we often do not bother to denote explicitly.

The boundary condition is that $\varphi_n = 0$ on all faces of the cube. We take the origin of the coordinate system at one corner of the cube. The wavefunctions that are the solutions of the wave equation have the form

$$\varphi_n = C \sin\left(\frac{n_x \pi x}{L}\right) \sin\left(\frac{n_y \pi y}{L}\right) \sin\left(\frac{n_z \pi z}{L}\right) , \tag{18}$$

with C a constant to be determined as before. The quantities n_x, n_y, n_z are nonzero integers and may all be taken as positive.

A negative integer does not give a new orbital, but just repeats an orbital already included among the positive integers:

$$\sin\left(\frac{-n_x \pi x}{L}\right) = -\sin\left(\frac{n_x \pi x}{L}\right) . \tag{19}$$

The minus sign on the right-hand side is equivalent to a change of sign of the

---

[2] Quantum mechanics requires fermion wavefunctions to be antisymmetrical—that is, to change sign when particles a and b are interchanged. The fermion ground state is

$$\psi_0(x_a, x_b) = \frac{2}{L}\left(\sin\frac{\pi x_a}{L} \sin\frac{\pi x_b}{L}\right)\frac{1}{\sqrt{2}}(\alpha_a \beta_b - \beta_a \alpha_b) \; . \tag{16a}$$

The orbital occupancies are $n_1\uparrow = 1$; $n_1\downarrow = 1$ and all others are zero. It is shown in texts on quantum mechanics that the antisymmetry requirement is equivalent to the statement of the Pauli principle in terms of occupancies $n_l = 0, 1$.

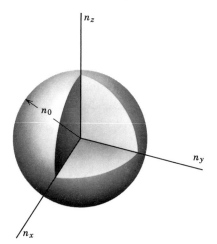

Figure 2 The positive octant of a sphere in the space defined by the quantum numbers n_x, n_y, n_z for free particle orbitals. There is one orbital (for each spin orientation) associated with every unit volume $\Delta n_x \, \Delta n_y \, \Delta n_z = 1$. Only positive values of n_x, n_y, n_z are allowed, as negative values do not give independent states. The energy of an orbital on the surface of the sphere of radius n_0 in $\mathbf{n}$ space is $(\hbar^2/2M) \cdot (\pi n_0/L)^2 = \epsilon_0$, for a particle in a box of side L. The number of orbitals in the allowed octant of a spherical shell of thickness Δn is $\tfrac{1}{8} \cdot 4\pi n^2 \, \Delta n$, not including a factor for the spin.

constant C. But only $|C|$ has physical significance, according to (8), so that the orbital with $-n_x$ is identical to the orbital with n_x.

When we substitute the orbital $\varphi_{\mathbf{n}}$ in (17) we find that the energy of an electron in the orbital $\mathbf{n}$ is given by

$$\epsilon_{\mathbf{n}} = \frac{\hbar^2}{2M} \left(\frac{\pi n}{L} \right)^2 , \tag{20}$$

with

$$n^2 = n_x{}^2 + n_y{}^2 + n_z{}^2 . \tag{21}$$

ENUMERATION OF ORBITALS[3]

Apart from spin, the number of orbitals with the magnitude n of the quantum number n less than some particular value n_0 is very closely equal to the volume of an octant of a sphere of radius n_0 in the space of the n_x, n_y, n_z, as in Fig. 2. The limitation to an octant occurs because the independent orbitals are described by the positive integers alone. The relation with the volume arises because the density of integers is the space defined by n_x, n_y, n_z is unity. To each triplet of integers there belongs a cube of unit volume.

In the allowed octant the number of orbitals within some radius n is

$$\gamma \cdot \frac{1}{8} \cdot \frac{4\pi}{3} n^3 = \frac{1}{6} \gamma \pi n^3 . \tag{22}$$

---

[3] This section is of central importance to many physical applications of the FD and BE distribution functions.

Here γ, the Greek letter gamma, denotes the number of independent spin orientations for a given value of n_x, n_y, n_z. Each spin orientation at a given $\mathbf{n}$ is assigned a distinct orbital. For particles of spin I, we know from quantum mechanics that $\gamma = 2I + 1$. For electrons, $I = \frac{1}{2}$ and $\gamma = 2$.

We shall often need an expression for the number of orbitals with the magnitude n of the quantum number $\mathbf{n} \equiv n_x$, n_y, n_z in the range Δn about n. This number is equal to $\Delta n \dfrac{d}{dn}$ of the result (22):

$$\Delta n \frac{d}{dn} (\tfrac{1}{6}\gamma\pi n^3) = \tfrac{1}{2}\gamma\pi n^2 \, \Delta n \; . \tag{23}$$

Equations (22) and (23) are only approximate, but are valid asymptotically as the volume of the sphere in $\mathbf{n}$ space increases. For small spheres the corrections have been discussed by Morse and Bolt[4] in connection with sound waves in rooms. It may be shown that the result (23) is essentially independent of the shape of the volume of the specimen, provided that the maximum dimension is much larger than the average wavelength of the orbitals of interest.

Problem 1. Fermion gas in ground state. Consider the ground state of a system of N free electrons in a volume V. Show that the kinetic energy U_0 of the system is

$$\frac{3}{10} \cdot \frac{\hbar^2}{m} \left(\frac{3\pi^2 N}{V} \right)^{\frac{2}{3}} \; .$$

Problem 2. Orbitals in a rectangular parallelepiped. Consider a particle confined within a rectangular parallelopiped of edges a, b, c such that $a = b = \eta c$. Show that the energy of an orbital is given by

$$\epsilon_{\mathbf{n}} = \frac{\pi^2\hbar^2}{2MV^{\frac{2}{3}}\eta^{\frac{2}{3}}} (n_x^2 + n_y^2 + \eta^2 n_z^2) \; ,$$

where V is the volume and n_x, n_y, n_z are positive integers. For a long square pipe, $\eta \ll 1$; for a square pancake, $\eta \gg 1$.

---

[4] P. M. Morse and R. H. Bolt, Reviews of Modern Physics **16**, 70 (1964) (see their Section 14 particularly); R. H. Bolt, Journal of the Acoustical Society of America **10**, 228 (1939).

The Monatomic Ideal Gas

CLASSICAL REGIME

The **classical regime** of a gas is defined as a condition of temperature and concentration such that the average number of atoms in any orbital is very much less than one. A gas at room temperature and atmospheric pressure is comfortably in the classical regime. In this regime the equilibrium properties of both fermions and bosons are identical, as we show below. The **quantum regime** is the opposite of the classical regime: the occupancy of an orbital may be comparable to one or larger. Here the properties of a fermion gas differ drastically from those of a boson gas, because for fermions the maximum occupancy is one, whereas for bosons it is unlimited. The characteristic features are summarized in the table below.

An ideal gas is defined as a system of free noninteracting atoms in the classical regime. By free we mean confined in a box with no restrictions on the motion within the box. Many of the traditional applications of thermal physics make the approximation that the working substance is an ideal gas. In this chapter we explore rather carefully the properties of an ideal monatomic gas.

We show that the Fermi-Dirac and Bose-Einstein distribution functions in the classical limit lead to the identical result for the average number of atoms in an orbital. First we have to clarify a notational difficulty: we have used $n(\epsilon)$ to denote the average occupancy of an orbital and we have also used n to denote the magnitude of the quantum number **n**. We cannot continue this double usage because both quantities will appear in the same equation. Therefore, instead of $n(\epsilon)$ or $\langle n(\epsilon) \rangle$ in this chapter we write $f(\epsilon)$ for the average occupancy of an orbital at energy ϵ:

$$\boxed{f(\epsilon_l) \equiv n(\epsilon_l) \ .}$$

Remember that ϵ_l is now the energy of an orbital and not the energy of the system of N particles.

Regime	Class of particle	Occupancies of an orbital	
Classical	Fermion } Boson	All $n(\epsilon) \ll 1$	
Quantum	Fermion	$n(\epsilon) \simeq 1$	for N orbitals
	Boson	$n(\epsilon) \gg 1$	for orbital of lowest energy

The Fermi-Dirac (FD) and Bose-Einstein (BE) distribution functions are

$$f(\epsilon) = \frac{1}{e^{(\epsilon - \mu)/T} \pm 1} \;, \tag{1}$$

where the plus sign is for the FD and the minus sign for the BE. If $f(\epsilon)$ is to be much smaller than unity for all states, we must have

$$e^{(\epsilon - \mu)/T} \gg 1 \;, \tag{2}$$

for all ϵ. When this is satisfied we are in the classical limit and we may neglect the term ± 1 in the denominator of (1). We then have, for either fermions or bosons,

$$\boxed{f(\epsilon) \cong e^{(\mu - \epsilon)/T} = \lambda e^{-\epsilon/T} \;,} \tag{3}$$

with $\lambda \equiv e^{\mu/T}$. The assumption (2) assures that $f(\epsilon) \ll 1$.

The result is called the **classical distribution function**, although its only significance is as the limit of the Fermi-Dirac or Bose-Einstein distribution function when the average occupancy $f(\epsilon)$ is very small in comparison with unity. Equation (3) is still basically a result for particles described by quantum mechanics: we shall see in (15) that the expression for the activity λ always involves the quantum constant $\hbar$, even in the classical regime. Any theory with $\hbar$ is not a classical theory. We get into the terrible difficulties shown at the end of Chapter 18 if we try to develop a sincerely classical statistical mechanics. The classical regime of quantum statistical mechanics, however, is important and useful.

We can use the classical distribution function (3) to explore the thermal properties of the monatomic ideal gas. There are many topics of importance: the entropy, chemical potential, heat capacity, pressure-volume-temperature relation, and the distribution of atomic velocities. To obtain results from the classical distribution function we need first to relate the chemical potential to the concentration of atoms. From the chemical potential we find the energy, then the entropy, and finally the pressure. We then examine fluctuations in the ideal gas as a test of the validity of our statistical approach.

CHEMICAL POTENTIAL

The chemical potential is usually found by demanding that the total number of atoms come out equal to the correct preassigned value N. The total number of atoms in the gas is related to the distribution function $f(\epsilon_l)$ by a sum over all orbitals:

$$N = \sum_l f(\epsilon_l) \;, \tag{4}$$

where l is the quantum number of an orbital of energy ϵ_l. Equation (4) says that the total number of particles is the sum of the average number in each orbital. We convert the sum into an integral by using the result $\frac{1}{2}\gamma\pi n^2\,dn$ for the number of orbitals with the translational quantum number n between n and $n + dn$. This is the result of Chapter 10.

Thus

$$\sum_n(\cdots) \to \tfrac{1}{2}\gamma\pi\int_0^\infty dn\,n^2\,(\cdots) \ . \tag{5}$$

The factor γ denotes the number of independent spin orientations $2I + 1$, which is the number of orbitals for each value of $n \equiv n_x$, n_y, n_z. **We shall for simplicity set $\gamma = 1$, which means zero spin, in this chapter except where specifically noted.** The table that follows (32) includes the effect of a general value of the spin.

The total number of particles may be written as

$$N = \int (\text{number of orbitals in } dn \text{ at } n) \times (\text{average occupancy of an orbital at } n) \ ,$$

whence in the classical regime

$$N = \int_0^\infty (\tfrac{1}{2}\pi n^2\,dn)(\lambda e^{-\epsilon/T}) = \tfrac{1}{2}\pi\lambda\int_0^\infty dn\,n^2 e^{-\epsilon/T} \ . \tag{6}$$

Please note again that n here denotes $|\mathbf{n}|$ and not the occupancy. Our immediate object is to evaluate the integral in (6) and solve for λ.

The energy of a free atom of mass M confined to a cube of volume $V = L^3$ is related to the quantum number n by

$$\epsilon = \frac{\hbar^2}{2M}\left(\frac{\pi n}{L}\right)^2 \ ; \qquad n^2 = (2M\epsilon)\left(\frac{L}{\pi\hbar}\right)^2 \ , \tag{7}$$

according to (10.20). Thus the expression for N becomes

$$N = \tfrac{1}{2}\pi\lambda\int_0^\infty dn\,n^2 \exp\left[-(\pi^2\hbar^2/2ML^2T)n^2\right] \ . \tag{8}$$

We introduce

$$x^2 \equiv \frac{\pi^2\hbar^2}{2ML^2T}\,n^2 \ , \tag{9}$$

whence

$$N = \tfrac{1}{2}\pi\lambda\left(\frac{2ML^2T}{\pi^2\hbar^2}\right)^{\frac{3}{2}}\int_0^\infty dx\,x^2 e^{-x^2} \ . \tag{10}$$

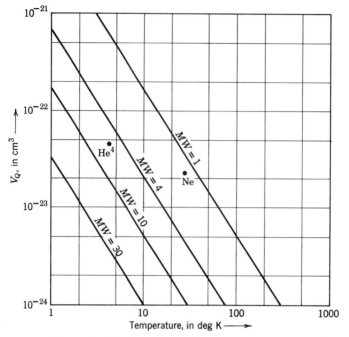

Figure 1 Plot of quantum volume versus temperature, for several values of the molecular weight. Here

$$V_Q = \frac{5.32 \times 10^{-21}}{[(MW)(T)]^{\frac{3}{2}}} \; \text{cm}^3 \; ,$$

where MW is the molecular weight and T is in deg K. Points shown are atomic volumes $V_A = V/N$ for liquid He⁴ and liquid Ne at their boiling points under a pressure of one atmosphere. For neon $V_A \gg V_Q$ at the boiling point, but $V_A \approx V_Q$ for helium.

The definite integral has the value $\frac{1}{4}\pi^{\frac{1}{2}}$, by (2.48). We note that $L^3 = V$, the volume, so that

$$N = \frac{V\lambda}{(2\pi\hbar^2/MT)^{\frac{3}{2}}} = \frac{V\lambda}{V_Q} \; , \tag{11}$$

where the **quantum volume** V_Q is defined as

$$V_Q \equiv \left(\frac{2\pi\hbar^2}{MT}\right)^{\frac{3}{2}} . \tag{12}$$

A plot of the quantum volume versus temperature is given in Fig. 1.

What is the physical significance of the quantum volume? The dimensions of MT are

$$[\text{mass}] \times [\text{energy}] = [\text{mass} \times \text{velocity}]^2 = [\text{momentum}]^2 \; .$$

By the de Broglie relation the quantum wavelength of a particle of momentum p is $2\pi\hbar/p$, so that $(2\pi\hbar^2/MT)^{\frac{3}{2}}$ is the cube of a wavelength. Now T is of the order of the kinetic energy $p^2/2M$ of an atom of an ideal gas, as we find below. Thus

$$\langle p^2 \rangle \approx 2MT \; ,$$

and

$$\frac{\hbar}{(MT)^{\frac{1}{2}}} \approx \text{thermal average wavelength, roughly} \; . \tag{13}$$

The cube of this quantity (apart from numerical factors) is the volume V_Q associated with the quantum wavelength of the particle. This volume turns up time and again in free particle problems, even in chemical reactions. It is a fundamental unit in thermal physics.

The number of atoms divided by the total volume is the concentration:

$$c \equiv \frac{N}{V} \; . \tag{14}$$

The result (11) for λ may be written as

$$\boxed{\lambda = e^{\mu/T} = \frac{N}{V}\left(\frac{2\pi\hbar^2}{MT}\right)^{\frac{3}{2}} = cV_Q \; .} \tag{15}$$

The result $\lambda = cV_Q$ is a compact, useful, and memorable expression for the absolute activity of an ideal monatomic gas. The activity is equal to the concentration times the quantum volume. It is applicable only if $\lambda \ll 1$ or $e^{\mu/T} \ll 1$, so that in the classical regime the value of the chemical potential must be negative and less than $-T$.

The integration of (6) was carried out with (7) for ϵ in terms of n^2. Thus (15) is based on an energy scale with zero, or very nearly zero, as the energy of the lowest orbital ($n = 1$). Another choice is used in (74) below.

The classical distribution function is now given by

$$f(\epsilon) = cV_Q e^{-\epsilon/T} \; . \tag{16}$$

The occupancy will be much less than one if

$$cV_Q \ll 1 \; , \tag{17}$$

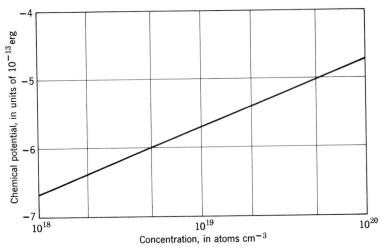

Figure 2   Chemical potential of He⁴ at 300 K, in the ideal gas approximation.

that is, if the average number of atoms in a quantum volume is much less than one. The inequality (17) and thus the ideal gas law fail at high concentration, low temperature, and low molecular weight.

   A plot of V_Q versus temperature for several values of the atomic weight is given in Fig. 1. For a gas at room temperature and atmospheric pressure, the concentration is of the order of 3×10^{19} atoms cm⁻³. We make an estimate of the value of the quantum volume for helium:

$$V_Q = \left(\frac{2\pi\hbar^2}{MT}\right)^{\frac{3}{2}} \approx \left[\frac{6 \times 10^{-54}}{(6 \times 10^{-24})(4 \times 10^{-14})}\right]^{\frac{3}{2}} \approx 10^{-25} \text{ cm}^3 \; ,$$

whence $cV_Q \approx 10^{-6}$ at room temperature and atmospheric pressure. This value is $\ll 1$, so that the gas is in the classical regime. The atmosphere around us is securely in the classical regime.

   The chemical potential is $\mu = T \log \lambda$, by the definition of λ. With the result (15) for λ, we have for the chemical potential of a monatomic ideal gas

$$\frac{\mu}{T} = \log cV_Q = \log c + \log V_Q = \log \frac{N}{V} + \tfrac{3}{2} \log \frac{2\pi\hbar^2}{MT} \; . \qquad (18)$$

This result is plotted for helium in Fig. 2.

   Notice that high values of the concentration mean high values of the chemical potential: the expression for μ contains the logarithm of the concentration. This dependence on concentration is an important intuitive aspect of the chemical potential. High chemical potential is associated with high concentration. Particles tend to move from regions of high concentration to regions of low concentration.

ENERGY

The total energy of the ideal gas is

$$U = \sum_l \epsilon_l f(\epsilon_l) = \lambda \sum_l \epsilon_l e^{-\epsilon_l/T} , \tag{19}$$

but we will evaluate the energy another way.[1] The partition function for one atom is

$$Z = \sum_l e^{-\epsilon_l/T} = \tfrac{1}{2}\pi \int_0^\infty dn\, n^2\, e^{-(\hbar\pi n)^2/2ML^2T} ; \tag{20}$$

with the new variable

$$x \equiv \frac{\hbar\pi n}{L} \cdot \frac{1}{(2MT)^{\frac{1}{2}}} ,$$

we have

$$Z = T^{\frac{3}{2}} \cdot \tfrac{1}{2}\pi \cdot (2M)^{\frac{3}{2}} \left(\frac{L}{\hbar\pi}\right)^3 \int_0^\infty dx\, x^2\, e^{-x^2} = \text{constant} \times T^{\frac{3}{2}} . \tag{21}$$

We obtain the energy from the partition function by (6.36):

$$U = T^2 \frac{\partial}{\partial T} \log Z ,$$

whence

$$U = T^2 \frac{\partial}{\partial T} (\log T^{\frac{3}{2}} + \text{constant}) . \tag{22}$$

It follows that the energy of the ideal gas is

$$\boxed{U = \tfrac{3}{2}T,} \tag{23}$$

per atom. This is an example of a famous result known as the principle of equipartition of energy: **the average kinetic energy of translational motion in the classical limit is equal to $\tfrac{1}{2}T$ or $\tfrac{1}{2}k_B T$ per translational degree of freedom of an atom.** The atoms move in three dimensions: each dimension of motion for each atom is called a **degree of freedom.**

---

[1] In the direct method we evaluate the energy as

$$U = \tfrac{1}{2}\pi\lambda \int_0^\infty dn\, n^2 \epsilon e^{-\epsilon/T} . \tag{19a}$$

With $x^2 \equiv \pi^2\hbar^2 n^2/2ML^2T$ as in (9), we have

$$U = \tfrac{1}{2}\pi\lambda \left(\frac{\pi^2\hbar^2}{2ML^2T}\right)^{-\frac{3}{2}} T \int_0^\infty dx\, x^4 e^{-x^2} , \tag{19b}$$

where by (2.51) the definite integral has the value $\tfrac{3}{8}\sqrt{\pi}$. Now substitute $N = V\lambda/V_Q$ to obtain the result $U = \tfrac{3}{2}NT$.

A more general form of the principle of equipartition of energy applies to a harmonic oscillator in the classical limit. In (6.28) we found that the energy of a harmonic oscillator in one dimension was equal to $\mathcal{T}$ in the high temperature limit $\mathcal{T} \gg \hbar\omega$. This result may be interpreted by classical statistical mechanics, Appendix E. Of the energy $\mathcal{T}$, the quantity $\frac{1}{2}\mathcal{T}$ is the thermal average kinetic energy and the other $\frac{1}{2}\mathcal{T}$ is the thermal average potential energy. This value for the thermal average potential energy applies only to a harmonic oscillator; the actual value depends on the details of the potential energy function. The result is different for an anharmonic oscillator. A polyatomic molecule has rotational degrees of freedom, and the average energy of each rotational degree of freedom is $\frac{1}{2}\mathcal{T}$ when the temperature is high in comparison with the energy difference between the rotational energy levels of the molecule. The rotational energy is kinetic (see Problem 9). A linear molecule has two degrees of rotational freedom which can be excited; a nonlinear molecule has three degrees of rotational freedom.

Problem 1. _Energy of gas of extreme relativistic particles._ Extreme relativistic particles have momenta p such that $pc \gg Mc^2$, where M is the rest mass of the particle. The de Broglie relation $\lambda = h/p$ for the quantum wavelength continues to apply. Show that the mean energy per particle of a nondegenerate extreme relativistic gas is $3\mathcal{T}$ if $\epsilon \cong pc$, in contrast to $\frac{3}{2}\mathcal{T}$ for the nonrelativistic problem. (An interesting variety of relativistic problems are discussed by E. Fermi in _Notes on thermodynamics and statistics_, University of Chicago Press, 1966, paperback.)

ENTROPY

Chemical potential was defined in Chapter 5 in terms of the derivative of the entropy with respect to N, taken at constant U. To find an expression for the entropy of the ideal gas, we first state the chemical potential as a function of N and U, instead of N and $\mathcal{T}$. To do this we use the result $U = \frac{3}{2}N\mathcal{T}$ or $\mathcal{T} = 2U/3N$. Then the result (18) for the chemical potential becomes

$$\frac{\mu}{\mathcal{T}} = \log \frac{N}{V} + \tfrac{3}{2} \log \frac{3\pi\hbar^2 N}{MU} \ . \tag{24}$$

It is convenient to rearrange (24) to group together the terms that contain the number of particles N:

$$\frac{\mu}{\mathcal{T}} = \tfrac{3}{2} \log \left(\frac{3\pi\hbar^2}{MUV^{\frac{2}{3}}} \right) + \tfrac{5}{2} \log N \ . \tag{25}$$

By the definition of the chemical potential μ,

$$\left(\frac{\partial \sigma}{\partial N}\right)_{U, V} = -\frac{\mu}{T} \ . \tag{26}$$

We obtain the entropy by integrating (26) at constant U and V to obtain

$$\int d\sigma = \int dN \left(-\frac{\mu}{T}\right) , \tag{27}$$

or

$$\sigma(N, U, V) = \int dN \left(-\frac{\mu}{T}\right) . \tag{28}$$

We substitute (25) in (28) and make use of the integral

$$\int_0^N dN \log N = N \log N - N \ . \tag{29}$$

We find

$$\sigma(N, U, V) = \tfrac{3}{2}N \log \left(\frac{MUV^{\frac{2}{3}}}{3\pi\hbar^2}\right) - \tfrac{3}{2}N \log N + \tfrac{5}{2}N \ . \tag{30}$$

This is the desired result. But for many purposes we want to reintroduce the temperature.

By use of $U = \tfrac{3}{2}NT$ we may rearrange (30) to obtain the entropy as a function of the temperature instead of the energy:

$$\boxed{\sigma(N, T, V) = N \log \left[\left(\frac{MT}{2\pi\hbar^2}\right)^{\frac{3}{2}} \left(\frac{V}{N}\right)\right] + \tfrac{5}{2}N} \tag{31}$$

or

$$\sigma(N, T, V) = N(-\log cV_Q + \tfrac{5}{2}) \ . \tag{32}$$

In the classical regime $cV_Q \ll 1$, so that $-\log cV_Q$ is positive.

This result is known as the **Sackur-Tetrode equation** for the entropy of a monatomic ideal gas.[2] It is very important historically and is essential in chemical thermodynamics. Even though the equation contains $\hbar$, the basic result was inferred from experiments on vapor pressure and on equilibrium in chemical reactions long before the quantum-mechanical basis was fully known or understood. It was a great challenge to theoretical physicists to explain the result, and many vain attempts were made in the early years of this century. We shall encounter applications of (30) and (31) in later chapters.

[2] O. Sackur, "Die Anwendung der kinetischen Theorie der Gase auf chemische Probleme," Annalen der Physik **36**, 958–980 (1911); see also O. Sackur, Annalen der Physik **40**, 67 (1913), and H. Tetrode, Annalen der Physik **38**, 434 (1912); **39**, 255 (1912).

Electron spin of atom, in units of $\hbar$	Spin of nucleus, in units of $\hbar$	Total number of independent spin states	Total spin entropy
0	I	$(2I + 1)^N$	$N \log (2I + 1)$
S	0	$(2S + 1)^N$	$N \log (2S + 1)$
S	I	$(2S + 1)^N(2I + 1)^N$	$N \log (2S + 1) + N \log (2I + 1)$

We have omitted the spin contribution $N \log (2I + 1)$ to the entropy of the ideal gas. This arises as the logarithm of the $(2I + 1)^N$ independent spin states that may be formed from N atoms of spin I. The contribution is called the **spin entropy**. If there are no unpaired electrons, the symbol I will refer to the nuclear spin alone. The spin entropy is given in the table below for an electronic system of spin S and a nuclear system of spin I. There may be both electronic and nuclear contributions to the spin entropy.

The entropy of the ideal gas is directly proportional to the number of particles N if the concentration N/V is constant, as we see from (32). When two identical gases are placed side-by-side, each system having entropy σ_1, the total entropy is $2\sigma_1$. We see that the entropy scales as the size of the system: the entropy is linear in the number of particles.

EXPERIMENTAL TESTS OF THE SACKUR-TETRODE EQUATION

We have seen that we can calculate the entropy of a monatomic ideal gas by use of the Sackur-Tetrode equation (31). The value thus calculated for one mole of a monatomic gas at a selected temperature and pressure may be compared with the experimental value of the entropy of the gas. The experimental value may be found by summing up several contributions, which typically may include the following:

1. Entropy increase on heating solid from absolute zero to the melting point.
2. Entropy increase in the solid to liquid transformation.
3. Entropy increase on heating liquid from melting point to the boiling point.
4. Entropy increase in the liquid to gas transformation.
5. Entropy change on taking gas from the boiling point to the selected temperature and pressure.

There may further be a slight correction for the nonideality of the gas.

Comparisons of experimental and theoretical values have now been carried out for many gases, and very satisfactory agreement is found between the two sets of values.[3]

We give details of the comparison for neon, after the measurements of Clusius[4]:

1. The heat capacity of the solid was measured from 12.3 K to the melting point 24.55 K under one atmosphere of pressure. The heat capacity of the solid below 12.3 K was estimated by a Debye law (Chapter 16) extrapolation to absolute zero of the measurements above 12.3 K. The entropy of the solid at the melting point is found from numerical integration of $\int dT(C_p/T)$ to be

$$S_{solid} = 14.29 \text{ J mol}^{-1} \text{ deg}^{-1} .$$

2. The heat input required to melt the solid at 24.55 K is observed to 335 J mol^{-1}. The associated entropy of melting is

$$\Delta S_{melting} = \frac{335 \text{ J mol}^{-1}}{24.55 \text{ deg}} = 13.64 \text{ J mol}^{-1} \text{ deg}^{-1} .$$

3. The heat capacity of the liquid was measured from the melting point to the boiling point of 27.2 K under one atmosphere of pressure. The entropy increase was found to be

$$\Delta S_{liquid} = 3.85 \text{ J mol}^{-1} \text{ deg}^{-1} .$$

4. The heat input required to vaporize the liquid at 27.2 K was observed to be 1761 J mol^{-1}. The associated entropy of vaporization is

$$\Delta S_{vaporization} = \frac{1761 \text{ J mol}^{-1}}{27.2 \text{ deg}} = 64.62 \text{ J mol}^{-1} \text{ deg}^{-1} .$$

The experimental value of the entropy of one mole of gas at 27.2 K at a pressure of one atmosphere is

$$S_{gas} = S_{solid} + \Delta S_{melting} + \Delta S_{liquid} + \Delta S_{vaporization} = 96.40 \text{ J mol}^{-1} \text{ deg}^{-1} .$$

The calculated value of the entropy of neon under the same conditions is

$$S_{gas} = 96.45 \text{ J mol}^{-1} \text{ deg}^{-1} ,$$

from the Sackur-Tetrode equation. The excellent agreement with the experimental value gives us confidence in the basis of the entire theoretical apparatus

---

[3] A classic study is "The heat capacity of oxygen from 12 K to its boiling point and its heat of vaporization. The entropy from spectroscopic data," W. F. Giauque and H. L. Johnston, Journal of the American Chemical Society **51**, 2300 (1929).
[4] K. Clusius, Zeitschrift für Physikalische Chemie **B31**, 459 (1936).

Table 1 *Comparison of Experimental and Calculated Values of the Entropy at the Boiling Point under One Atmosphere*

Gas	$T_{b.p.}$, in deg	Entropy in J mol⁻¹ deg⁻¹	
		Experimental	Calculated
Ne	27.2	96.40	96.45
Ar	87.29	129.75	129.24
Kr	119.93	144.56	145.06

From *Landolt Börnstein* tables, 6th ed., Vol. 2, Part 4, pp. 394–399.

that led to the Sackur-Tetrode equation. The result (31) is hardly one that we could have just guessed; to find it verified by observation is a real experience. Results for argon and krypton are given in Table 1.

Problem 2. *Integration of the thermodynamic identity for an ideal gas.* From the thermodynamic identity at constant number of particles we have

$$dS = \frac{dU}{T} + \frac{p\,dV}{T} = \frac{1}{T}\left(\frac{\partial U}{\partial T}\right)_V dT + \frac{1}{T}\left(\frac{\partial U}{\partial V}\right)_T dV + \frac{p\,dV}{T} \ .$$

Show that for an ideal gas the entropy is

$$S = C_V \log T + Nk_B \log V + S_1 \ , \tag{33}$$

where S_1 is a constant, independent of T and V. We use the fact that $(\partial U/\partial V)_T = 0$, for an ideal gas. Note that (31) and (32) are more powerful results than (33), because they give the value of S_1 in terms of fundamental constants; that is, they give the absolute entropy.

Problem 3. *Entropy of mixing.* Suppose that a system of N atoms of type A is placed in diffusive contact with a system of N atoms of type B at the same temperature and volume. Show that after diffusive equilibrium is reached the total entropy is increased by $2N \log 2$. If the atoms are identical ($A \equiv B$), show that there is no increase in entropy when diffusive contact is established. The entropy increase $2N \log 2$ is known as the **entropy of mixing.**

PRESSURE

The entropy of an ideal gas depends on the volume as

$$\sigma(V) = N \log V + \text{constant} ,\qquad (34)$$

as we see from (31) or (33). The pressure of the gas is obtained by use of the relation

$$\frac{p}{T} = \left(\frac{\partial \sigma}{\partial V}\right)_{N, U} ,\qquad (35)$$

whence

$$\frac{p}{T} = \frac{N}{V} .\qquad (36)$$

We may write (36) in the form

$$\boxed{\begin{aligned} pV &= NT ; \\ pV &= Nk_B T . \end{aligned}}\qquad (37)$$

This is called the **ideal gas law.**

A relationship between pressure, volume, and temperature for a gas, liquid, or solid is called the **equation of state;** thus (37) is the equation of state of an ideal gas.

One mole of gas contains N_0 molecules, and for one mole we have

$$pV = N_0 k_B T = RT ,\qquad (38)$$

where the **gas constant** R is defined by

$$R \equiv N_0 k_B = (6.02252 \times 10^{23} \text{ atoms mol}^{-1})(1.38054 \times 10^{-16} \text{ erg deg}^{-1})$$
$$= 8.31434 \times 10^7 \text{ ergs mol}^{-1} \text{ deg}^{-1} .\qquad (39)$$

Here N_0 is the **Avogadro number,** defined as the number of molecules in one mole. Its value is 6.02252×10^{23} molecules per mole. (The recent (1969) review of the values of the physical constants gives $N_0 = 6.02217 \times 10^{23}$ and $k_B = 1.38062 \times 10^{-16}$.)

The ideal gas law (37) is one of the most important results in the statistical theory of gases. The present derivation was achieved by use only of the fundamental assumptions of statistical mechanics. The ideal gas law can be derived also by quite elementary kinetic arguments if the result $U = \frac{3}{2}NT$ for the energy is assumed. We give an elementary argument in Chapter 13.

Yet another form of the ideal gas law is often used, particularly in astrophysics. The mass density ρ is defined by

$$\rho = \frac{NM}{V} , \tag{40}$$

where M is the mass of an atom. Thus $N/V = \rho/M$, or

$$p = \frac{T}{M}\rho ; \qquad \rho = \frac{M}{T}p . \tag{41}$$

The right-hand version of this relation has the form

$$\text{response} = \text{constant} \times \text{force} , \tag{42}$$

if the density ρ is the response and the pressure p is the force. The law of paramagnetism (4.39) has a similar form, $\mathcal{M} = (N\mu_0^2/T)H$, where the magnetic moment $\mathcal{M}$ is the response and the magnetic field H is the force.

Problem 4. *Relation of pressure and energy density.* (a) Show that the average pressure in a system in thermal contact with a heat reservoir is given by

$$p = -\frac{\sum_l \left(\dfrac{d\epsilon_l}{dV}\right) e^{-\epsilon_l/T}}{Z} , \tag{43}$$

where the sum is over all states of the system. (b) From the result (10.20) show for a gas of free particles that

$$\left(\frac{\partial \epsilon_l}{\partial V}\right)_N = -\frac{2}{3}\frac{\epsilon_l}{V} , \tag{44}$$

as a result of the boundary conditions of the problem. The result (44) holds equally whether ϵ_l refers to a state of N noninteracting particles or to an orbital. (c) Show that for a gas of free nonrelativistic particles

$$p = \frac{2U}{3V} , \tag{45}$$

where U is the thermal average energy of the system. This result is not limited to the classical regime; it holds equally for fermion and boson particles, as long as they have a nonzero rest mass.

HEAT CAPACITY

The heat capacity of a system held at constant volume is defined as

$$C_V = T\left(\frac{\partial S}{\partial T}\right)_V .$$ (46)

It is understood that the number of atoms is held constant in the differentiation. We may use the thermodynamic identity to express C_V as

$$C_V \equiv \left(\frac{\partial U}{\partial T}\right)_V = k_B\left(\frac{\partial U}{\partial \tau}\right)_V ,$$ (47)

because $dU = T\,dS$ in a change in which $dN = 0$ and $dV = 0$. For the ideal monatomic gas $U = \frac{3}{2}N\tau$, from (23), so that

$$C_V = \frac{3}{2}Nk_B ,$$ (48)

or $\frac{1}{2}k_B$ per degree of freedom. For one mole, $C_V = \frac{3}{2}R$.

The heat capacity at constant pressure is defined by

$$C_p = T\left(\frac{\partial S}{\partial T}\right)_p ,$$ (49)

with the number of particles understood to be constant. We use the thermodynamic identity to obtain

$$T\left(\frac{\partial S}{\partial T}\right)_p = \left(\frac{\partial U}{\partial T}\right)_p + p\left(\frac{\partial V}{\partial T}\right)_p .$$ (50)

We expect C_p to be larger than C_V, because the gas will do no external work when heated at constant volume, but when heated at constant pressure the gas expands and does work against the external pressure. The work done by the gas appears as a contribution to C_p.

We want to find an expression for the difference between C_p and C_V. Let us first consider C_V for a *polyatomic* ideal gas. To a good approximation the energy is the sum of translational contributions plus vibrational and rotational contributions, grouped together as internal contributions. Thus

$$U = U_{\text{translational}} + U_{\text{internal}} ,$$ (51)

or

$$U = \frac{3}{2}N\tau + Nu_{\text{internal}} ,$$ (52)

where u_{internal} is the vibrational and rotational energy of one molecule. The heat capacity at constant volume is

$$C_V = \left(\frac{\partial U}{\partial T}\right)_V = \frac{3}{2}Nk_B + Nk_B\frac{\partial u_{\text{internal}}}{\partial T} ,$$ (53)

a generalization of (48). From (52) we also have

$$\left(\frac{\partial U}{\partial T}\right)_p = \frac{3}{2} Nk_B + Nk_B \frac{\partial u_{\text{internal}}}{\partial T} , \qquad (54)$$

which is identical with (53) for $(\partial U/\partial T)_V$. (This result is limited to the ideal gas.)

Now

$$pV = Nk_B T ; \qquad p\left(\frac{\partial V}{\partial T}\right)_p = Nk_B , \qquad (55)$$

so that (50) becomes

$$C_p = C_V + Nk_B . \qquad (56)$$

A general expression for $C_p - C_V$ applicable to solids, liquids, and gases is given as an example in Chapter 19. For an ideal monatomic gas we have

$$C_p = \tfrac{3}{2}Nk_B + Nk_B = \tfrac{5}{2}Nk_B . \qquad (57)$$

For one mole of gas

$$\boxed{C_p - C_V = R ,} \qquad (58)$$

from (56).

Problem 5. Alternate evaluation of C_V for monatomic ideal gas. (a) Show from (31) that

$$S(T) = Nk_B \log T^{\frac{3}{2}} + Nk_B \log V + S_1 . \qquad (59)$$

(b) Verify from this result that

$$C_V = \tfrac{3}{2}Nk_B . \qquad (60)$$

This is an alternate derivation to (47) of the result for C_V.

FLUCTUATIONS IN NUMBER OF PARTICLES

We calculate for an ideal gas the mean square deviation of N from $\langle N \rangle$, in order to see if the total number of particles in a system in diffusive contact with a reservoir is well-defined. The mean square deviation is

$$\langle (\Delta N)^2 \rangle \equiv \langle (N - \langle N \rangle)^2 \rangle = \langle N^2 \rangle - 2\langle N \rangle\langle N \rangle + \langle N \rangle^2 = \langle N^2 \rangle - \langle N \rangle^2 .$$

(61)

We saw in (6.62) that

$$\langle (\Delta N)^2 \rangle = T \frac{\partial \langle N \rangle}{\partial \mu} .$$

(62)

For an ideal gas the number of particles is related to the chemical potential by

$$\langle N \rangle = e^{\mu/T}(V/V_Q) ,$$

(63)

from (15). Then the mean square fluctuation is

$$\langle (\Delta N)^2 \rangle = T \frac{\partial \langle N \rangle}{\partial \mu} = \langle N \rangle ,$$

(64)

and the mean square fractional fluctuation is

$$\boxed{\frac{\langle (\Delta N)^2 \rangle}{\langle N \rangle^2} = \frac{1}{\langle N \rangle} .}$$

(65)

The fractional fluctuations are extremely small if there is a macroscopic number of particles in the volume under consideration. If $\langle N \rangle = 10^{20}$, the root mean square fractional fluctuation is

$$\left[\frac{\langle (\Delta N)^2 \rangle}{\langle N \rangle^2} \right]^{\frac{1}{2}} = 10^{-10} .$$

(66)

This demonstrates the accuracy with which a system in diffusive contact simulates a system with a fixed total number of particles.

Problem 6. *Fluctuations in a Fermi gas.* Show for a single orbital of a fermion system that

$$\langle (\Delta n)^2 \rangle = \langle n \rangle(1 - \langle n \rangle) ,$$

(67)

if $\langle n \rangle$ is the average number of fermions in that orbital. Notice that the fluctuation vanishes for orbitals with energies deep enough below the Fermi energy so that $\langle n \rangle = 1$.

Problem 7. *Fluctuations in a Bose gas.* Show that if $\langle n \rangle$ is the occupancy of a single orbital of a boson system, then

$$\langle (\Delta n)^2 \rangle = \langle n \rangle (1 + \langle n \rangle) \ . \tag{68}$$

This shows that if the occupancy is large, with $\langle n \rangle \gg 1$, then the fractional fluctuations are of the order of unity:

$$\frac{\langle (\Delta n)^2 \rangle}{\langle n \rangle^2} \approx 1 \ , \tag{69}$$

so that the actual fluctuations can be enormous.[5] The physical basis is discussed in Chapter 15.

FLUCTUATIONS IN ENERGY

A system in thermal contact with a reservoir does not have a precisely constant energy. Even the act of defining the temperature of a system by placing it in contact with a thermometer or a heat reservoir leads to an uncertainty in the value of the energy. In Chapter 6 we found a famous expression for the mean square fluctuation of the energy ϵ of the system:

$$\langle (\epsilon - \langle \epsilon \rangle)^2 \rangle = k_B T^2 C_V \ , \tag{70}$$

where C_V is the heat capacity at constant volume. For an ideal monatomic gas

$$C_V = \tfrac{3}{2} N k_B \ ; \qquad U = \tfrac{3}{2} N k_B T = \langle \epsilon \rangle \ ,$$

so that

$$\frac{\langle (\epsilon - \langle \epsilon \rangle)^2 \rangle}{\langle \epsilon \rangle^2} = \frac{2}{3N} \ . \tag{71}$$

Thus the root mean square fractional fluctuation in energy of an ideal gas is of the order of $1/\sqrt{N}$. For $N \approx 10^{20}$ the fractional fluctuation is of the order of 10^{-10}, which is negligibly small. For systems of usual laboratory dimensions the energy for all practical purposes is just as well-defined when the system is in thermal contact with a reservoir as when the system is perfectly insulated or closed.

---

[5] Fluctuations may be reduced, however, if the total number of particles in the system is held constant; see M. J. Klein and L. Tisza, Physical Review **76**, 1861 (1949), and references cited there.

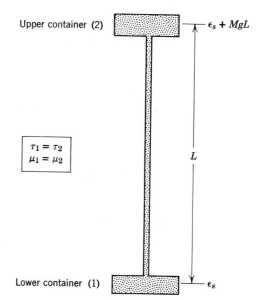

Upper container (2)

$\epsilon_s + MgL$

$$\tau_1 = \tau_2$$
$$\mu_1 = \mu_2$$

L

Lower container (1)

ϵ_s

Figure 3   Two containers of gas in thermal and diffusive contact in a uniform gravitational field. If an orbital s in the lower container has energy ϵ_s, the corresponding orbital in the upper container has energy $\epsilon_s + MgL$. Here g is the acceleration of gravity. Because the two containers are in contact, $\mu_1 = \mu_2$ and $\tau_1 = \tau_2$.

FLUCTUATIONS IN PRESSURE

As was noted by Gibbs in his treatise, fluctuations in pressure are more difficult to discuss than fluctuations in energy. It is meaningless to speak of the instantaneous pressure, because of the instantaneous nature of molecular impacts with the boundary. Hence we cannot discuss pressure fluctuations in a gas without bringing in the idea of the frequency spectrum of the fluctuations. The concept of the power spectrum of a noise source or of fluctuations is valuable in physics, but it lies beyond the modest limits we have set for this book, except for the discussion in Appendix H of noise in electrical circuits.

EQUILIBRIUM IN A GRAVITATIONAL FIELD

We treat an ideal gas in static equilibrium in a uniform gravitational field of acceleration g. We want to find the variation of pressure as a function of height. The particular geometry is displayed in Fig. 3 and is chosen for convenience. The two containers, which might be cut out from a larger volume, are in thermal and diffusive contact, so that $\tau_1 = \tau_2$ and $\mu_1 = \mu_2$. The behavior of the chemical potential is the key to the problem.

The grand sum for the lower container is

$$\mathfrak{Z}_1 = \sum_N \sum_l e^{[N\mu_1 - \epsilon_l(N)]/\tau} \ . \tag{72}$$

where $\epsilon_l(N)$ is the energy of an N-particle state in the lower container. The corresponding N-particle state l in the upper container has its energy increased by the gravitational potential energy $NMgL$ of N particles of mass M displaced by a distance L in the gravitational field g. The energy of this state in the upper container is

$$\epsilon_l(N) + NMgL , \tag{73}$$

so that the grand sum for the upper container may be written as

$$\mathfrak{Z}_2 = \sum_N \sum_l e^{[N\mu_2 - \epsilon_l(N) - NMgL]/\mathcal{T}} = \sum_N \sum_l e^{[N(\mu_2 - MgL) - \epsilon_l(N)]/\mathcal{T}} . \tag{74}$$

Wherever μ_1 appears in $\mathfrak{Z}_1$, there $\mu_2 - MgL$ appears in $\mathfrak{Z}_2$. For the lower container we have, from (18),

$$\mu_1 = \mathcal{T} \log c_1 V_Q . \tag{75}$$

Here c_1 is the concentration in the lower container. The integral (6) which led to (11) will be modified for the upper container because

$$\lambda_2 e^{-MgL/\mathcal{T}}$$

will appear in every line in place of λ_1; therefore for the upper container

$$\mu_2 - MgL = \mathcal{T} \log c_2 V_Q , \tag{76}$$

or

$$\mu_2 = \mathcal{T} \log c_2 V_Q + MgL . \tag{77}$$

Observe that μ_2 has been increased by the potential energy MgL of one particle.

The two chemical potentials, μ_1 and μ_2, are equal in diffusive equilibrium. We must decrease the concentration in the upper container, for we must have

$$\mu_1 = \mathcal{T} \log c_1 V_Q = \mathcal{T} \log c_2 V_Q + MgL = \mu_2 , \tag{78}$$

or

$$\mathcal{T} \log \frac{c_1}{c_2} = MgL , \tag{79}$$

whence

$$\boxed{c_2 = c_1 e^{-MgL/\mathcal{T}} .} \tag{80}$$

This gives the dependence of the concentration on the height L.

The pressure of an ideal gas is proportional to the concentration; therefore

$$\boxed{p_2 = p_1 e^{-MgL/\mathcal{T}} .} \tag{81}$$

This is called the **barometric pressure equation**. It gives the dependence of the

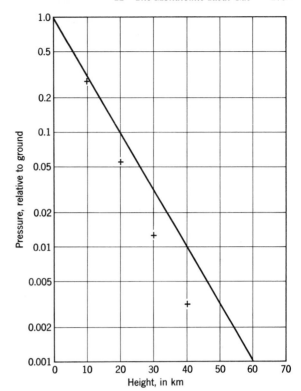

Figure 4   Decrease of pressure with altitude, for atmosphere of N_2 at 290 K. Crosses represent the average atmosphere as sampled on rocket flights. The actual atmosphere is not isothermal.

pressure on the height L in an isothermal atmosphere of a single chemical species.

We see from (81) that there is a characteristic height T/Mg at which the atmospheric pressure decreases by the fraction $e^{-1} \cong 0.37$. To estimate the characteristic height, let us consider an isothermal atmosphere composed of nitrogen molecules with a molecular weight of 28. The mass of a molecule is $(28)(1.66 \times 10^{-24}$ gm$) \cong 48 \times 10^{-24}$ gm. At a temperature of 290 K the value of T is $(290$ K$)(1.38 \times 10^{-16}$ erg deg$^{-1} \cong 4.0 \times 10^{-14}$ erg. With g, the acceleration of gravity, as 980 cm^2 sec^{-1}, we have for the characteristic height

$$\frac{T}{Mg} \cong \frac{4.0 \times 10^{-14} \text{ erg}}{(48 \times 10^{-24} \text{ gm})(980 \text{ cm}^2 \text{ sec}^{-1})} \cong 0.85 \times 10^6 \text{ cm} = 8.5 \text{ km} . \quad (82)$$

This is approximately 5 miles. Lighter molecules such as H_2 and He will extend farther up, but these have largely escaped from the atmosphere in the course of time.

The pressure calculated for N_2 at 290 K is plotted against height in Fig. 4. Experimental values of the total pressure as found from rocket observations are also shown. (The actual atmosphere of the earth is not isothermal.)

CHEMICAL POTENTIAL IN A FORCE FIELD

We obtained in (15) the expression

$$\mu = T \log c V_Q \tag{83}$$

for the chemical potential of a monatomic ideal gas with zero spin; here c is the concentration and V_Q is the quantum volume. This result was calculated with zero as the energy of the lowest orbital. In a gravitational field the energy of all orbitals is increased by the gravitational potential energy MgL of one particle at height L above the reference level. We found for the chemical potential at L the expression

$$\mu(L) = T \log c V_Q + MgL \ , \tag{84}$$

where c is now the concentration at L. This equation has two terms typical for the chemical potential, one that is

$$T \log \textbf{(fractional concentration)}$$

and one that is the

potential energy of a particle.

In (84) we notice that $c V_Q$ is the probability that there is a particle in the quantum volume V_Q.

The result is easily generalized to a system of charged particles in an electrostatic potential $\varphi(\mathbf{r})$. If q is the charge of a particle, then $q\varphi(\mathbf{r})$ is the potential energy of the particle. It follows that

$$\mu(\mathbf{r}) = T \log c(\mathbf{r}) V_Q + q\varphi(\mathbf{r}) \ . \tag{85}$$

But for a system in diffusive equilibrium the chemical potential is constant and independent of $\mathbf{r}$. Thus changes in the electrostatic potential from $\mathbf{r}_1$ to $\mathbf{r}_2$ must be compensated by changes in the particle concentration from $\mathbf{r}_1$ to $\mathbf{r}_2$:

$$T \log c(\mathbf{r}_1) V_Q + q\varphi(\mathbf{r}_1) = T \log c(\mathbf{r}_2) V_Q + q\varphi(\mathbf{r}_2) \ , \tag{86}$$

or

$$\frac{c(\mathbf{r}_2)}{c(\mathbf{r}_1)} = e^{q[\varphi(\mathbf{r}_1) - \varphi(\mathbf{r}_2)]/T} \ . \tag{87}$$

This result relates the ratio of the particle concentrations to the potential difference. The result is of special importance for semiconductor devices: a potential difference across a pn junction causes a difference in electron concentration across the junction.

CHEMICAL POTENTIAL OF IDEAL GAS
WITH INTERNAL DEGREES OF FREEDOM*

A gas need not be monatomic to be ideal. We consider here an ideal gas of identical polyatomic molecules. Each molecule has rotational and vibrational motions: these are internal degrees of freedom of the molecule as distinguished from the translational degrees of freedom. We suppose that the total energy ϵ_t of an orbital t of the molecule is the sum of two parts,

$$\epsilon_t = \epsilon_i + \epsilon_{\mathbf{n}} , \tag{88}$$

where ϵ_i refers to the internal degrees of freedom and $\epsilon_{\mathbf{n}}$ to the translational motion of the center of mass of the molecule. For the translational motion we know that

$$\epsilon_{\mathbf{n}} = \frac{\hbar^2}{2M} \left(\frac{\pi n}{L} \right)^2 , \tag{89}$$

where $\mathbf{n}$ is the quantum number of the translational orbital as in Chapter 10. The dependence of the vibrational energy ϵ_i of the molecule on the force constant was treated in Chapter 7; the rotational energy is the subject of Problem 9.

We have assumed in writing (88) that the orbital φ_t of the molecule can be factored into internal and external parts,

$$\varphi_t = \varphi_i \varphi_{\mathbf{n}} , \tag{90}$$

where φ_i relates to rotational and vibrational excitation and $\varphi_{\mathbf{n}}$ describes the translational motion.

In the classical regime the probability that a given translational orbital $\mathbf{n}$ be occupied is always very small in comparison with one. When we write the grand sum in the classical regime for the system that consists of this orbital, we neglect terms in λ^2 and higher powers of λ, because such terms correspond to occupancy of the orbital by more than one molecule. This approximation is a matter of the classical regime as defined at the beginning of this section. For a fermion system the terms in λ^2 and higher do not enter in any event.

Accordingly, the grand sum for the system of all orbitals t for which the translational quantum number is precisely $\mathbf{n}$ and for which the internal quantum number i assumes all possible values is

$$\mathfrak{Z} = 1 + \lambda \sum_i e^{-(\epsilon_i + \epsilon_{\mathbf{n}})/T} , \tag{91}$$

in the classical regime. We may factor $e^{-\epsilon_{\mathbf{n}}/T}$ to obtain

$$\mathfrak{Z} = 1 + \lambda \left(\sum_i e^{-\epsilon_i/T} \right) e^{-\epsilon_{\mathbf{n}}/T} . \tag{92}$$

° This section may be omitted on the first reading.

We introduce the partition function Z_{int} of the internal degrees of freedom:

$$Z_{int} \equiv \sum_i e^{-\epsilon_i/T} , \tag{93}$$

whence (81) becomes

$$\mathcal{Z} = 1 + \lambda Z_{int} e^{-\epsilon_n/T} . \tag{94}$$

The probability that the translational orbital **n** is occupied, irrespective of the state i of internal motion of the molecule, is given by the ratio of the term in λ to the grand sum $\mathcal{Z}$:

$$f(\epsilon_n) = \frac{\lambda Z_{int} e^{-\epsilon_n/T}}{1 + \lambda Z_{int} e^{-\epsilon_n/T}} \cong \lambda Z_{int} e^{-\epsilon_n/T} . \tag{95}$$

The classical regime is defined as $f(\epsilon_n) \ll 1$. The result (95) is entirely analogous to (3) for the monatomic case, but λZ_{int} now plays the role of λ. The quantity λ in (95) is still the absolute activity and is still related to the chemical potential by $\lambda \equiv e^{\mu/T}$.

Several of the results derived for the monatomic ideal gas are different for the polyatomic ideal gas:

(a) Equation (15) for λ is replaced by

$$\lambda = \frac{cV_Q}{Z_{int}} . \tag{96a}$$

Thus we must add to the chemical potential as given by (18) a new term $-T \log Z_{int}$:

$$\mu = T(\log cV_Q - \log Z_{int}) . \tag{96b}$$

(b) Equation (20) for the energy assumes the form

$$\frac{U}{N} = T^2 \frac{\partial}{\partial T} \log \left(Z_{int} \sum_n e^{-\epsilon_n/T} \right) = T^2 \frac{\partial}{\partial T} \log \sum_n e^{-\epsilon_n/T} + T^2 \frac{\partial}{\partial T} \log Z_{int} . \tag{97}$$

The energy is thereby increased by a term

$$U_{int} = NT^2 \frac{\partial}{\partial T} \log Z_{int} . \tag{98}$$

The former result $U = \frac{3}{2}NT$ applies to the translational energy alone:

$$U_{translational} = \frac{3}{2}NT . \tag{99}$$

(c) The heat capacity is increased because the energy is increased.
(d) The entropy is also increased. An explicit expression for the increase of entropy is given in (101).

(e) The addition to the entropy is independent of the volume, because Z_{int} is independent of the volume. The pressure is

$$\frac{p}{T} = \left(\frac{\partial \sigma}{\partial V}\right)_{N,\, U} , \tag{100}$$

so that the pressure at a given temperature is not changed by the addition of the internal degrees of freedom. We have $pV = NT$, as for the ideal monatomic gas.

(f) Although C_V is changed, the relation $C_p - C_V = R$ is unchanged.

Problem 8. *Entropy of the internal degrees of motion.* Show that the entropy σ_{int} associated with the internal degrees of freedom is given by

$$\sigma_{int} = N \log Z_{int} + NT \frac{\partial}{\partial T} \log Z_{int} = N \frac{\partial}{\partial T} T \log Z_{int} . \tag{101}$$

Hint. Form

$$\sigma_{int} = \int_0^T \frac{\partial U_{int}}{\partial T} \frac{dT}{T} , \tag{102}$$

where U_{int} is given by (98). The result (101) follows on integration of (102) by parts.

Problem 9. *Rotation of molecules.* We consider the rotational states of a diatomic molecule such as CO. The energy of each state is

$$\frac{\hbar^2}{2I} J(J + 1) ,$$

where I is the moment of inertia about an axis through the center of mass and normal to the line connecting the two atoms. The rotational quantum number J may assume the values 0, 1, 2, 3, ... For each J, there are $2J + 1$ states of equal rotational energy; each state is distinguished by the projection of J on an arbitrary direction. (a) Write down the partition function for the rotational states and evaluate it approximately, but explicitly, for temperatures high in comparison with $\hbar^2/I$. *Hint:* Replace the sum by an integral. (b) Write out exact and approximate expressions in the same approximation as (a) for the rotational contributions to the energy, entropy, heat capacity, and free energy. (c) Give the approximate low temperature form of the rotational energy. Sketch roughly the energy versus temperature, using appropriate units for the two axes. Be sure the limiting behaviors are shown correctly.

Summary of Steps Leading to the Ideal Gas Law

(a) $f(\epsilon) = \lambda e^{-\epsilon/T}$

Occupancy of an orbital in the classical limit of $f(\epsilon) \ll 1$.

(b) $\lambda = \dfrac{N}{\sum_n e^{-\epsilon_n/T}}$

Given N, this equation determines λ in the classical limit.

(c) $\epsilon_n = \dfrac{\hbar^2}{2M}\left(\dfrac{\pi n}{V^{\frac{1}{3}}}\right)^2$

Energy of a free particle orbital of quantum number n in a cube of volume V.

(d) $\sum_n e^{-\epsilon_n/T} = \frac{1}{2}\pi \int dn\, n^2 e^{-\epsilon/T}$

Transformation of the sum to an integral.

(e) $\lambda = \dfrac{NV_Q}{V}$

Result of the integration (d) after substitution in (b).

(f) $V_Q = \left(\dfrac{2\pi\hbar^2}{MT}\right)^{\frac{3}{2}}$

Definition of the quantum volume.

(g) $\mu = -T \log V +$ terms independent of the volume

(h) $\sigma = -\int dN \dfrac{\mu}{T} = N \log V +$ terms independent of the volume

(i) $\dfrac{p}{T} = \left(\dfrac{\partial \sigma}{\partial V}\right)_{U,\,N} = \dfrac{N}{V}$

Ideal gas law.

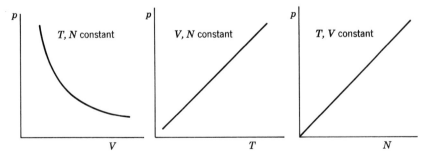

Figure 1 Dependence of pressure of ideal gas on V, T, and N. The gas will not be in the classical regime at very low temperature. The effects of interatomic interactions will be important at low T and high concentration (high N/V).

CHAPTER 12

NUMERICAL CALCULATIONS FOR AN IDEAL MONATOMIC GAS

In this chapter we show how to make numerical calculations for the central thermodynamic properties of an ideal monatomic gas. We consider also the effects on these properties of reversible and irreversible changes in the volume occupied by the gas.

In our model example we consider 1×10^{22} atoms of He^4 at an initial volume of 10^3 cm³ at 300 K. Under these conditions helium approximates quite well to an ideal monatomic gas, and the results of Chapter 11 may be applied with high accuracy.

What is the pressure of the gas?

According to the ideal gas law

$$pV = Nk_BT \; ; \qquad p = Nk_BT/V \; . \tag{1}$$

The variation of pressure with N, T, and V is sketched in Fig. 1. With the Boltzmann constant $k_B = 1.3805 \times 10^{-16}$ erg deg⁻¹, we have

$$p = \frac{(1 \times 10^{22} \text{ atoms})(1.38 \times 10^{-16} \text{ erg deg}^{-1})(300 \text{ K})}{(1 \times 10^3 \text{ cm}^3)}$$

$$= 4.14 \times 10^5 \text{ dynes cm}^{-2} \; . \tag{2}$$

We may express a pressure given in dynes per square centimeter in terms of pressure in standard atmospheres or in terms of bars, another unit of pressure.[1] A pressure of one **standard atmosphere** is defined as

$$1 \text{ atm} \equiv 1.01325 \times 10^6 \text{ dynes cm}^{-2} \; . \tag{3a}$$

This is the pressure exerted by a column of mercury 760 mm high at 0 C at the standard acceleration of gravity, defined as 980.665 cm sec⁻². For the example above

$$p = \frac{4.14 \times 10^5 \text{ dynes cm}^{-2}}{1.0133 \times 10^6 \text{ dynes cm}^{-2} \text{ atm}^{-1}} = 0.408 \text{ atm} \; . \tag{3b}$$

A **bar** is a unit of pressure defined as

$$1 \text{ bar} \equiv 1 \times 10^6 \text{ dynes cm}^{-2} \; . \tag{3c}$$

In the example $p = 0.414$ bar.

---

[1] There is much to be said for the nonproliferation of units. Our standard unit of pressure is the dyne cm⁻², but to understand the literature you will need numerous other units.

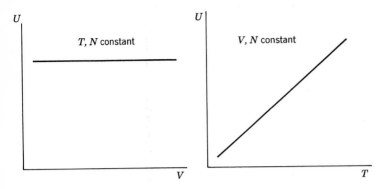

Figure 2 Dependence of energy of ideal gas on V and T.

What is the energy of the gas?

For an ideal monatomic gas we know that

$$U = \tfrac{3}{2}Nk_BT \ . \tag{4}$$

The variation of energy with temperature is sketched in Fig. 2. From (4) we have for the standard example

$$U = \tfrac{3}{2}(1 \times 10^{22})(1.38 \times 10^{-16} \text{ erg deg}^{-1})(300 \text{ K}) = 6.21 \times 10^{8} \text{ ergs} \ . \tag{5}$$

We may express this result in joules and in calories by the conversions

$$10^{7} \text{ ergs} \equiv 1 \text{ joule}$$

and

$$4.184 \times 10^{7} \text{ ergs} \equiv 1 \text{ calorie} \equiv 4.184 \text{ joules} \ . \tag{6}$$

Thus in the model example

$$U \cong 62 \text{ joules} \cong 15 \text{ cal} \ .$$

The abbreviation of joule is J, and the abbreviation of calorie is cal.

A **calorie** is a unit of energy defined as 4.184 J. The original definition of the calorie was the quantity of heat required to raise the temperature of one gram of water by one degree Celsius. Later the specification was added in some countries that the temperature should be raised from 14.5 to 15.5 C at atmospheric pressure. Because there is little need for another unit of energy, the use of the calorie as a unit is sensibly discouraged by international bodies. *Beware:* one Calorie spelled with a capital C is equal to 1000 calories spelled with a lowercase c, or to one kilocalorie. Food Calories are kilocalories.

In the chemical literature the unit of energy **kilocalorie per mole** is used. This unit is 1000 calories per mole of molecules. A mole is $N_0 = 6.0225 \times 10^{23}$ molecules. We may convert kilocalories per mole to electron volts per molecule by the relation

$$23.061 \text{ kcal mol}^{-1} = 1 \text{ eV} \ ,$$

where

$$1 \text{ eV} = 1.6021 \times 10^{-12} \text{ erg} \ .$$

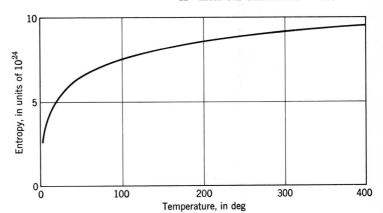

Figure 3a   Entropy of one mole of He⁴ at a pressure of one atmosphere, as a function of the temperature.

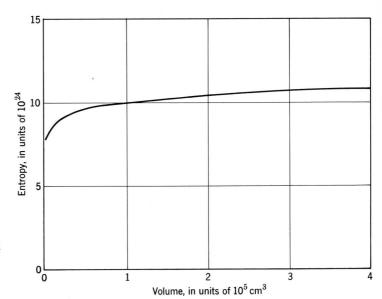

Figure 3b   Entropy of one mole of He³ at 300 K, as a function of the volume.

The value of the gas constant $R \equiv N_0 k_B$ is

$$R = 8.3143 \times 10^7 \text{ erg deg}^{-1} \text{ mol}^{-1}$$

or 1.987 cal deg^{-1} mol^{-1}.

What is the entropy of the gas?

By (11.31) the entropy of an ideal gas of spin of atoms of spin zero is

$$\sigma = N \left[\log \left(\frac{M k_B T}{2\pi \hbar^2} \right)^{\frac{3}{2}} - \log \left(\frac{N}{V} \right) + \frac{5}{2} \right] . \tag{7}$$

The dependence of the entropy on temperature and on volume is shown in Figs. 3a and 3b.

The mass of an atom of He⁴ is found as the product of the atomic weight times the atomic mass unit:

$$M = (4.003)(1.66 \times 10^{-24} \text{ gm}) = 6.64 \times 10^{-24} \text{ gm} ,$$

where we have used the definition

$$1 \text{ unified atomic mass unit} \equiv 1.66042 \times 10^{-24} \text{ gm} , \tag{8}$$

which is $\frac{1}{12}$ the mass of an atom of C^{12}. The atomic weight of He⁴ is 4.003.

With $\hbar = 1.05449 \times 10^{-27}$ erg-sec, we have

$$\frac{Mk_BT}{2\pi\hbar^2} = \frac{(6.64 \times 10^{-24} \text{ gm})(1.38 \times 10^{-16} \text{ erg deg}^{-1})(300 \text{ K})}{(6.28)(1.11 \times 10^{-54} \text{ erg}^2\text{-sec}^2)}$$

$$= 3.94 \times 10^{16} \text{ cm}^{-2} , \tag{9}$$

and

$$\left(\frac{Mk_BT}{2\pi\hbar^2}\right)^{\frac{3}{2}} = 7.8 \times 10^{24} \text{ cm}^{-3} . \tag{10}$$

This quantity is the reciprocal of what we denoted as V_Q in Chapter 11. We have for the quantum volume of He⁴ at 300 K:

$$V_Q = 1.28 \times 10^{-25} \text{ cm}^2 . \tag{11}$$

Under the prescribed conditions of 1×10^{22} atoms in 1×10^3 cm³, we have for the concentration

$$c = \frac{N}{V} = \frac{1 \times 10^{22}}{1 \times 10^3 \text{ cm}^3} = 1 \times 10^{19} \text{ cm}^{-3} . \tag{12}$$

Thus $cV_Q \cong 1.28 \times 10^{-6} \ll 1$, so that the gas is in the classical regime.
From (10) and (12) we form

$$\log\left[\left(\frac{Mk_BT}{2\pi\hbar^2}\right)^{\frac{3}{2}}\left(\frac{V}{N}\right)\right] = \log\left[(7.8 \times 10^{24})(1 \times 10^{-19})\right]$$

$$= 2.30 \log_{10}(7.8 \times 10^5) = (2.30)(5.89) = 13.55 . \tag{13}$$

Here we have used the relation

$$\log x = (\log 10)(\log_{10} x) ,$$

with $\log 10 = 2.303$. This follows on taking the natural logarithm of both sides of the identity

$$x = 10^{\log_{10} x} .$$

From (7) and (13) we have for the entropy of the gas of 10^{22} atoms the value

$$\sigma = (1 \times 10^{22})(13.55 + 2.50) = 1.60 \times 10^{23} , \qquad (14)$$

or

$$S = k_B\sigma = (1.38 \times 10^{-16} \text{ erg deg}^{-1})(1.60 \times 10^{23}) = 2.21 \times 10^7 \text{ ergs deg}^{-1} . \qquad (15)$$

The entropy σ is dimensionless.

If the gas were He^3, which has a nuclear spin $I = \frac{1}{2}$, the entropy would be increased by the spin entropy $N \log (2I + 1) = N \log 2$. Also, the value of V_Q for He^3 is increased with respect to He^4 because of the decrease in the mass.

SLOW ISOTHERMAL EXPANSION

Let the gas expand slowly at constant temperature until the volume is 2×10^3 cm^3. A slow expansion is a **reversible process,** because at any instant the system is in its most probable configuration.

What is the pressure after expansion?

The final volume is twice the initial volume; the final temperature is equal to the initial temperature. Thus by $pV = N\mathcal{T}$ we see that the final pressure is one-half the initial pressure.

What is the entropy after expansion?

The entropy of an ideal gas at constant temperature depends on volume as

$$\sigma(V) = N \log V + \text{constant} , \qquad (16)$$

whence

$$\sigma_2 - \sigma_1 = N \log (V_2/V_1) = N \log 2 = (1 \times 10^{22})(0.693) = 0.069 \times 10^{23} . \qquad (17)$$

The value of the entropy σ_1 before the expansion is given by (14); from it the entropy at a volume $V = 2 \times 10^3$ cm^3 is

$$\sigma_2 = (1.60 + 0.07) \times 10^{23} = 1.67 \times 10^{23} . \qquad (18)$$

Notice that the entropy is larger at the larger volume, because the system has

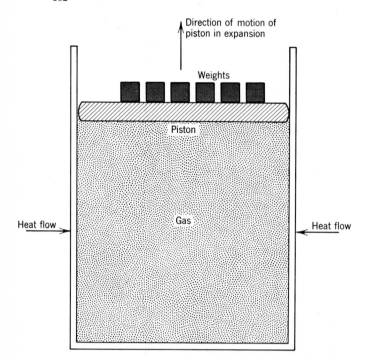

Direction of motion of piston in expansion

Weights

Piston

Gas

Heat flow

Heat flow

Figure 4 Work is done by the gas in an isothermal expansion. Here the gas does work by raising the weights. Under isothermal conditions pV is constant for an ideal gas, so that the pressure must be reduced to allow the volume to expand. The pressure is reduced by removing the load of weights a little at a time.

more accessible states in the larger volume than in the smaller volume at the same temperature.

How much work is done by the gas in the expansion?

When the gas expands isothermally, it does work against a piston, as in Fig. 4. The work done on the piston when the volume is doubled is

$$\int_{V_1}^{V_2} p \, dV = \int_{V_1}^{V_2} dV \frac{N k_B T}{V} = N k_B T \log \left(\frac{V_2}{V_1} \right) = N k_B T \log 2 \ . \quad (19)$$

We may evaluate $N k_B T$ directly, or we may take it as the value of the product $p_1 V_1$ in the initial condition, which is

$$(4.14 \times 10^5 \text{ dynes cm}^{-2})(1 \times 10^3 \text{ cm}^3) = 4.14 \times 10^8 \text{ ergs} \ .$$

Thus the work done on the piston is, from (19),

$$N k_B T \log 2 = (4.14 \times 10^8 \text{ ergs})(0.693) = 2.87 \times 10^8 \text{ ergs} \ . \quad (20)$$

We have defined W as the work done *on* the gas by external agencies. This is the negative of the work done by the gas on the piston, so that

$$W = -\int p \, dV = -2.87 \times 10^8 \text{ ergs} = -28.7 \text{ J} \quad (21)$$

from (19).

What is the change of energy in the expansion?

The energy of an ideal monatomic gas is $U = \frac{3}{2}Nk_BT$ and thus does not change in an expansion at constant temperature.

How much heat flowed into the gas from the reservoir?

We have seen that the energy of the ideal gas remained constant when the gas did work on the piston. By a decent respect for the conservation of energy it is necessary that a flow of energy or heat into the gas occur from the reservoir through the walls of the container. The quantity Q of heat added to the gas must be equal, but be opposite in sign, to the work done by the piston, so that

$$Q + W = 0 , \tag{22}$$

or

$$Q = 2.87 \times 10^8 \text{ ergs} = 28.7 \text{ J} , \tag{23}$$

from the result (20).

We can use this result for Q to calculate the entropy change by a second method, the first method being (17). We have

$$\sigma_2 - \sigma_1 = \frac{Q}{T} = \frac{Q}{k_BT} \tag{24}$$

in a reversible process. We calculated Q in (23), and the temperature is 300 K. Thus

$$\sigma_2 - \sigma_1 = \frac{Q}{k_BT} = \frac{2.87 \times 10^8 \text{ ergs}}{(1.38 \times 10^{-16} \text{ erg deg}^{-1})(300 \text{ K})} = 0.693 \times 10^{22} , \tag{25}$$

in agreement with result of the direct calculation in (17).

SLOW EXPANSION AT CONSTANT ENTROPY

We have considered an expansion at constant temperature. Suppose instead that the gas expands slowly from 1×10^3 to 2×10^3 cm^3 in an insulated container. No heat flow to or from the gas is permitted, so that $DQ = 0$. The entropy is constant, because $DQ = T \, d\sigma$ in a quasistatic process with a constant number of particles.

A process at constant entropy is called **isentropic.** Any thermodynamic process with no heat flow is called **adiabatic,** so that a reversible adiabatic process is isentropic.

What is the temperature of the gas after expansion?

In Chapter 11 we found that the entropy of an ideal monatomic gas depends on the volume and the temperature as

$$S(T, V) = N(\log T^{\frac{3}{2}} + \log V + \text{constant}) , \qquad (26)$$

so that the entropy remains constant if

$$\log T^{\frac{3}{2}} V = \text{constant} ; \qquad T^{\frac{3}{2}} V = \text{constant} . \qquad (27)$$

In an expansion at constant entropy from V_1 to V_2 we have

$$\boxed{T_1^{\frac{3}{2}} V_1 = T_2^{\frac{3}{2}} V_2} \qquad (28)$$

for an ideal monatomic gas.

We use the relation $pV = Nk_BT$ to obtain an alternative form:

$$\frac{p_1V_1}{T_1} = \frac{p_2V_2}{T_2} ; \qquad p_1V_1\left(\frac{T_2}{T_1}\right) = p_2V_2 ; \qquad p_1V_1\left(\frac{V_1}{V_2}\right)^{\frac{2}{3}} = p_2V_2 , \qquad (29)$$

or

$$\boxed{p_1V_1^{\frac{5}{3}} = p_2V_2^{\frac{5}{3}} ,} \qquad (30)$$

for an ideal monatomic gas. It can be shown for a polyatomic gas that $p_1 V_1{}^\gamma = p_2 V_2{}^\gamma$ in an expansion at constant entropy; here $\gamma \equiv C_p/C_V$ is the ratio of the heat capacities.

With $T_1 = 300$ K and $V_1/V_2 = \frac{1}{2}$ we find from (28):

$$T_2 = (\tfrac{1}{2})^{\frac{2}{3}}(300 \text{ K}) = 189 \text{ K} . \qquad (31)$$

This is the final temperature after the expansion at constant entropy. The gas is cooled in the expansion process by

$$T_1 - T_2 = 300 - 189 = 111 \text{ K} . \qquad (32)$$

Expansion at constant entropy is an important method of refrigeration.

What is the change in energy in the expansion?

The energy change is calculated from the temperature change (32). We have for an ideal monatomic gas

$$U_2 - U_1 = C_V(T_2 - T_1) , \qquad (33)$$

or,

$$U_2 - U_1 = \tfrac{3}{2}Nk_B(T_2 - T_1) = \tfrac{3}{2}(1 \times 10^{22})(1.38 \times 10^{-16} \text{ erg deg}^{-1})(-111 \text{ K})$$
$$= -2.3 \times 10^8 \text{ ergs} . \qquad (34)$$

The energy decreases in an expansion at constant entropy. The work done on the gas is equal to $U_2 - U_1$. The work done by the gas is $U_1 - U_2 = 2.3 \times 10^8$ ergs.

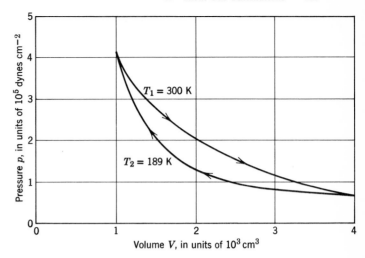

Figure 5 Carnot cycle calculated for an ideal monatomic gas with 10^{22} atoms. Two legs are at constant temperature, and two are at constant entropy.

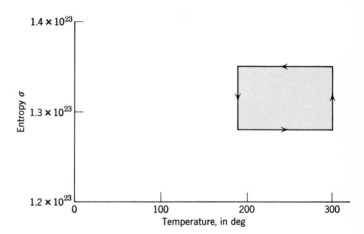

Figure 6 The Carnot cycle of Fig. 5, but plotted in the entropy-temperature plane.

The Carnot cycle of Chapter 8 as calculated for our system is shown in Fig. 5 in the p-V plane, and in Fig. 6 in the σ-T plane.

SUDDEN EXPANSION INTO A VACUUM

Let the gas expand suddenly into a vacuum from an initial volume of 1 liter to a final volume of 2 liters. This is an irreversible process. When a hole is opened in the partition to permit the expansion, the first atoms rush through the hole and strike the opposite wall. If no heat flow through the walls is permitted, there is no way for the atoms to lose their kinetic energy. The subsequent flow may be turbulent (irreversible), with different parts of the gas at different values of the energy density. Irreversible energy flow between regions will eventually equalize conditions throughout the gas.

Table 1 Summary of Ideal Gas Expansion Experiments

	$U_2 - U_1$	$\sigma_2 - \sigma_1$	W	Q
Quasistatic isothermal expansion	0	$N \log \dfrac{V_2}{V_1}$	$-N k_B T \log \dfrac{V_2}{V_1}$	$N k_B T \log \dfrac{V_2}{V_1}$
Quasistatic isentropic expansion	$-\frac{3}{2} N k_B T_1 \left[1 - \left(\dfrac{V_1}{V_2} \right)^{\frac{2}{3}} \right]$	0	$-\frac{3}{2} N k_B T_1 \left[1 - \left(\dfrac{V_1}{V_2} \right)^{\frac{2}{3}} \right]$	0
Irreversible expansion into vacuum	0	$N \log \dfrac{V_2}{V_1}$	0	0

How much work is done in the expansion?

No means of doing external work is provided, so that the work done is zero. Zero work is not necessarily a characteristic of all irreversible processes, but the work is zero for expansion into a vacuum.

What is the temperature after expansion?

No work is done and no heat is added in the expansion: $DW = 0$ and $DQ = 0$. Because $dU = DQ + DW$, we have $dU = 0$. The energy is unchanged, so that the temperature of the ideal gas is unchanged. The energy of a real gas will change in general in the process because the atoms are moved further apart, which affects their interaction energy. This effect is treated in thermodynamics textbooks.

What is the change of entropy in the expansion?

The increase of entropy when the volume is doubled at constant temperature is given by the result of (17). We have for the change of entropy

$$\Delta \sigma = \sigma_2 - \sigma_1 = N \log 2 = 0.069 \times 10^{23} . \tag{35}$$

In an expansion into a vacuum $DQ = 0$. Thus the entropy change (35) in the irreversible expansion is larger than DQ/T:

$$T \, \Delta \sigma > DQ . \tag{36}$$

This inequality is always satisfied in processes which are not reversible, as we established in Chapter 7.

Expansion into a vacuum is not a reversible process: the system is not in the most probable configuration at every stage of the expansion. Only the configuration before removal of the partition and the final configuration after equilibration are most probable configurations. At intermediate stages the distribution of atoms between the two regions into which the system is divided does not correspond to an equilibrium distribution. When isothermal or isentropic expansions are carried out reversibly, we have

$$T \, \Delta \sigma = DQ \ . \tag{37}$$

The central results of this chapter are summarized in Table 1.

Problem 1. *Energy of imperfect gas.* (a) With the help of the Maxwell relation

$$\left(\frac{\partial \sigma}{\partial V} \right)_T = \left(\frac{\partial p}{\partial T} \right)_V , \tag{38}$$

of which a proof is given in (18.16), show that

$$\left(\frac{\partial U}{\partial V} \right)_T = T \left(\frac{\partial p}{\partial T} \right)_V - p \ . \tag{39}$$

We integrate this to obtain

$$U(T, V) - U(T, \infty) = \int_\infty^V dV \left[T \left(\frac{\partial p}{\partial T} \right)_V - p \right] . \tag{40}$$

(b) Use the **virial equation of state** defined by

$$pV = RT \left[1 + \frac{B(T)}{V} + \frac{C(T)}{V^2} + \frac{D(T)}{V^3} + \cdots \right] \tag{41}$$

for one mole of a real gas to obtain the result

$$U(T, V) - U(T, \infty) = - \frac{RT^2}{V} \left[\frac{dB}{dT} + \frac{1}{2V} \frac{dC}{dT} + \cdots \right] . \tag{42}$$

Here $B(T)$ is called the second virial coefficient; $C(T)$ is the third virial coefficient. The energy $U(T, \infty)$ at infinite volume is that of an ideal gas. The energy of an ideal gas is independent of the volume.

(c) For argon $B(T) = -178$ cm^3 mol^{-1} at 100 K and -15 cm^3 mol^{-1} at 300 K. Make a rough plot of $U(T, V)/U(T, \infty)$ versus V at $T = 200$ K. Neglect the third virial coefficient.

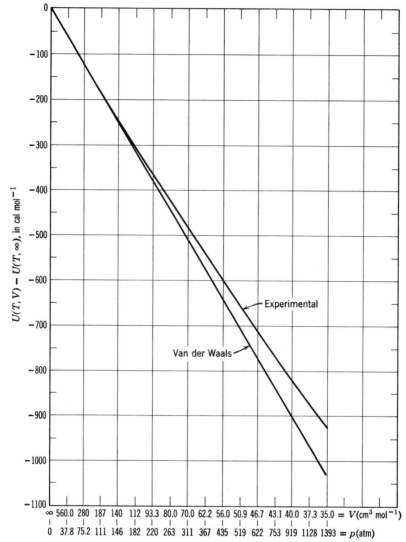

Figure 7 The experimental energy of one mole of argon at 0 C, and the energy as inferred from the van der Waals equation of state. The value of $U(T, \infty)$ is the ideal gas value, which at 0 C is

$$\tfrac{3}{2}(1.987 \text{ cal deg}^{-1} \text{ mol}^{-1})(273 \text{ K}) = 814 \text{ cal mol}^{-1} .$$

At a pressure of 1000 atm the interatomic interactions lower the energy by an amount comparable with this value.

Problem 2. *Van der Waals equation of state.* The van der Waals equation of state is

$$\left(p + \frac{a}{V^2}\right)(V - b) = RT , \tag{43}$$

written for one mole of gas. This is an empirical[2] equation with two constants, a and b. We can rewrite (43) for n moles of gas if we substitute n^2a for a; nb for b; and nR for R. A dimensionless form of (43) is introduced in (20.38). (a) Use the virial equation of state (41) to show that

$$B(T) = b - \frac{a}{RT} ; \tag{44}$$

$$C(T) = b^2 ; \tag{45}$$

$$U(T, V) - U(T, \infty) = -\frac{a}{V} . \tag{46}$$

Notice that the term b (that is related to a molar volume) does not enter the energy $U(T, V)$; only the intermolecular force term a enters the energy. An experimental curve of $U(T, V)$ versus V for argon is given in Fig. 7.

(b) Write a brief account (about 300 words) of the physical basis of the van der Waals equation of state. The subject is treated in standard thermodynamics textbooks; see, for example, J. C. Slater, *Introduction to chemical physics*, McGraw-Hill, 1939, Chapter 12.

(c) Equation (46) shows that the energy is lowered as the volume is decreased; explain how it is that the pressure still tends to expand the gas.

Problem 3. *Time for a large fluctuation.* In footnote 3 of Chapter 4 we quoted Boltzmann to the effect that two gases in a 0.1 liter container will unmix only in a time enormously long compared to $10^{(10^{10})}$ years. We shall investigate a related problem: we let a gas of atoms of He^4 occupy a container of volume of 0.1 liter under standard conditions of temperature and pressure, and we ask how long it will be before the atoms assume a configuration in which all are in one-half of the container.

(a) Estimate the number of states accessible to the system in this initial condition.

(b) The gas is compressed isothermally to a volume of 0.05 liter. How many states are accessible now?

---

[2] Empirical means "guided by experiment, without knowledge of general principles." Actually, one can give simple arguments for the terms in a and b: that in a represents the effect of $1/R^6$ attractive forces, and that in b represents the volume excluded by the size of the atoms.

(c) For the system in the 0.1 liter container, estimate the value of the ratio

$$\frac{\text{number of states for which all atoms are in one-half of the volume}}{\text{number of states for which the atoms are anywhere in the volume}}.$$

(d) If the collision rate of an atom is $\approx 10^{10}$ sec^{-1}, what is the total number of collisions of all atoms in the system in a year? We use this as a crude estimate of the frequency with which the state of the system changes.

(e) Estimate the number of years you would expect to wait before all atoms are in one-half of the volume, starting from the equilibrium configuration.

CHAPTER 13

Kinetic Theory of Gases

"I am conscious of being only an individual struggling weakly against the stream of time. But it still remains in my power to contribute in such a way that, when the theory of gases is again revived, not too much will have to be rediscovered."   (L. Boltzmann, 1898)

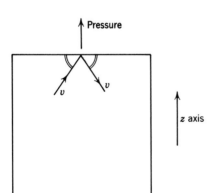

Figure 1 The change of momentum of a molecule of velocity **v** which is reflected in a mirror-like fashion from the wall of the container is $2M|v_z|$.

In this chapter we give a kinetic calculation of the ideal gas law, and we derive the distribution of velocities of molecules in a gas. We then introduce the effects of collisions between atoms of a classical gas. We discuss in an elementary way various transport processes in gases: diffusion, thermal conductivity, and viscosity. These are all controlled by collisions. A **transport process** is the result of a nonequilibrium condition in the system, as a result of which there is a net flux of particles, energy, momentum, or charge between two parts of the system. When a system is in equilibrium, there is no net transport.

The treatment will be given entirely in the language of classical mechanics, because the classical theory of transport processes is clearer and simpler than the quantum theory. We shall not talk of states, but of velocities. This circumstance marks a curious division of the subject, for quantum theory has enormous advantages in the basic formulation of the concepts of statistical physics and is absolutely essential to the calculation of the equilibrium properties of matter. This chapter on nonequilibrium processes is an island of classical physics.[1]

We first illustrate the kinetic method by means of an elementary derivation of the ideal gas law, $pV = N\mathcal{T}$.

KINETIC ARGUMENT FOR IDEAL GAS LAW

Consider molecules that strike the wall of a container. Let v_z denote the velocity component of a molecule normal to the plane of the wall, as in Fig. 1. If the molecule strikes the wall and is reflected specularly (mirror-like reflection) from the wall, the change of momentum of the molecule is

$$2M|v_z| \ , \tag{1}$$

where M is the mass of the molecule.

There is a net force on the wall from the change of momentum of the molecules that strike it. The pressure p on the wall is equal to the change of

---

[1] However, linear transport coefficients are related to fluctuations in the equilibrium system by a theorem known as the fluctuation-dissipation theorem. This circumstance does not often simplify the calculation of the transport coefficients. The simplest example of the theorem concerns the electrical resistivity and is called the Nyquist theorem. A short derivation is found in Appendix H; see also H. B. Callen, *Thermodynamics*, Wiley, 1960, Part III.

momentum per unit area per unit time as a consequence of Newton's second law of motion:

$$p = (\text{momentum change per molecule})(\text{number of molecules striking unit area per unit time}) \ . \tag{2}$$

The number of molecules that strike a unit area in unit time is one-half the number in a cylinder of length $|v_z|$. This number is $\frac{1}{2}n|v_z|$, where n is the concentration of molecules. The factor one-half enters because at a given instant only half the molecules are moving toward the wall; the other half are moving away from the wall.

We combine (1) and (2) to obtain

$$p = (2M|v_z|)(\tfrac{1}{2}n|v_z|) = nM|v_z|^2 = nM\langle v_z^2\rangle \ . \tag{3}$$

Now the average value of $\frac{1}{2}Mv_z^2$ is $\frac{1}{2}T$, by the result of Chapter 11. We recall that $\langle\frac{1}{2}Mv_z^2\rangle$ is the translational kinetic energy of one degree of freedom. Here the degree of freedom is the motion of the molecule along the axis normal to the wall. Thus the pressure is given by

$$p = nT = (N/V)T \ ; \qquad pV = NT \ . \tag{4}$$

This is the ideal gas law. The kinetic argument given is essentially correct, although we might well have taken more care[2] in obtaining (3).

MAXWELL DISTRIBUTION OF VELOCITIES

Our next enterprise is to translate the energy distribution function of an ideal gas into a classical velocity distribution function. (Often when we mean *speed* we shall say *velocity*, as this is the oral tradition in physics when no confusion is caused.) In Chapter 11 we found the result

$$f(\epsilon_{\mathbf{n}}) = \lambda e^{-\epsilon_{\mathbf{n}}/T} \tag{5}$$

for the probability of occupancy of an orbital $\mathbf{n}$ of energy

$$\epsilon_{\mathbf{n}} = \frac{\hbar^2}{2M}\left(\frac{\pi n}{L}\right)^2 \ . \tag{6}$$

Here L is the edge of a cube of volume $V = L^3$. The probability that there is one atom somewhere among the set of orbitals with the magnitude $n \equiv |\mathbf{n}|$ of the quantum number between n and $n + dn$ is given by the product of the

---

[2] Let $a(v_z)\,dv_z$ be the number of molecules per unit volume with the z component of the velocity between v_z and $v_z + dv_z$. The number of molecules with velocity in this range which strike a unit area of the wall in unit time is $a(v_z)v_z\,dv_z$. The momentum change of the molecules in this group is $(2Mv_z) \cdot a(v_z)v_z\,dv_z$. Thus the total pressure is

$$p = \int_0^{\infty} 2Mv_z^2 a(v_z)\,dv_z = M\int_{-\infty}^{\infty} v_z^2 a(v_z)\,dv_z = Mn\langle v_z^2\rangle \ . \tag{3a}$$

number of orbitals in the range dn times the probability an orbital is occupied. From (5) and (10.23):

$$(\tfrac{1}{2}\pi n^2 \, dn)f(\epsilon_n) = \tfrac{1}{2}\pi\lambda n^2 e^{-\epsilon_n/T} \, dn \ , \tag{7}$$

where we have taken the spin of the atom as zero.

We want the probability distribution of the classical velocity, so that we must find a connection between the quantum number **n** and the classical velocity of a particle in the state **n**. The results we give are exact for the square of the velocity, and they are valid for the velocity itself in the classical limit of quantum mechanics.

The classical kinetic energy $\tfrac{1}{2}Mv^2$ is related to the quantum energy (6) by

$$\tfrac{1}{2}Mv^2 = \frac{\hbar^2}{2M}\left(\frac{\pi n}{L}\right)^2 \ ; \qquad v = \frac{\hbar\pi}{ML}n \ ; \qquad n = \frac{ML}{\hbar\pi}v \ . \tag{8}$$

Let

$P(v) \, dv$ = probability that an atom has velocity magnitude, or speed, in the range dv at v .

This is evaluated from (7) by replacing dn by $(dn/dv) \, dv$, which by (8) is equal to $(ML/\hbar\pi) \, dv$. Thus

$$P(v) \, dv = \tfrac{1}{2}\pi\lambda n^2 e^{-\epsilon_n/T}\frac{dn}{dv} \, dv = \tfrac{1}{2}\pi\lambda\left(\frac{ML}{\hbar\pi}\right)^3 v^2 e^{-Mv^2/2T} \, dv \ . \tag{9}$$

It is convenient to write (9) simply as

$$P(v) \, dv = Av^2 e^{-Mv^2/2T} \, dv \ , \tag{10}$$

where A is a constant determined by the normalization condition

$$\int_0^\infty dv \, P(v) = 1 = A\int_0^\infty dv \, v^2 e^{-Mv^2/2T} \ . \tag{11}$$

With

$$y^2 \equiv \frac{Mv^2}{2T} \ ; \qquad v^2 = \frac{2T}{M}y^2 \ , \tag{12}$$

we have from (11):

$$1 = A\left(\frac{2T}{M}\right)^{\frac{3}{2}}\int_0^\infty dy \, y^2 e^{-y^2} = A\left(\frac{2T}{M}\right)^{\frac{3}{2}}\frac{\pi^{\frac{1}{2}}}{4} \ . \tag{13}$$

Thus

$$A = 4\pi\left(\frac{M}{2\pi T}\right)^{\frac{3}{2}} \tag{14}$$

and, from (10),

$$\boxed{P(v) = 4\pi\left(\frac{M}{2\pi T}\right)^{\frac{3}{2}} v^2 e^{-Mv^2/2T} \ .} \tag{15}$$

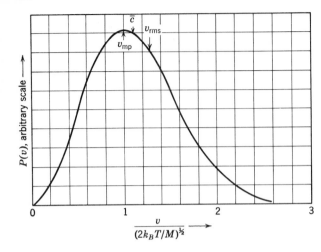

Figure 2 Maxwell velocity distribution as a function of the speed in units of the most probable speed $v_{mp} = (2k_BT/M)^{\frac{1}{2}}$. Also shown are the mean speed $\bar{c}$ and the root mean square velocity v_{rms}.

This is the **Maxwell velocity distribution.** It is plotted in Fig. 2. The quantity $P(v)\,dv$ is the probability that an atom has its speed in dv at v. Of this distribution Boltzmann wrote:

"The self-regulating most probable [configuration]—which we call the Maxwell velocity distribution since Maxwell first found its mathematical expression in a special case—is not some kind of special singular state which is contrasted to infinitely many more non-Maxwellian distributions. Rather it is, on the contrary, characterized by the fact that by far the largest number of possible states have the characteristic properties of the Maxwell distribution, and compared to this number, the number of possible velocity distributions which significantly deviate from the Maxwellian is vanishingly small."

Table 1   Molecular Velocities at 0 C = 273 K

(As calculated from $v_{rms} = (3T/M)^{\frac{1}{2}}$.)

Gas	v_{rms}, in 10^4 cm sec^{-1}	Gas	v_{rms}, in 10^4 cm sec^{-1}
H_2	18.4	Ar	4.3
He	13.1	Kr	2.86
H_2O	6.2	Xe	2.27
Ne	5.8	Hg	1.85
N_2	4.9	Free electron	1100.
O_2	4.6		

Problem 1. *Mean speeds in a Maxwellian distribution.* (a) Show by use of (15) and by integration that the root mean square velocity v_{rms} is

$$v_{\text{rms}} = \langle v^2 \rangle^{\frac{1}{2}} = \left(\frac{3T}{M} \right)^{\frac{1}{2}} . \tag{16}$$

Because $\langle v^2 \rangle = \langle v_x{}^2 \rangle + \langle v_y{}^2 \rangle + \langle v_z{}^2 \rangle$ and $\langle v_x{}^2 \rangle = \langle v_y{}^2 \rangle = \langle v_z{}^2 \rangle$, it follows that

$$\langle v_x{}^2 \rangle^{\frac{1}{2}} = \left(\frac{T}{M} \right)^{\frac{1}{2}} = \frac{v_{\text{rms}}}{\sqrt{3}} . \tag{17}$$

The results (16) and (17) can also be obtained directly from the expression in Chapter 11 for the average kinetic energy of an ideal gas. Values of v_{rms} are given in Table 1.

(b) Show that the most probable value of the speed v_{mp} is

$$v_{\text{mp}} = \left(\frac{2T}{M} \right)^{\frac{1}{2}} . \tag{18}$$

By most probable value of the speed we mean the maximum of the Maxwell distribution as a function of v. Notice that $v_{\text{mp}} < v_{\text{rms}}$.

(c) Show that the mean speed $\bar{c}$ is

$$\bar{c} = \int_0^\infty dv\, vP(v) = \left(\frac{8T}{\pi M} \right)^{\frac{1}{2}} . \tag{19}$$

The mean speed may also be written as $\langle |v| \rangle$. The ratio

$$\frac{v_{\text{rms}}}{\bar{c}} = 1.086 . \tag{20}$$

(d) Show that $\bar{c}_z$, the mean of the absolute value of the z component of the velocity of an atom, is

$$\bar{c}_z \equiv \langle |v_z| \rangle = \tfrac{1}{2}\bar{c} = \left(\frac{2T}{\pi M} \right)^{\frac{1}{2}} . \tag{21}$$

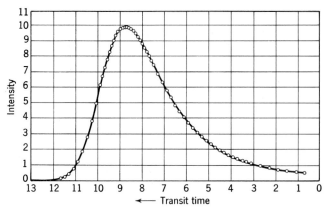

Figure 3 Measured transmission points and calculated Maxwell transmission curve for potassium atoms that exit from an oven at a temperature 157 C. The horizontal axis is the transit time of the atoms transmitted. The intensity is in arbitrary units; the curve and the points are normalized to the same maximum value. The oven pressure is 0.84×10^{-3} mm Hg. (After Marcus and McFee.)

EXPERIMENTAL VERIFICATION OF THE VELOCITY DISTRIBUTION IN A BEAM

A very careful study of the velocity distribution of atoms of potassium which effuse from the slit of an oven has been reported by P. M. Marcus and J. H. McFee.[3] The curve in Fig. 3 compares the experimental results with the prediction of (22). The agreement is excellent at the oven pressure indicated. Some details of the experiments are shown in Figs. 4 and 5.

VELOCITY DISTRIBUTION OF ATOMS EXITING FROM AN OVEN

To present the experimental evidence for the Maxwell distribution (15) we need an expression for the distribution of speeds of atoms that exit in a fixed direction from a small hole[4] in an oven. This distribution is different from the velocity distribution within the oven, because the flux through the hole involves an extra factor, the velocity component normal to the wall. (In the ex-

[3] P. M. Marcus and J. H. McFee, *Recent research in molecular beams*, ed. I. Esterman, Academic Press, 1959.

[4] In such experiments a round hole is said to be small if the diameter is less than a mean free path of an atom in the oven. If the hole is not small in this sense, the flow of gas from it will be governed by the laws of hydrodynamic flow and not by gas kinetics. Flow of gas through a small hole is called **effusion**.

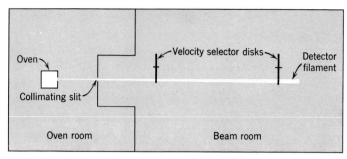

Figure 4 Top view of beam apparatus. The oven, velocity selector, and detector are mounted on an optical bench and aligned optically, then inserted in a cylindrical vacuum chamber. The chamber is divided into the separate oven room and beam room. (After Marcus and McFee.)

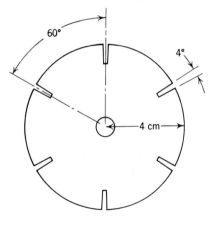

Figure 5 Detail of a selector disk. The velocity selector consists of two identical slotted disks of aluminum. Each disk is turned by an 8000 rpm synchronous motor. A phase shifter varies the phase of the voltage fed to one motor relative to the other. A light beam is used to determine when the velocity selector is set for zero transit time. This beam is produced by a small light source near one disk. The beam then passes through the velocity selector and is detected by a photocell near the other disk. A pressure of $\leq 5 \times 10^{-7}$ mm Hg can be maintained indefinitely in the beam room. (After Marcus and McFee.)

periments a second hole is used to select atoms that exit in a narrowly defined direction.) The exit beam is weighted in favor of atoms of high velocity at the expense of those at low velocity.

The weight factor is the velocity in the direction normal to the plane of the hole. This factor is proportional to the velocity v. The probability that an atom which leaves the hole will have a velocity between v and $v + dv$ defines the quantity $P_{\text{beam}}(v)\, dv$, where

$$P_{\text{beam}}(v) \propto v P_{\text{Maxwell}} \propto v^3 e^{-Mv^2/2T} \, , \qquad (22)$$

with P_{Maxwell} given by (15). The distribution (22) is often called the **Maxwell transmission distribution** and refers to the transmission through a hole. The experimental results were shown in Fig. 3.

COLLISION CROSS SECTIONS AND MEAN FREE PATHS

Atoms are not rigid spheres. At large separations the potential energy between two atoms is the van der Waals interaction. This interaction is attractive and varies approximately as $1/r^6$, where r is the internuclear separation. At small separations there is a repulsive interaction with a range dependence roughly as $1/r^{12}$. Two atoms that collide with a high relative velocity will usually exhibit a smaller collision cross section than if they had collided with low relative velocity. The collision cross sections depend on velocity and therefore depend on the temperature of the gas.

These features are details that may be neglected in a first look at the effects of interactions in gases. We shall see what happens if we treat atoms as rigid spheres of diameter d. The value to be taken for d is not likely to be very far wrong, in spite of the crudeness of the model. We might take the atomic diameter $d \cong 2.2$ Å for helium and 3.6 Å for argon; both are values quoted in the literature.

Two atoms of diameter d will collide with each other if their centers pass within a distance d of each other (Fig. 6). By the argument of the figure we see that one collision will occur when an atom has traversed an average distance

$$\ell = \frac{1}{\pi d^2 n} , \tag{23}$$

where n is the number of atoms per unit volume. The length ℓ is called the **mean free path:** it is the average distance traveled by an atom between collisions. Our result (23) did not consider the velocity of the other atoms and some smaller effects.[5] If the velocity of the other atoms is taken into account, ℓ will be reduced by the factor $1/\sqrt{2}$. (A slow molecule has a short mean free path, for it will get hit by faster molecules not originally in the volume swept out by the slow molecule.)

We can easily estimate the order of magnitude of the mean free path. If the atomic diameter d is taken as 2.2 Å, which may roughly apply to helium, then the cross section is

$$\pi d^2 = (3.14)(2.2 \times 10^{-8} \text{ cm})^2 = 15.2 \times 10^{-16} \text{ cm}^2 . \tag{24}$$

At a temperature of 0 C and a pressure of 760 mm of mercury, the concentration of atoms of an ideal gas is given by the **Loschmidt number,**

$$n_0 = 2.69 \times 10^{19} \text{ atoms cm}^{-3} . \tag{25}$$

---

[5] J. H. Jeans, *Introduction to the kinetic theory of gases,* Cambridge University Press, 1940, Chapter 5.

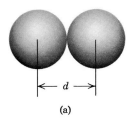

(a)

Figure 6   (a) Two rigid spheres will collide if their centers pass within a distance d of each other. (b) An atom of diameter d which travels a long distance L will sweep out a volume $\pi d^2 L$, in the sense that it will collide with any atom whose center lies within the volume. If n is the concentration of atoms, the average number of atoms in this volume is $n\pi d^2 L$. This is the number of collisions. The average distance between collisions is

$$\ell = \frac{L}{n\pi d^2 L} = \frac{1}{n\pi d^2} \ .$$

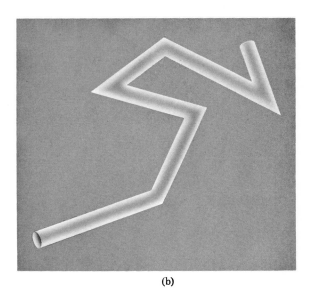

(b)

We combine these two results to obtain the mean free path under standard conditions:

$$\ell = \frac{1}{\pi d^2 n_0} = \frac{1}{(15.2 \times 10^{-16} \text{ cm}^2)(2.69 \times 10^{19} \text{ cm}^{-3})} = 2.5 \times 10^{-5} \text{ cm} \ .$$

$$(26)$$

This is about 1000 times larger than the diameter of an atom. The associated collision frequency is

$$\frac{v_{rms}}{\ell} \approx \frac{10^5 \text{ cm sec}^{-1}}{10^{-5} \text{ cm}} \approx 10^{10} \text{ sec}^{-1} \ .$$

At a pressure of 10^{-6} atm or 1 dyne cm^{-2} the concentration of atoms is reduced by the factor 10^{-6} and the mean free path is increased to 25 cm. Thus at a pressure of 10^{-6} atm the mean free path may not be small in comparison with the dimensions of any particular experimental apparatus. This will depend on the apparatus, but the answer is relevant to the validity of the calculations that follow. We assume in this chapter that the mean free path is

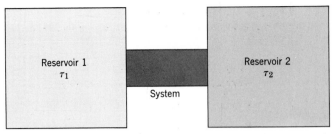

Figure 7 Opposite ends of the system are in thermal contact with reservoirs at temperatures T_1 and T_2.

small in comparison with any relevant dimension of the apparatus concerned, except as otherwise specified. The region of mean free path long in comparison with the apparatus is known as the Knudsen region and is treated in texts on the kinetic theory of gases and on high vacuum systems.

We give only approximate physical treatments of the topics that follow in this chapter. We bring out all the essential physics, but neglect such matters as careful averages over solid angles. The results may consequently be in error by numerical factors such as $\frac{2}{3}$ or $\frac{1}{2}$.

TRANSPORT PROCESSES

We consider a system which is not in thermal equilibrium. It is of practical interest to consider a nonequilibrium steady state, which means that the system is maintained in same nonequilibrium condition for the time interval of the experiment, however long. For example, we may create a steady state nonequilibrium condition of temperature in a system by placing opposite ends in thermal contact with large reservoirs at two different temperatures, as in Fig. 7.

If reservoir 1 is at the higher temperature, energy will flow through the system from reservoir 1 to reservoir 2. This direction of energy flow will increase the total entropy of (reservoir 1 + reservoir 2 + system). The temperature difference or, more precisely, the temperature difference per unit length of the system is the driving force for the process. The physical quantity that is transported through the specimen in this experiment is energy.

The result of experiment and of both rudimentary and sophisticated calculations is that the rate of energy transport is directly proportional to the temperature gradient. The standard form in which we express transport results is

$$\text{flux} = (\text{coefficient})(\text{driving force}) . \tag{27}$$

The definition of the flux of a physical quantity A is:

$$\textbf{flux of } A = \text{net amount of } A \text{ transported across unit area in unit time} . \tag{28}$$

Table 2   *Summary of Phenomenological Transport Laws*

Effect	Flux of particle property	Gradient	Coefficient	Law	Approximate expression for coefficient
Diffusion	Number	$\dfrac{dn}{dz}$	Diffusivity D	$\mathbf{J}_n = -D \text{ grad } n$	$D = \frac{1}{3}\bar{c}l$
Thermal conductivity	Energy	$\dfrac{du}{dz} = \hat{C}_V \dfrac{dT}{dz}$	Thermal conductivity K	$\mathbf{J}_U = -K \text{ grad } T$	$K = \frac{1}{3}\hat{C}_V \bar{c}l$
Viscosity	Transverse momentum	$M\dfrac{dv_x}{dz}$	Viscosity η	$\dfrac{F_x}{A} = J_p{}^x = -\eta\dfrac{dv_x}{dz}$	$\eta = \frac{1}{3}\rho\bar{c}l$
Electrical conductivity	Charge	$-\dfrac{d\varphi}{dz} = E_z$	Conductivity σ	$\mathbf{J}_q = \sigma\mathbf{E}$	$\sigma = \dfrac{nq^2 l}{M\bar{c}}$

Symbols: n = number of particles per unit volume
$\quad\bar{c}$ = mean thermal speed = $\langle|v|\rangle$
$\quad l$ = mean free path
$\quad\hat{C}_V$ = heat capacity per unit volume
$\quad u$ = mean thermal energy per unit volume
$\quad F_x/A$ = shear force per unit area

φ = electrostatic potential
E = electric field intensity
q = electric charge
M = mass of particle
ρ = mass per unit volume

The flux of A is a vector and will be denoted by $\mathbf{J}_A$. The transport laws are summarized in Table 2.

If the quantity transported through the system is the energy U, then we write

$$\mathbf{J}_U = -K \text{ grad } T \ , \tag{29}$$

where $\mathbf{J}_U$ is the flux of energy; the negative gradient of the temperature acts as the driving force; and K is the **coefficient of thermal conductivity**.[6]

Before showing how to calculate transport coefficients, let us show how not to calculate them. We can derive an energy flux as the product of the energy density times the velocity of a particle. We might guess that the net energy flux in the z direction is, with ρ_U as the energy density,

$$J_U \overset{?}{=} [\rho_U(2) - \rho_U(1)]\bar{c}_z \overset{?}{=} \hat{C}_V(T_2 - T_1)\bar{c}_z \ , \tag{30}$$

where $\hat{C}_V$ is the heat capacity per unit volume and $\bar{c}_z$ is the average of the absolute value of the z component of the molecular velocity (21). Our reason for suggesting (30) is that $\rho_U(2)\bar{c}_z$ is a flux of energy characteristic of the end of the system at temperature T_2, and $\rho(1)\bar{c}_z$ is a flux of energy characteristic of the end of the system at temperature T_1. The difference of the two fluxes gives the net flux (30).

---

[6] The energy flux is not necessarily equal to the heat flux, although the two are often confused. Thus K_U, which is what we calculate, is not exactly equal to K_Q for the heat flux. A good discussion of this point is given by Callen, Chapter 17.

This would indeed be the correct result if the molecules traveled from one end of the system to the other without collision, because the energy in this situation is transported with the velocity $\bar{c}_z$, as defined by (21). But in almost every transport problem the collisions limit severely the rate of transport of energy, and the expression for the energy flux is modified by collisions in an important way. The molecules travel freely only over distances of the order of the mean free path ℓ, after which they collide.[7] We assume that in a collision at position z the molecules come into a new local equilibrium condition at a local temperature $T(z)$ and local energy density $\rho_U(z)$ which is characteristic of the position z.

Thus across a plane at z there is a flux of energy in the positive z direction equal to $\rho_U(z)\bar{c}_z$ and a flux in the negative z direction equal to $\rho_U(z + \ell)\bar{c}_z$, for the molecules transport energy without collision only over the distance ℓ. The molecules that reach z from $z + \ell$ have an energy density characteristic of the local temperature at $z + \ell$. We have for the net flux

$$J_U = [\rho_U(z) - \rho_U(z + \ell)]\bar{c}_z = -\ell \frac{\partial \rho_U}{\partial z}\bar{c}_z = -\ell \frac{\partial \rho_U}{\partial T}\frac{dT}{dz}\bar{c}_z = -\ell \hat{C}_V \bar{c}_z \frac{dT}{dz} \, ,$$

(31)

where

$$\hat{C}_V \equiv \frac{\partial \rho_U}{\partial T}$$

(32)

is the heat capacity per unit volume. The form of the result (31) is quite different from that of (30). In (30) we described a free flow of energy; in (31) we describe a random scattering or diffusion of energy (Fig. 8) over the distance of a mean free path ℓ.

In (31) the quantity ℓ should really be ℓ_z, the projection of the mean free path on the z axis. When we properly consider the distribution of molecular directions[8] we find that the spatial average of $\ell_z\bar{c}_z$ is $\frac{1}{3}\ell\bar{c}$, so that the correct expression for the energy flux is

$$J_U = -\tfrac{1}{3}\hat{C}_V \ell\bar{c} \frac{dT}{dz} \, .$$

(33)

[7] The discussion assumes that ℓ and c are uncorrelated. But the mean free path for a fast molecule is longer than for a slow molecule. Thus the numerical factor $\frac{1}{3}$ in the result (34) is not meaningful, and we include it only as a matter of convention.

[8] We are concerned with the average value of $\ell_z\bar{c}_z$, where $\ell_z = \ell \cos\theta$ is the projection of the mean free path on the z axis, and $\bar{c}_z = \bar{c}\cos\theta$ is the projection of the speed. The polar angle with the z axis is θ, and the average is taken over the surface of a hemisphere, because all directions in space are equally likely. The element of surface area is $2\pi \sin\theta\, d\theta$. Thus

$$\langle \ell_z\bar{c}_z \rangle = \ell\bar{c} \frac{2\pi \int_0^{\frac{1}{2}\pi} \cos^2\theta \sin\theta\, d\theta}{2\pi} = \tfrac{1}{3}\ell\bar{c} \, .$$

(32a)

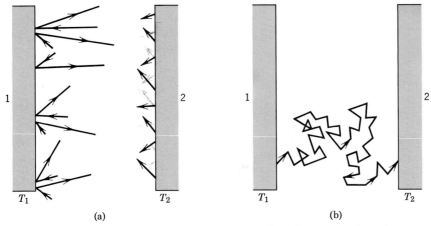

(a)   (b)

Figure 8 (a) Energy transfer by molecules in the collisionless regime (very low pressure) between a high temperature reservoir 1 and a low temperature reservoir 2. Here the lengths of the arrows are proportional to the velocity of the molecules. (b) Energy transfer by diffusion of energy in regime where collisions control the transfer (higher pressure). Here the arrows trace the path followed by a single molecule in its motion between reservoirs 1 and 2; the breaks shown are due to collisions with other molecules.

By comparison with (29) the coefficient of thermal conductivity is

$$K = \tfrac{1}{3}\hat{C}_V \bar{c} l \ .$$
(34)

The thermal conductivity increases directly with the mean free path, the heat capacity, and the speed of the particles. An improved value of K given by Kennard is $K = 1.23\hat{C}_V \bar{c} l$, quite a bit higher than (34).

DIFFUSION

We consider a system with one end in diffusive contact with a reservoir at chemical potential μ_1 and the other end in diffusive contact with a reservoir at chemical potential μ_2, as in Fig. 9. The temperature is taken to be constant everywhere.

If reservoir 1 is at the higher chemical potential, then particles will flow through the system from reservoir 1 to reservoir 2. This direction of particle flow will increase the total entropy of (reservoir 1 + reservoir 2 + system).

Most often the diffusion process is discussed for situations where the difference of chemical potential is caused simply by a difference in particle concentration. The driving force of diffusion is taken as the gradient of the particle concentration, and the flux $\mathbf{J}_n$ is the number of particles passing through a unit area in unit time:

$$\mathbf{J}_n = -D \operatorname{grad} n \ .$$
(35)

Here D is the diffusion constant or **diffusivity.**

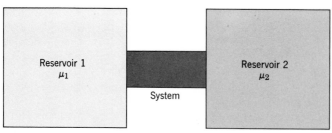

Figure 9 Opposite ends of the system are in diffusive contact with reservoirs at chemical potentials μ_1 and μ_2. The temperature is taken to be constant everywhere.

By direct analogy with the derivation of (31) the particle flux in the z direction is given by

$$J_n = [n(z) - n(z + \ell_z)]\bar{c}_z = -\ell_z \frac{dn}{dz}\bar{c}_z \ . \tag{36}$$

By use of (32a) for the spatial average of $\ell_z \bar{c}_z$ we have

$$J_n = -\tfrac{1}{3}\bar{c}\ell \frac{dn}{dz} \ . \tag{37}$$

On comparison with (35) we see that the diffusivity is given by

$$\boxed{D = \tfrac{1}{3}\bar{c}\ell \ .} \tag{38}$$

We observe that the coefficient of thermal conductivity (34) can be written as

$$K = \hat{C}_V D \ . \tag{39}$$

In $\mathbf{J}_U$ we transport energy; in $\mathbf{J}_n$ we transport particles.

VISCOSITY

"Experiment 26 . . . . We observ'd also that when the Receiver was full of Air, the included Pendulum continu'd its Recursions about fifteen minutes (or a quarter of an hour) before it left off swinging; and that after the exsuction of the Air, the Vibration of the same Pendulum (being fresh put into motion) appear'd not (by a minutes Watch) to last sensibly longer. So that the event of this Experiment being other than we expected, scarce afforded us any other satisfaction, than that of our not having omitted to try it." (Robert Boyle, 1660.)

This early experiment by Boyle shows that the viscosity of a gas is independent of the pressure. The result is "other than we expected," for we might expect the denser medium to be the more viscous.

The **coefficient of viscosity** η is defined by

$$Z_x = -\eta \frac{dv_x}{dz} . \tag{40}$$

Here η is the Greek letter eta; v_x is the x component of the flow velocity of the gas; and Z_x is the x component of the shear stress exerted by the gas on a unit area of the xy plane.

By Newton's second law of mechanics there is a shear stress Z_x on the xy plane if the plane receives a net flux of momentum in the x direction, for this is a rate of change of the momentum of the plane. There will be such a flux of momentum if the gas has a flow velocity parallel to the plane and in the x direction.

By direct analogy with (31) the net momentum flux is

$$J_p{}^x = Mn[v_x(z) - v_x(z + l_z)]\bar{c}_z , \tag{41}$$

where Mn is the mass density of the gas; $v_x(z)$ is the x component of the flow velocity at z; and $\bar{c}_z$ is the z component of the average molecular speed. We note that $Mnv_x(z)$ is the x component of the momentum density at z. The z component of the mean free path is l_z. Then the shear stress component is

$$Z_x = J_p{}^x = -Mn\bar{c}_z l_z \frac{dv_x}{dz} , \tag{42}$$

or, because the spatial average $\langle c_z l_z \rangle = \frac{1}{3}\bar{c}l$, we have

$$Z_x = -\frac{1}{3}Mn\bar{c}l \frac{dv_x}{dz} . \tag{43}$$

On comparison with (40) we have for the coefficient of viscosity

$$\boxed{\eta = \frac{1}{3}Mn\bar{c}l .} \tag{44}$$

Now by (23) the mean free path is given by $l = 1/\pi d^2 n$, where d is the molecular diameter and n is the concentration. Thus the viscosity may be expressed as

$$\eta = \frac{M\bar{c}}{3\pi d^2} , \tag{45}$$

so that the viscosity is independent of the pressure of the gas.

Jeans makes the following comment on this circumstance: "In spite of its apparent improbability, this law was predicted by Maxwell on purely theoretical grounds, and its subsequent experimental confirmation has constituted one of the most striking triumphs of the kinetic theory." The result that the viscosity

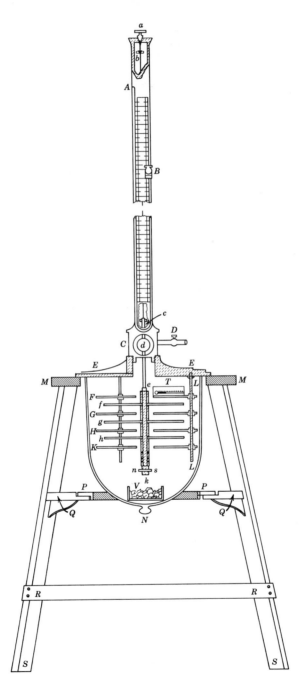

Figure 10 Maxwell's apparatus for the study of the pressure dependence of the coefficient of viscosity of gases. A set of three disks suspended by a long wire b oscillates with respect to a fixed set of parallel disks. The extent and duration of the vibrations are observed by a telescope directed on the mirror at d. [After J. C. Maxwell, Philosophical Transactions of the Royal Society of London **156**, 249 (1866).]

Table 3   Experimental Values of K, η, and Kρ/ηC_V at 0 C and One Atmosphere of Pressure. (After Jeans.)

Gas	K, in 10^{-4} cal cm^{-1} sec^{-1} deg^{-1}	η, in 10^{-4} gm cm^{-1} sec^{-1}	$K\rho/\eta\hat{C}_V$
He	3.36	1.88	2.40
Ar	0.389	2.10	2.49
H_2	3.97	0.857	1.91
N_2	0.54	1.67	1.91
O_2	0.57	1.92	1.90

For experimental values of η, see pp. 562 and 577 of Hirschfelder, Curtis, and Bird as cited at the end of this chapter; for K, see pp. 573 and 577; for the diffusivity, see p. 581.

is independent of the pressure fails when at very high pressures the molecules are nearly always in contact, or when at very low pressures the mean free path becomes longer than the dimensions of the apparatus.

The apparatus used by Maxwell to verify the pressure-independence of the viscosity is illustrated in Fig. 10. His experiments were carried out at pressures between 0.02 and 1 atm, and over this range of pressure there was no perceptible change in the damping of the oscillations of the suspended disks.

Comparison of our earlier result (34) for the thermal conductivity K and the result (44) for the viscosity η gives the relation

$$K = \frac{\eta \hat{C}_V}{\rho} , \tag{46}$$

where $\hat{C}_V/\rho$ is the heat capacity per unit mass of the gas; ρ is the mass density. Observed values of the ratio

$$\frac{K\rho}{\eta\hat{C}_V}$$

are given in Table 3. The observed values are somewhat higher than the value unity predicted by the approximate calculations we have given above. Many improved calculations of the kinetic coefficients K, D, η have been carried out which take account of effects we have neglected, but the results are not changed to a major extent.

Comparison of the expressions for the diffusivity and the viscosity gives the ratio

$$D = \eta/\rho \tag{47}$$

where ρ is the mass density. A careful calculation by Chapman for a hard sphere model of a gas gives

$$D = 1.200 \, \eta/\rho \, , \tag{48}$$

which is not very far from our result (47).

REFERENCES

L. Loeb, *Kinetic theory of gases*, 2nd ed., Dover, 1934. This classical textbook for a generation of physicists remains of great value.

M. Knudsen, *Kinetic theory of gases*, Methuen, London, 1946. Elementary account of phenomena at very low pressures.

J. H. Jeans, *Introduction to the kinetic theory of gases*, Cambridge University Press, 1940.

S. Chapman and T. G. Cowling, *Mathematical theory of nonuniform gases*, Cambridge, 1952. Very detailed treatment.

J. O. Hirschfelder, C. F. Curtis, and R. B. Bird, *Molecular theory of gases and liquids*, Wiley, 1954. An encyclopedic work.

L. Boltzmann, *Lectures on gas theory*, English translation published by University of California Press, 1964; German original published by J. A. Barth, Leipzig, Part I, 1896; Part II, 1898. Great historical interest.

Applications of the Fermi-Dirac Distribution: Metals and White Dwarfs

CHAPTER 14   APPLICATIONS OF THE FERMI-DIRAC DISTRIBUTION:
METALS AND WHITE DWARFS

By a **degenerate**[1] **Fermi gas** we mean a gas of noninteracting or weakly interacting fermions under conditions of concentration and temperature such that the temperature τ is low in comparison with the Fermi energy:

$$\tau \ll \epsilon_F .$$

The orbitals of energy lower than the Fermi energy will be almost entirely occupied when this inequality is satisfied, and the orbitals of higher energy will be almost entirely vacant. An orbital is occupied fully when it contains one particle.

A Fermi gas is **nondegenerate** if $\tau \gg \epsilon_F$; that is, if the temperature is high in comparison with the Fermi energy. Chapters 11 through 13 were concerned with a gas in the nondegenerate limit. In this chapter we treat degenerate Fermi gases, with $\tau \ll \epsilon_F$.

The most important applications of the theory of degenerate Fermi gases are to conduction electrons in metals; to the interiors of white dwarf stars; to liquid He^3; and to nuclear matter.

GROUND STATE OF FERMI GAS IN ONE DIMENSION

Suppose that we want to accommodate N free electrons in a length L in a one-dimensional world. What orbitals will be occupied in the ground state of the system? By the Pauli exclusion principle each orbital can be occupied by at most one electron. This principle applies, of course, to electrons in solids as well as to electrons in atoms. In our one-dimensional world the quantum numbers of an electron in the free electron orbitals of form $\sin(n\pi x/L)$ are n and the **spin quantum number**[2] m_s. With each value of the integer n in the wavefunction there is an orbital for each of the two possible orientations of the electron spin, along the $+z$ or along the $-z$ direction, with $m_s = \pm\frac{1}{2}$. Each energy level labeled by the quantum number n has two orbitals, one with the spin up and one with the spin down.

---

[1] This is the second distinct usage of the term *degenerate* in statistical physics. The first usage was introduced in Chapter 1, where we called an energy level degenerate if more than one state has the same energy.

[2] The spin quantum number is the projection along the z axis of the spin of the particle, expressed in units of $\hbar$. The spin of the electron is $\frac{1}{2}\hbar$, and the values of the projections are restricted by quantum mechanics to $\pm\frac{1}{2}\hbar$. Thus $m_s = \pm\frac{1}{2}$ for an electron.

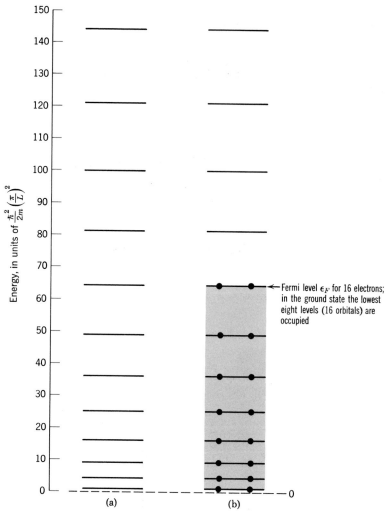

Figure 1 (a) The orbital energies of the orbitals $n = 1, 2, \ldots, 12$ for an electron confined to a line of length L. Each level corresponds to two orbitals, one for spin up and one for spin down. (b) The ground state of a system of 16 electrons. The orbitals above the shaded region are vacant in the ground state. *Note.* In one dimension the orbitals become less dense as the energy increases; in three dimensions they become more dense, as in Fig. 1.2.

If the system has 8 electrons, then in the ground state the orbitals with $n = 1, 2, 3,$ and 4 are filled and the orbitals of higher n are empty. Any other arrangement gives a higher energy. Let n_F denote the quantum number of the topmost filled orbital in the ground state of a system of N noninteracting electrons. We fill the orbitals starting from $n = 1$ at the bottom, and we continue filling the higher orbitals with electrons until all N electrons are accommodated. It will be convenient in this chapter to take N as an even number. The filled orbitals for the ground state of a system of 16 electrons are shown in Fig. 1.

GROUND STATE OF FERMI GAS IN THREE DIMENSIONS

The Fermi energy ϵ_F is determined by the requirement that the system in the ground state hold N electrons, with each orbital filled with one electron if the energy of the orbital is less than

$$\epsilon_F = \frac{\hbar^2}{2m}\left(\frac{\pi n_F}{L}\right)^2 . \tag{1}$$

The system is in a cube of side L. If the system holds N electrons the orbitals must be filled up to the quantum number n_F determined by (10.22):

$$N = \frac{\pi}{3} n_F^3 \; ; \qquad n_F^2 = \left(\frac{3N}{\pi}\right)^{\frac{2}{3}} . \tag{2}$$

In using (10.22) we have set $\gamma = 2$ because an electron has two possible spin orientations.

With $L^3 = V$, we may write (1) as

$$\boxed{\epsilon_F = \frac{\hbar^2}{2m}\left(\frac{3\pi^2 N}{V}\right)^{\frac{2}{3}} .} \tag{3}$$

The total energy of the system in the ground state is, with the familiar result $\pi n^2 \, \Delta n$ from (10.23) for the number of orbitals in the range Δn at n,

$$U_0 = 2 \sum_{|n| < n_F} \epsilon_n = \pi \int_0^{n_F} dn \, n^2 \epsilon_n = \frac{\pi^3}{2m}\left(\frac{\hbar}{L}\right)^2 \int_0^{n_F} dn \, n^4 , \tag{4}$$

with $\epsilon_n = (\hbar^2/2m)(\pi n/L)^2$. We have let

$$2 \sum_{n} \to \pi \int dn \, n^2 \tag{5}$$

in the conversion of the sum into an integral over an octant of a spherical shell. On integration of (4) we have

$$U_0 = \frac{\pi^3}{10m}\left(\frac{\hbar}{L}\right)^2 n_F^5 = \frac{3\hbar^2}{10m}\left(\frac{\pi n_F}{L}\right)^2 N = \tfrac{3}{5} N\epsilon_F , \tag{6}$$

using (1) and (2). The average kinetic energy in the ground state is U_0/N per particle; this is $\tfrac{3}{5}$ of the Fermi energy. The result (6) is plotted as a function of volume in Fig. 2. Notice that at constant N the energy increases as the volume decreases, so that the Fermi energy of an electron gas gives a repulsive contribution to the binding of a metal.

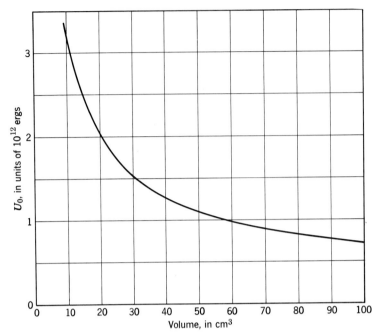

Figure 2 Total ground state energy U_0 of one mole of electrons, versus volume. (Courtesy of R. Cahn.)

Problem 1. *Energy of relativistic Fermi gas.* Find in the extreme relativistic limit the total energy in the ground state of a relativistic Fermi gas of N electrons in volume V. The energy is given by $E^2 = m^2c^4 + p^2c^2$, but assume that $E \cong pc$. *Hint.* The momentum $p = \hbar n\pi/L$ exactly as for the non-relativistic limit; here $n = (n_x{}^2 + n_y{}^2 + n_z{}^2)^{\frac{1}{2}}$, with positive integral values of n_x, n_y, n_z. The approximate value of the total energy is therefore obtained on summing pc over all orbitals up to $n = n_F$. The general problem is treated by F. Juttner, Zeitschrift für Physik **47**, 542 (1928).

DENSITY OF ORBITALS OF FREE PARTICLES

We saw in (10.23) that for a free particle the number of orbitals having the magnitude of their quantum number $\mathbf{n} \equiv n_x, n_y, n_z$ in the range Δn at n is

$$\tfrac{1}{2}\gamma\pi n^2 \, \Delta n \ , \tag{7}$$

where γ is the number of independent spin orientations. If the energy of an orbital is a function of n alone, then we can always calculate thermal averages by use of (7), as we did for the ground state energy U_0 of the Fermi gas.

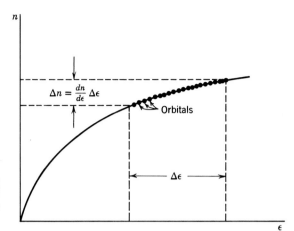

Figure 3 The number of orbitals in the energy range $\Delta\epsilon$ is equal to the number of orbitals in the quantum number range $\Delta n = (dn/d\epsilon)\,\Delta\epsilon$.

It often seems more convenient to express the thermal averages as integrals over the orbital energy ϵ in place of integrals over the orbital quantum number n. To do this we transform each integral over n by use of a function called the density of orbitals. The **density of orbitals** is defined as the number of orbitals per unit energy range and is denoted by $\mathcal{D}(\epsilon)$. In the literature $\mathcal{D}(\epsilon)$ is nearly always referred to as the density of states, but we wish to preserve the usage of orbitals for one particle and states for many particles.

We want to find the number of orbitals with energy between ϵ and $\epsilon + d\epsilon$. This number is denoted by $\mathcal{D}(\epsilon)\,d\epsilon$ and is given by

$$\mathcal{D}(\epsilon)\,d\epsilon = \tfrac{1}{2}\gamma\pi n^2\,\frac{dn}{d\epsilon}\,d\epsilon\ , \tag{8}$$

if the energy ϵ is a function only of n. The result (8) follows directly from (7), as we see from Fig. 3.

For free particles of mass M the energy is related to the quantum number by

$$\epsilon = \frac{\hbar^2}{2M}\left(\frac{\pi n}{L}\right)^2\ ; \qquad n = \left(\frac{2M\epsilon L^2}{\hbar^2\pi^2}\right)^{\frac{1}{2}}\ , \tag{9}$$

whence

$$\frac{dn}{d\epsilon} = \left(\frac{ML^2}{2\hbar^2\pi^2\epsilon}\right)^{\frac{1}{2}}\ ; \qquad n^2\,\frac{dn}{d\epsilon} = \frac{L^3(2M)^{\frac{3}{2}}}{2\pi^3\hbar^3}\epsilon^{\frac{1}{2}}\ . \tag{10}$$

From (8) the density of orbitals is

$$\boxed{\mathcal{D}(\epsilon) = \frac{\gamma V}{4\pi^2}\left(\frac{2M}{\hbar^2}\right)^{\frac{3}{2}}\epsilon^{\frac{1}{2}}\ .} \tag{11}$$

This is the density-of-orbitals function for free particles in the volume V.

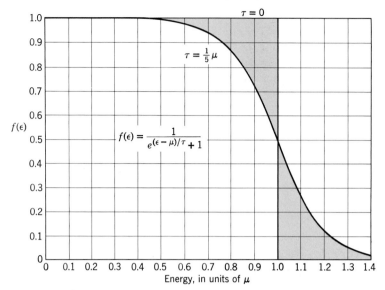

Figure 4 Plot of the Fermi-Dirac distribution function $f(\epsilon)$ versus ϵ/μ, for zero temperature and for a temperature $\mathcal{T} = \frac{1}{5}\mu$. The value of $f(\epsilon)$ gives the fraction of orbitals at a given energy which are occupied when the system is in thermal equilibrium. When the system is heated from absolute zero, electrons are transferred from the shaded region at $\epsilon/\mu < 1$ to the shaded region at $\epsilon/\mu > 1$. For a metal μ might correspond to a temperature of 50 000 K.

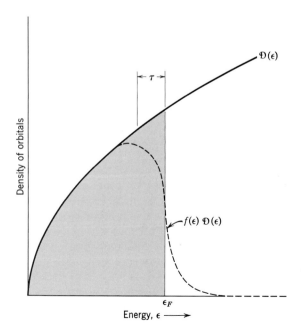

Figure 5 Density of orbitals as a function of energy, for a free electron gas in three dimensions. The dashed curve represents the density $f(\epsilon)\,\mathcal{D}(\epsilon)$ of occupied orbitals at a finite temperature, but such that $\mathcal{T}$ is small in comparison with ϵ_F. The shaded area represents the occupied orbitals at absolute zero.

When multiplied by the Fermi-Dirac function of Fig. 4, the density of orbitals $\mathcal{D}(\epsilon)$ becomes $\mathcal{D}(\epsilon) f(\epsilon)$, the density of occupied orbitals (Fig. 5).

As examples of the use of (11), the total number of electrons in a system may now be written as

$$N = \int_0^\infty d\epsilon \, \mathcal{D}(\epsilon) f(\epsilon) \ , \tag{12}$$

where $f(\epsilon)$ is the FD distribution function. The product $\mathcal{D}(\epsilon) f(\epsilon)$ gives the density of occupied orbitals. The total energy is

$$U = \int_0^\infty d\epsilon \, \epsilon \, \mathcal{D}(\epsilon) f(\epsilon) \ . \tag{13}$$

If the system is in the ground state, all orbitals are filled up to the energy ϵ_F, above which they are vacant. The number of electrons is equal to

$$N = \int_0^{\epsilon_F} d\epsilon \, \mathcal{D}(\epsilon) \ , \tag{14}$$

and the energy is

$$U_0 = \int_0^{\epsilon_F} d\epsilon \, \epsilon \, \mathcal{D}(\epsilon) \ . \tag{15}$$

Problem 2. *Density of orbitals at the Fermi energy.* Consider a free electron gas having N electrons. Show that the density of orbitals at the Fermi energy ϵ_F is

$$\mathcal{D}(\epsilon_F) = \frac{3N}{2\epsilon_F} \ . \tag{16}$$

Hint: Take the logarithm of both sides of (3) and form $dN/d\epsilon_F$.

Problem 3. *Density of orbitals in one and two dimensions.* (a) Show that the density of orbitals of a free particle in one dimension is

$$\mathcal{D}_1(\epsilon) = \frac{\gamma L}{\pi} \left(\frac{M}{2\epsilon\hbar^2} \right)^{\frac{1}{2}} \ , \tag{17}$$

where L is the length of the line.

(b) Show that in two dimensions

$$\mathcal{D}_2(\epsilon) = \frac{\gamma M L^2}{2\pi\hbar^2} \ , \tag{18}$$

independent of ϵ. Here L^2 is the area of the surface. The result in three dimensions is given in (11).

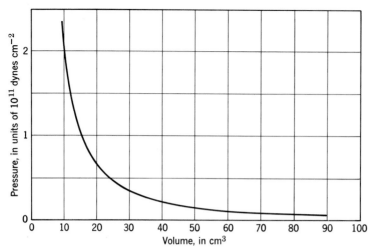

Figure 6 Pressure of one mole of electrons versus volume, for the ground state. (Courtesy of R. Cahn.)

Problem 4. Pressure of degenerate Fermi gas. (a) Show that a Fermi electron gas in the ground state exerts a pressure

$$p = \frac{(3\pi^2)^{\frac{2}{3}}}{5} \cdot \frac{\hbar^2}{m} \left(\frac{N}{V}\right)^{\frac{5}{3}}, \tag{19}$$

because in a uniform increase of the volume of a cube every orbital has its energy lowered: the energy of an orbital is proportional to $1/L^2$ or to $1/V^{\frac{2}{3}}$. *Hint.* Use (6). (b) Evaluate the pressure for electrons in metallic sodium, with use of the data in Table 2. A plot of pressure versus volume is given in Fig. 6, for the ground state of a Fermi gas.

ENERGY AND HEAT CAPACITY OF THE ELECTRON GAS

We derive a quantitative expression for the heat capacity of a degenerate Fermi gas of electrons in three dimensions. The calculation is perhaps the most impressive accomplishment of the theory of the degenerate Fermi gas. For an ideal monatomic gas the heat capacity is $\frac{3}{2}Nk_B$, but for electrons in a metal very much lower values are found. The calculation that follows gives an excellent account of the experimental results.

The increase in the total energy of a system of N electrons when heated from 0 to T is denoted by $\Delta U \equiv U(T) - U(0)$, whence

$$\Delta U = \int_0^\infty d\epsilon \, \epsilon \, \mathfrak{D}(\epsilon) \, f(\epsilon) - \int_0^{\epsilon_F} d\epsilon \, \epsilon \, \mathfrak{D}(\epsilon) \tag{20}$$

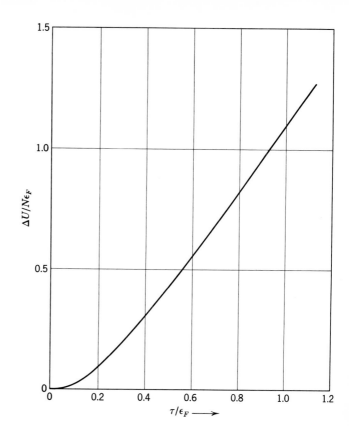

Figure 7 Temperature dependence of the energy of a noninteracting fermion gas in three dimensions. The energy is plotted in normalized form as $\Delta U/N\epsilon_F$, where N is the number of electrons. The temperature is plotted as τ/ϵ_F.

Here $f(\epsilon)$ is the Fermi-Dirac function of Chapter 9, and $\mathcal{D}(\epsilon)$ is the number of orbitals per unit energy range. We multiply the identity

$$N = \int_0^\infty d\epsilon\, f(\epsilon)\, \mathcal{D}(\epsilon) = \int_0^{\epsilon_F} d\epsilon\, \mathcal{D}(\epsilon) \tag{21}$$

by ϵ_F to obtain

$$\left(\int_0^{\epsilon_F} + \int_{\epsilon_F}^\infty\right) d\epsilon\, \epsilon_F\, f(\epsilon)\, \mathcal{D}(\epsilon) = \int_0^{\epsilon_F} d\epsilon\, \epsilon_F\, \mathcal{D}(\epsilon)\ , \tag{22}$$

which we use to rewrite (20) as

$$\Delta U = \int_{\epsilon_F}^\infty d\epsilon\, (\epsilon - \epsilon_F)\, f(\epsilon)\, \mathcal{D}(\epsilon) + \int_0^{\epsilon_F} d\epsilon\, (\epsilon_F - \epsilon)[1 - f(\epsilon)]\, \mathcal{D}(\epsilon)\ . \tag{23}$$

The first integral on the right-hand side of (23) gives the energy needed to take electrons from ϵ_F to the orbitals of energy $\epsilon > \epsilon_F$, and the second integral gives the energy needed to bring the same electrons to ϵ_F from orbitals below ϵ_F. The product $f(\epsilon)\, \mathcal{D}(\epsilon)\, d\epsilon$ in the first integral is the number of electrons elevated to orbitals in the energy range $d\epsilon$ at an energy ϵ. The factor $[1 - f(\epsilon)]$ in the second integral is the probability that an electron has been removed from an orbital ϵ. The function ΔU is plotted in Fig. 7. We made the plot with the aid of the numerical results of Appendix C.

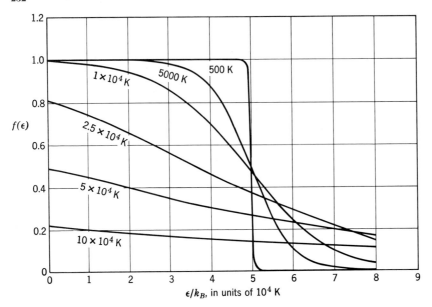

Figure 8 Fermi-Dirac distribution function at various temperatures, for $T_F \equiv \epsilon_F/k_B = 50\,000$ K. The results apply to a gas in three dimensions. The total number of particles is constant, independent of temperature. (Courtesy of B. Feldman.)

Table 1 Values of the Chemical Potential and Absolute Activity of a Fermi Gas at Several Temperatures

(The gas is chosen such that $\epsilon_F/k_B = 50\,000$ K.)

Temperature T, in deg K	Chemical potential μ/k_B, in deg K	Absolute activity $\lambda = e^{\mu/k_B T}$
0	50,000	∞
500	50,000	e^{100}
5,000	49,700	21,000
10,000	48,200	124
25,000	36,200	4.3
50,000	$-100°$	0.98
100,000	$-128,000$	0.28

° The chemical potential is close to zero, but is not exactly zero, at the temperature $k_B T = \epsilon_F$.

In Fig. 8 we plot the FD distribution function versus ϵ, for six values of the temperature. The electron concentration of the Fermi gas was taken such that $\epsilon_F/k_B = 50\,000$ K, characteristic of the conduction electrons in a metal. Values of the chemical potential μ and absolute activity $\lambda = e^{\mu/T}$ as a function of temperature are given in Table 1.

The heat capacity of the electron gas is found on differentiating the

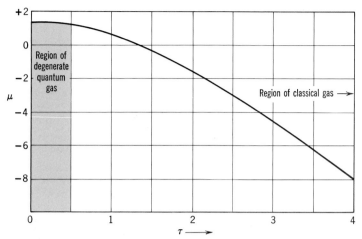

Figure 9   Plot of the chemical potential μ versus temperature T for a gas of noninteracting fermions in three dimensions. For convenience in plotting, the particle concentration has been chosen so that $\mu(0) \equiv \epsilon_F = (\frac{3}{2})^{\frac{2}{3}}$.

energy U or ΔU with respect to T. The only temperature-dependent term in (23) is $f(\epsilon)$, whence we can group terms to obtain

$$C_{\text{el}} = \frac{dU}{dT} = \int_0^\infty d\epsilon \, (\epsilon - \epsilon_F) \frac{df}{dT} \mathfrak{D}(\epsilon) \; . \tag{24}$$

At the temperatures of interest in metals $k_B T/\epsilon_F < 0.01$, so that the derivative df/dT is large only at energies near ϵ_F. It is a good approximation to evaluate the density of orbitals $\mathfrak{D}(\epsilon)$ at ϵ_F and take it outside of the integral:

$$C_{\text{el}} \cong \mathfrak{D}(\epsilon_F) \int_0^\infty d\epsilon \, (\epsilon - \epsilon_F) \frac{df}{dT} \; . \tag{25}$$

Examination of the data in Table 1 and of the graphs in Figs. 8 and 9 of the variation of μ with T suggests that when $T \ll \epsilon_F$ we may replace the chemical potential μ in the FD distribution function by the constant Fermi energy $\epsilon_F \equiv \mu(0)$. Then

$$\frac{df}{dT} = \frac{\epsilon - \epsilon_F}{k_B T^2} \cdot \frac{e^{(\epsilon - \epsilon_F)/k_B T}}{[e^{(\epsilon - \epsilon_F)/k_B T} + 1]^2} \; . \tag{26}$$

We set

$$x \equiv (\epsilon - \epsilon_F)/k_B T \; , \tag{27}$$

and it follows from (25) and (26) that

$$C_{\text{el}} = k_B^2 T \, \mathfrak{D}(\epsilon_F) \int_{-\epsilon_F/k_B T}^\infty dx \, x^2 \frac{e^x}{(e^x + 1)^2} \; . \tag{28}$$

We may safely replace the lower limit by $-\infty$ because the factor e^x in the integrand is already negligible at $x = -\epsilon_F/k_B T$ if we are concerned with low temperatures such that $\epsilon_F/k_B T \sim 100$ or more. The integral[3] becomes

$$\int_{-\infty}^{\infty} dx\, x^2 \frac{e^x}{(e^x + 1)^2} = \frac{\pi^2}{3} , \tag{29}$$

whence we have for the heat capacity of an electron gas

$$C_{\text{el}} = \tfrac{1}{3}\pi^2\, \mathfrak{D}(\epsilon_F) k_B^2 T . \tag{30}$$

We found in Problem 2 that the density of orbitals at the Fermi energy is

$$\mathfrak{D}(\epsilon_F) = \frac{3N}{2\epsilon_F} = \frac{3N}{2k_B T_F} \tag{31}$$

for a free electron gas, with $k_B T_F \equiv \epsilon_F$. Do not be deceived by the notation T_F: it is *not* the temperature of the Fermi gas, but only a convenient reference point. For $T \ll T_F$ the gas is degenerate; for $T \gg T_F$ the gas is in the classical regime. Thus (30) becomes

$$C_{\text{el}} = \tfrac{1}{2}\pi^2 N k_B \cdot \frac{k_B T}{\epsilon_F} = \tfrac{1}{2}\pi^2 N k_B \frac{T}{T_F} . \tag{32}$$

We can give a physical explanation of the form of the result (32). When the specimen is heated from absolute zero, chiefly those electrons in states within an energy range T of the Fermi level are excited thermally, because the FD distribution function is affected over a region of the order of T in width, as illustrated by Figs. 4 and 9. Thus the number of excited electrons is of the order of NT/ϵ_F, and each of these has its energy increased approximately by T. The total electronic thermal energy ΔU is therefore of the order of

$$\Delta U \approx \frac{NT^2}{\epsilon_F} . \tag{33}$$

The electronic contribution to the heat capacity is given by

$$C_{\text{el}} = \frac{d\,\Delta U}{dT} = k_B \frac{d\,\Delta U}{dT} \approx N k_B \frac{T}{\epsilon_F} \approx N k_B \frac{T}{T_F} , \tag{34}$$

which is directly proportional to T, in agreement with the exact result (30) and with the experimental results.

[3] The integral is not elementary, but may be evaluated from the more familiar result

$$\int_0^{\infty} dx \frac{x}{e^{ax} + 1} = \frac{\pi^2}{12a^2} \tag{29a}$$

on differentiation of both sides with respect to the parameter a.

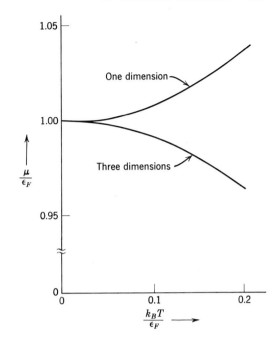

Figure 10 Variation with temperature of the chemical potential μ, for free electron Fermi gases in one and three dimensions. In common metals $k_B T/\epsilon_F \approx 0.01$ at room temperature, so that μ is closely equal to ϵ_F. These curves were calculated from series expansions of the integral (12) for the number of particles in the system.

Problem 5. *Chemical potential versus temperature.* Explain graphically why the initial curvature of μ versus T is upward for a gas in one dimension and downward in three dimensions (Fig. 10). *Hint.* The $\mathfrak{D}_1(\epsilon)$ and $\mathfrak{D}_3(\epsilon)$ curves are different, where $\mathfrak{D}_1$ is given by (17) and $\mathfrak{D}_3$ by (11). It will be found useful to set up the integral for N, the number of particles, and to consider the behavior of the integrand at two quite different temperatures.

FERMI GAS IN METALS

It is well-established that the alkali metals and copper, silver, and gold have one free electron, or conduction electron, per atom. These elements have one valence electron per atom, and the valence electron becomes the conduction electron in the metal. Thus the concentration of conduction electrons is equal to the concentration of atoms, which may be evaluated either from the density and the atomic weight or from the crystal lattice parameter.

If the conduction electrons in metals act as a free fermion gas, the value of the Fermi energy ϵ_F may be calculated from the relation

$$\epsilon_F = \frac{\hbar^2}{2m} \left(\frac{3\pi^2 N}{V} \right)^{\frac{2}{3}} . \tag{35}$$

Table 2   Calculated Fermi Energy Parameters for Free Electrons

	Electron concentration N/V, in cm^{-3}	Velocity v_F, in cm sec^{-1}	Fermi energy ϵ_F, in ev	Fermi temperature $T_F = \epsilon_F/k_B$, in deg K
Li	4.6×10^{22}	1.3×10^8	4.7	5.5×10^4
Na	2.5	1.1	3.1	3.7
K	1.34	0.85	2.1	2.4
Rb	1.08	0.79	1.8	2.1
Cs	0.86	0.73	1.5	1.8
Cu	8.50	1.56	7.0	8.2
Ag	5.76	1.38	5.5	6.4
Au	5.90	1.39	5.5	6.4

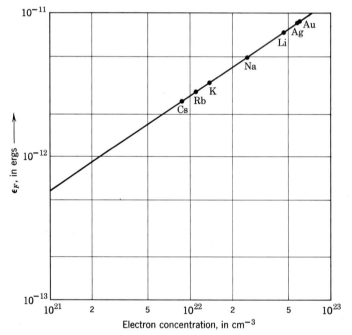

Figure 11   Fermi energy ϵ_F of a free electron gas as a function of the concentration. Calculated values are shown for several monovalent metals. The straight line is drawn for $\epsilon_F = 5.835 \times 10^{-27}$ $(N/V)^{\frac{2}{3}}$ ergs, with N/V in cm^{-3}.

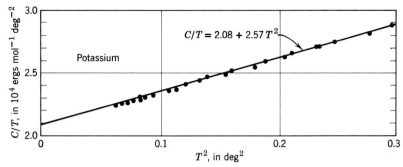

Figure 12 Experimental heat capacity values for potassium, plotted as C/T versus T^2. [After W. H. Lien and N. E. Phillips, Physical Review **133**, A1370 (1964).]

Values of N/V and of ϵ_F are given in Table 2 and in Fig. 11. The electron velocity v_F at the Fermi surface is also given in the table; it is defined so that the kinetic energy of an electron at the Fermi surface is equal to ϵ_F:

$$\tfrac{1}{2}mv_F{}^2 = \epsilon_F \ , \tag{36}$$

where m is the mass of the electron.

The values of the Fermi temperature $T_F \equiv \epsilon_F/k_B$ for ordinary metals[4] are of the order of 5×10^4 K, so that the assumption $T \ll T_F$ used in the derivation of (30) is an excellent approximation at room temperature and below.

The heat capacity of many metals at constant volume may be written as the sum of an electronic contribution and a lattice vibration contribution. At low temperatures the sum has the form

$$C_V = \gamma T + AT^3 \ , \tag{37}$$

where γ and A are constants characteristic of the material. Here $\gamma \equiv \tfrac{1}{2}\pi^2 Nk_B/T_F$ from (32), and the lattice vibration term AT^3 is discussed in Chapter 16. The electronic term is linear in T and is dominant at sufficiently low temperatures.

It is helpful to display the experimental values of the heat capacity for a given material as a plot of C_V/T versus T^2:

$$C_V/T = \gamma + AT^2 \ , \tag{38}$$

for then the points should lie on a straight line. The intercept at $T = 0$ gives the value of γ. Such a plot is shown for potassium in Fig. 12. Observed values of γ are given in Tables 3 and 4.

---

[4] It is possible to simulate metals with arbitrarily low values of T_F by the controlled addition of suitable impurities to semiconductor crystals. A discussion of semiconductors is given in ISSP, Chapter 10.

Table 3 Experimental and Free Electron Electronic Heat Capacities of Monovalent Metals

(The values of γ and γ_0 are in millijoules mol^{-1} deg^{-2}, or 10^4 ergs mol^{-1} deg^{-2}.)

Metal	γ (exp), mJ mol^{-1} deg^{-2}	γ_0 (free electron), mJ mol^{-1} deg^{-2}	γ/γ_0
Li	1.63	0.75	2.17
Na	1.38	1.14	1.21
K	2.08	1.69	1.23
Rb	2.41	1.97	1.22
Cs	3.20	2.36	1.35
Cu	0.695	0.50	1.39
Ag	0.646	0.65	1.00
Au	0.729	0.65	1.13

Courtesy of N. E. Phillips.

Table 4 Experimental Values of Electronic Heat Capacity Constant γ of Metals

(The value of γ in millijoules mol^{-1} deg^{-2}.)

Li	Be												B	C	N
1.63	0.17														
Na	Mg												Al	Si	P
1.38	1.3												1.35		
K	Ca	Sc	Ti	V	Cr	Mnγ	Fe	Co	Ni	Cu	Zn	Ga	Ge		As
2.08	2.9	10.7	3.35	9.26	1.40	9.20	4.98	4.73	7.02	0.695	0.64	0.596			0.19
Rb	Sr	Y	Zr	Nb	Mo	Tc	Ru	Rh	Pd	Ag	Cd	In	Sn		Sb
2.41	3.6	10.2	2.80	7.79	2.0	—	3.3	4.9	9.42	0.646	0.688	1.69	1.78		0.11
Cs	Ba	La	Hf	Ta	W	Re	Os	Ir	Pt	Au	Hg	Tl	Pb		Bi
3.20	2.7	10.	2.16	5.9	1.3	2.3	2.4	3.1	6.8	0.729	1.79	1.47	2.98		0.008

From compilations furnished by N. Phillips and N. Pearlman.

We compare the observed values of γ with the calculated values γ_0 for a free electron Fermi gas from (32):

$$\gamma_0 = \frac{\pi^2 N k_B^2}{2\epsilon_F} = \frac{5.668 \times 10^{-8}}{\epsilon_F} \text{ erg mol}^{-1} \text{ deg}^{-2}, \qquad (39)$$

with the Fermi energy ϵ_F in ergs. Or

$$\gamma_0 = \frac{3.538 \times 10^4}{\epsilon_F} \text{ ergs mol}^{-1} \text{ deg}^{-2}, \qquad (40)$$

with ϵ_F in electron volts as given in Table 2. It is helpful to remember that

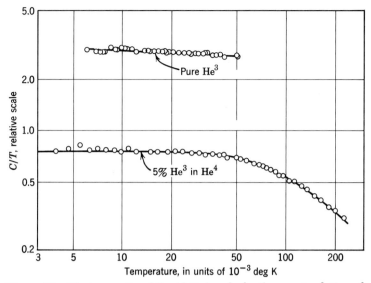

Figure 13   Heat capacity of liquid He^3 and of a 5 percent solution of He^3 in liquid He^4. The quantity plotted on the vertical axis is C/T, and the horizontal axis is T. Thus for a Fermi gas in the degenerate temperature region the theoretical curves of C/T at constant volume are horizontal. The curve for pure He^3 is taken at constant pressure, which accounts for the slight slope. The curve for the solution of He^3 in liquid He^4 indicates that the He^3 in solution acts as a Fermi gas; the degenerate region at low temperature goes over to the nondegenerate region at higher temperature. The solid line through the experimental points for the solution is drawn for $T_F = 0.331$ K, which agrees with the calculation for free atoms if the effective mass is taken as 2.38 times the mass of an atom of He^3. (Curves from the review by J. C. Wheatley.)

1 millijoule $= 1 \times 10^4$ ergs. The close agreement of γ with γ_0 shown in Table 3 is characteristic of the monovalent metals. The differences are attributed to electron-electron interactions and electron-lattice interactions; their calculation requires advanced techniques.

Problem 6. *Liquid He^3 as a Fermi gas.* The atom He^3 has spin $I = \frac{1}{2}$ and is a fermion. (a) Calculate as in Table 2 the Fermi sphere parameters v_F, ϵ_F, and T_F for He^3 at absolute zero, viewed as a gas of noninteracting fermions. The density of the liquid is 0.081 gm cm^{-3}. (b) Calculate the heat capacity at low temperatures $T \ll T_F$ and compare with the experimental value $C_V = 2.89\, Nk_BT$ as observed for $T < 0.1$ K by A. C. Anderson, W. Reese, and J. C. Wheatley, *Physical Review* **130**, 495 (1963); see also Fig. 13. Excellent surveys of the properties of liquid He^3 are given by J. Wilks, *The properties of liquid and solid helium,* Clarendon Press, 1967, and by J. C. Wheatley, "Dilute solutions of He^3 in He^4 at low temperatures," American Journal of Physics **36**, 181–210 (1968). The principles of refrigerators based on He^3-He^4 mixtures are reviewed by N. S. Betts, Contemporary Physics **9**, 97 (1968); such refrigerators produce steady temperatures down to 0.01 K in continuously acting operation.

WHITE DWARF STARS

White dwarf stars[5] have masses comparable to that of the Sun. The mass and radius of the Sun are

$$M_\odot = 1.99 \times 10^{33} \text{ gm} \; ; \qquad R_\odot = 7.0 \times 10^{10} \text{ cm} \; . \qquad (41)$$

The radii of white dwarfs are very small, perhaps 0.01 that of the Sun. The density of the Sun, which is a normal star, is of the order of 1 gm cm^{-3}. The densities of white dwarfs are exceedingly high, of the order of 10^4 to 10^7 gm cm^{-3}. What is the state of matter at such high densities? It is believed that atoms under the densities prevalent in white dwarfs are entirely ionized into nuclei and free electrons, and the electron gas is a degenerate gas.

The companion of Sirius was the first of these astral objects to be discovered. In 1844 Bessel observed that the path of the star Sirius oscillated slightly about a straight line, as if it had an invisible companion (Fig. 14). The companion, Sirius B, was discovered near its predicted position by Clark in 1862. The mass of Sirius B was determined to be 1.96 $\times$ 10^{33} gm by measurements on the orbits. The radius of Sirius B was estimated as 1.9 $\times$ 10^9 cm by a comparison of the surface temperature and the radiant energy flux, using the properties of thermal radiant energy as developed in Chapter 15.

The values of the mass and radius of Sirius B lead to the mean density

$$\rho = \frac{M}{V} = \frac{1.96 \times 10^{33} \text{ gm}}{\frac{4}{3}\pi(1.9 \times 10^9 \text{ cm})^3} \approx 0.69 \times 10^5 \text{ gm cm}^{-3} \; . \qquad (42)$$

The matter of which the white dwarf is composed must be very tightly packed. This extraordinarily high value of the density was appraised by Eddington in 1926 in the following words: "Apart from the incredibility of the result, there was no particular reason to view the calculation with suspicion." Other white dwarfs have higher densities: that named Van Maanen No. 2 has a mean density of 6.8 $\times$ 10^6 gm cm^{-3}. We have noted that the mean density of the Sun is only of the order of 1 gm cm^{-3}.

---

[5] For an introduction to the properties of white dwarf stars see pp. 84–90 of L. H. Aller, *Astrophysics, nuclear transformations, stellar interiors, and nebulae*, Ronald, 1954; see also Chapter 12 of D. H. Menzel, P. L. Bhatnagar, and H. K. Sen, *Stellar interiors*, Chapman and Hall, 1963. Perhaps 3 percent of the stars in the neighborhood of the sun are white dwarfs.

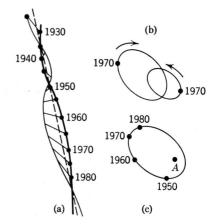

Figure 14 The visual orbit of Sirius. The heavy curve in (a) shows the sinusoidal motion of the primary star. The thin curve represents the sinusoidal motion of the white dwarf companion. The dashed curve is the motion of the center of gravity of the system. The curve in (b) shows the apparent orbits of the two components around their common center of gravity; and (c) is the apparent orbit of the companion about the primary. (After Struve, Lynds, and Pillans.)

Hydrogen atoms at a density of the order of 10^6 gm cm^{-3} have a volume per atom equal to

$$V_A \approx \frac{1}{(10^6 \text{ mol cm}^{-3})(6 \times 10^{23} \text{ atoms mol}^{-1})} \approx 2 \times 10^{-30} \text{ cm}^3 \text{ per atom },$$

$$(43)$$

or 2×10^{-6} Å^3 per atom. The average nearest-neighbor separation is then of the order of 0.01 Å, as compared with the internuclear separation of 0.74 Å in a molecule of hydrogen.

Under such conditions of high density the atomic electrons are no longer attached to individual nuclei. The electrons are ionized and form an electron gas, as suggested originally by Eddington[6] in 1924. The matter in the white dwarfs is held together by gravitational attraction, which is the binding force in all stars.

The condition of white dwarf stars is such that in the interior the electron gas is degenerate; the temperature is much less than the Fermi energy ϵ_F. The Fermi energy of an electron gas at a concentration of 1×10^{30} electrons cm^{-3} is given by

$$\epsilon_F = \frac{\hbar^2}{2m} \left(\frac{3\pi^2 N}{V} \right)^{\frac{2}{3}} \approx 0.5 \times 10^{-6} \text{ erg} \approx 3 \times 10^5 \text{ ev },\qquad(44)$$

which is about 10^5 higher than in a typical metal. The Fermi temperature $T_F \equiv \epsilon_F/k_B \approx 3 \times 10^9$ K. The temperature in the interior[7] of a star or a white dwarf is believed to be of the order of 10^7 K. This temperature is needed to maintain a thermonuclear reaction at a rate compatible with the

[6] A. S. Eddington, *Internal constitution of the stars,* Cambridge University Press, 1926.
[7] The surface temperatures of stars are very much lower because of cooling by radiation from the surface. The surface temperature of the Sun is about 6000 K.

rate of emission of radiant energy from the surface of the star. Thus in a white dwarf $T \ll T_F$. The electron gas in the interior of a white dwarf is highly degenerate in spite of the high temperature of the electrons; the thermal energy is much lower than the Fermi energy.

Are the electron energies in the relativistic regime? We are concerned with this question because our development of the theory of the Fermi gas has used the nonrelativistic expression $p^2/2m$ for the kinetic energy of an electron. The energy equivalence of the rest mass of an electron is

$$\epsilon_0 = mc^2 \simeq (1 \times 10^{-27} \text{ gm})(3 \times 10^{10} \text{ cm sec}^{-2})^2 \simeq 1 \times 10^{-6} \text{ erg} . \quad (45)$$

This energy is of the same order as our estimate (44) of the Fermi energy of the electrons in a white dwarf. Relativistic effects will be significant, but not dominant. At higher densities than that used in our estimate it is important to treat the Fermi gas as relativistic.[8]

Do the nuclei in a white dwarf star also form a degenerate Fermi gas? The Fermi energy is inversely proportional to the mass of the particle. The Fermi energy of nucleons, which are protons or neutrons, may be found from the Fermi energy of electrons at the same concentration if we multiply the Fermi energy of electrons by the mass ratio $m/M = 1/1840$. Under the conditions of the estimate above, the Fermi energy of a nucleon is of the order of 0.25×10^{-9} erg. This value is somewhat less than the value of the thermal energy $k_B T$ at 1×10^7 K, which is 1.4×10^{-9} erg. Thus the nucleon gas may not be degenerate, although close to degeneracy. Some types of white dwarfs are known that are believed to have a degenerate nucleon distribution.

Problem 7. Mass-radius relationship for white dwarfs. Consider a white dwarf of mass M and radius R. Let the electrons be degenerate but nonrelativistic; the protons are nondegenerate. (a) Show that the order of magnitude of the gravitational self-energy is

$$-\frac{GM^2}{R} , \quad (46)$$

where G is the gravitational constant. (If the mass density is constant within the sphere of radius R, the exact potential energy is $-3GM^2/5R$). (b) Show that the order of magnitude of the kinetic energy of the electrons is

$$\frac{\hbar^2 N^{\frac{5}{3}}}{mR^2} \simeq \frac{\hbar^2 M^{\frac{5}{3}}}{mM_H^{\frac{5}{3}}R^2} , \quad (47)$$

where m is the mass of an electron and M_H is the mass of a proton. (c) Show

[8] See the discussion in Chapters 2 and 12 of D. H. Menzel, P. L. Bhatnagar, and H. K. Sen, *Stellar interiors*, Wiley, 1963.

that if the gravitational and kinetic energies are of the same order of magnitude (as required by the virial theorem of mechanics, Appendix D) then

$$M^{\frac{1}{3}}R \simeq 10^{20} \text{ gm}^{\frac{1}{3}} \text{ cm} . \tag{48}$$

(d) If the mass is equal to that of the Sun (2×10^{33} gm), what is the density of the white dwarf? (e) It is believed that pulsars are stars composed of a cold degenerate gas of neutrons. Show that for a neutron star

$$M^{\frac{1}{3}}R \simeq 10^{17} \text{ gm}^{\frac{1}{3}} \text{ cm} . \tag{48a}$$

What is the value of the radius for a neutron star with a mass equal to that of the Sun? Express the result in km.

NUCLEAR MATTER

By nuclear matter we mean that state of matter which exists within nuclei. In nuclear matter the neutrons and protons of which nuclei are composed may be viewed as a degenerate fermion gas, at least qualitatively. We estimate here the Fermi energy of the nucleon gas in nuclear matter.[9]

The radius of a nucleus that contains A nucleons, where

$$A = Z \text{ protons} + (A - Z) \text{ neutrons} ,$$

is given by the empirical relation[10]

$$R \cong (1.3 \times 10^{-13} \text{ cm}) \times A^{\frac{1}{3}} . \tag{49}$$

According to this relation the average volume per particle is constant, for the volume goes as R^3, which is proportional to A. The concentration of nucleons in nuclear matter is

$$\frac{N}{V} \cong \frac{A}{\frac{4}{3}\pi(1.3 \times 10^{-13} \text{ cm})^3 A} \cong 0.11 \times 10^{39} \text{ cm}^{-3} , \tag{50}$$

about 10^8 times higher than the concentration of nucleons in a white dwarf star. Neutrons and protons are not identical particles. The Fermi energy of the neutrons need not equal the Fermi energy of the protons. The concentration of one or the other, but not both, enters the familiar relation

$$\epsilon_F = \frac{\hbar^2}{2M}\left(\frac{3\pi^2 N}{V}\right)^{\frac{2}{3}} . \tag{51}$$

---

[9] See J. M. Blatt and V. F. Weisskopf, *Theoretical nuclear physics*, Wiley, 1952; M. A. Preston, *Physics of the nucleus*, Addison-Wesley, 1962.

[10] This relation is based on various experiments on the scattering of particles by nuclei, and also on theoretical inferences from the binding energy of nuclei.

For simplicity let us suppose that the number of protons is equal to the number of neutrons. With $Z \approx \frac{1}{2}A$ the concentrations of the nucleons are

$$\left(\frac{N}{V}\right)_{\text{protons}} \approx \left(\frac{N}{V}\right)_{\text{neutrons}} \approx 0.05 \times 10^{39} \text{ cm}^{-3} , \tag{52}$$

as obtained from (50) on dividing by 2. We have for the Fermi energy

$$\epsilon_F \cong (3.17 \times 10^{-30})(N/V)^{\frac{2}{3}} \approx 0.43 \times 10^{-4} \text{ erg} \approx 27 \text{ Mev} . \tag{53}$$

The average kinetic energy of a particle in a degenerate Fermi gas is $\frac{3}{5}$ of the Fermi energy, so that in nuclear matter the average kinetic energy is 16 Mev per nucleon.

Problem 8. *Entropy of fermions.* In (19.30) the general result is established that at constant μ and V,

$$\sigma = \frac{\partial}{\partial \tau}(\tau \log \mathfrak{Z}) . \tag{54}$$

Consider an orbital which may be occupied by 0 or 1 particle, with energy 0 or ϵ, respectively. Show that the entropy for this problem may be written as

$$\sigma = -[f \log f + (1 - f) \log (1 - f)] , \tag{55}$$

where f is the Fermi-Dirac distribution function. Compare this result with (15.12) for bosons.

CHAPTER 15

Planck Distribution Function for Photons

━━━━━━━━━━━━━━━━━━━━━━━━━━━━━━━━━━━━━

"The expression . . . is simple: $S = f(U/\nu)$, where ν is the frequency of the oscillator. I will derive this on another occasion."

(M. Planck,[1] 19 October 1900.)

"[We consider] the distribution of the energy U among N oscillators of frequency ν. If U is viewed as divisible without limit, then an infinite number of distributions are possible. We consider however—and this is the essential point of the whole calculation—U as made up of an entirely determined number of finite equal parts, and we make use of the natural constant $h = 6.55 \times 10^{-27}$ erg-sec. This constant when multiplied by the common frequency ν of the oscillators gives the element of energy ϵ in ergs . . . ."

(M. Planck,[2] 14 December 1900.)

---

[1] Max Planck, *Physikalische Abhandlungen und Vortraege*, Vieweg, Braunschweig, 1958. Our two quotations are from precursors to the definitive paper "Über das Gesetz der Energieverteilung im Normalspektrum," Annalen der Physik (4), **4**, 553–563 (1901). The value of h is taken at present to be 6.626×10^{-27} erg-sec. An excellent brief historical review of Planck's work on quanta is given by M. J. Klein, Physics Today **19**, 23 (Nov. 1966); see also D. ter Haar, *Old quantum theory*, Pergamon, 1967.

[2] Planck said later that he made this assumption as an act of desperation.

What we now view as an application of the boson distribution function

$$\langle n \rangle = \frac{1}{e^{(\epsilon - \mu)/\tau} - 1} \tag{1}$$

for the occupancy of an orbital of energy ϵ was made by Planck in the first work in which the concept of a quantum of energy was ever used. Planck was concerned with the problem of thermal radiation: what is the distribution in frequency of electromagnetic energy in thermal equilibrium with the walls of a cavity that contains the radiation?

In agreement with experiment, he found the result

$$\langle n \rangle = \frac{1}{e^{\hbar\omega/\tau} - 1} , \tag{2}$$

where $\langle n \rangle$ is the number of photons in a cavity mode of angular frequency ω. Here $\hbar\omega = \epsilon$ is the energy of a mode or orbital when occupied by a single light quantum or photon. The term **photon** denotes a quantum of energy of the electromagnetic field.

The result (2) follows from (1) if the chemical potential is set equal to zero. We shall derive (2) from first principles, but we may remark that the origin of the difference between the Planck and the Bose-Einstein results is that the number of photons in a system is not a conserved quantity, whereas in the original derivation of (1) we assumed explicitly that the number of particles was conserved. There are many more photons in a cavity when the walls are at a high temperature than when the walls are at a low temperature. The total number of photons in the system plus reservoir is not conserved. Therefore it is no longer appropriate to consider the entropy variation $(\partial\sigma/\partial N)\,\delta N$ which previously led us to introduce the chemical potential. We shall now see how to avoid this step when treating a system of photons.

PHOTON DISTRIBUTION

We need to modify slightly the derivation of the Bose-Einstein distribution function that was given in Chapter 9. There we considered an orbital of energy ϵ which might be occupied by $0, 1, 2, \ldots, n, \ldots$ particles. But here we imagine as the system a single particle with the properties of a harmonic oscillator.[3] Such an oscillator on quantum theory has orbitals of energy $0\epsilon, 1\epsilon, 2\epsilon, \ldots, n\epsilon, \ldots$, as in Fig. 1. The oscillator is treated as a system in thermal

---

[3] Planck introduced the concept of a harmonic oscillator of frequency ω to represent a mode of frequency ω of the electromagnetic field in a cavity. The oscillator is associated with the electromagnetic field and not with the walls of the cavity. Our ϵ is equal to $\hbar\omega$.

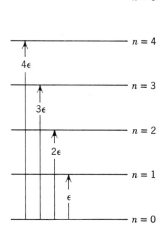

$n = 5$

$n = 4$

4ϵ

$n = 3$

3ϵ

$n = 2$

2ϵ

$n = 1$

ϵ

$n = 0$

Figure 1 Orbitals of an oscillator that represents a mode of frequency ω of an electromagnetic field. When the oscillator is in the orbital of energy $n\epsilon = n\hbar\omega$, this is equivalent to n photons in the mode.

contact with a reservoir, but the oscillator itself is not allowed to diffuse into the reservoir. An oscillator is equivalent to a mode of the electromagnetic field in the cavity.

In the quantum theory of a harmonic oscillator it is shown that the ground state energy is $\frac{1}{2}\hbar\omega$ above the energy of a classical harmonic oscillator at rest. (The quantum oscillator is not at rest in the ground state.) The energy of the nth orbital of a quantum harmonic oscillator is $(n + \frac{1}{2})\epsilon$, where $\frac{1}{2}\epsilon$ is the **zero-point energy** of the oscillator. The zero-point quantum motion of a harmonic oscillator has definite physical consequences; for example, the Lamb shift of the energy levels of the hydrogen atom is caused by the zero-point amplitude of the electromagnetic field. Also, the inelastic scattering of x-rays by a crystal at absolute zero is caused by the zero-point vibrations of the atoms. We shall, however, omit the zero-point energy from our entire discussion in this chapter and the next. To do so will simplify the appearance of the equations without altering their thermodynamic consequences, for the entropy of a system of harmonic oscillators at a given temperature is not affected by the omission or inclusion of the zero-point energy.

When the oscillator is excited to occupy the orbital of energy $n\epsilon_l$, the system is equivalent in every respect to the real system with n photons in the state l. A change in the orbital that is occupied in the oscillator is equivalent to a change in the number of photons in the real system. But the oscillator system is made up of one and only one oscillator. We conserve our oscillator. We do not allow the system to exchange oscillators with the reservoir, but only to

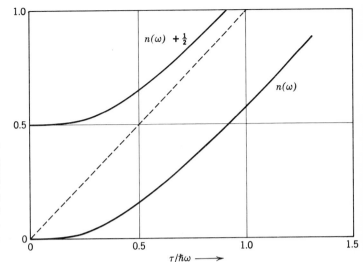

Figure 2 Planck distribution as function of the reduced temperature $T/\hbar\omega$. Here $\langle n(\omega) \rangle$ is the thermal average of the number of photons in the mode of frequency ω. A plot of $\langle n(\omega) \rangle + \frac{1}{2}$ is also given, where $\frac{1}{2}$ is the effective zero-point occupancy of the mode. The dashed line is the classical limit.

exchange energy. By this conservation device the chemical potential does not appear in the problem. We use the Boltzmann factor as obtained in Chapter 6 from the Gibbs factor by holding the number of particles constant. Therefore we work with the partition function instead of the grand sum.

For the system of one oscillator the partition function is

$$Z = \sum_{n=0}^{\infty} e^{-n\epsilon/T} , \tag{3}$$

where n is the number of photons having energy ϵ. The thermal average value of the occupancy is

$$\langle n \rangle = \frac{\Sigma n e^{-n\epsilon/T}}{\Sigma e^{-n\epsilon/T}} = \frac{\Sigma n x^n}{\Sigma x^n} , \tag{4}$$

where $x \equiv e^{-\epsilon/T}$. We may express (4) as

$$\langle n \rangle = x \frac{d}{dx} \log \Sigma x^n = -x \frac{d}{dx} \log (1 - x)$$

$$= \frac{x}{1-x} = \frac{e^{-\epsilon/T}}{1 - e^{-\epsilon/T}} = \frac{1}{e^{\epsilon/T} - 1} , \tag{5}$$

which is the promised result (2).

The thermal average number of photons in a mode of frequency ω is

$$\boxed{\langle n \rangle = \frac{1}{e^{\hbar\omega/T} - 1} ,} \tag{6}$$

where $\hbar\omega \equiv \epsilon$. This is called the **Planck distribution function.** It is plotted in Fig. 2 as a function of temperature. The Bose-Einstein distribution (1) re-

Table 1 Functions Related to the Planck Distribution

x	$\dfrac{x^2 e^x}{(e^x - 1)^2}$	$\dfrac{x}{e^x - 1}$	$\ln(1 - e^{-x})$	$\dfrac{\dfrac{x}{e^x - 1}}{-\ln(1 - e^{-x})}$
0.00	1.00000	1.00000	$-\infty$	∞
0.05	0.99979	0.97521	-3.02063	3.99584
0.10	0.99917	0.95083	-2.35217	3.30300
0.15	0.99813	0.92687	-1.97118	2.89806
0.20	0.99667	0.90333	-1.70777	2.61110
0.25	0.99481	0.88020	-1.50869	2.38888
0.30	0.99253	0.85749	-1.35023	2.20771
0.35	0.98985	0.83519	-1.21972	2.05491
0.40	0.98677	0.81330	-1.10963	1.92293
0.45	0.98329	0.79182	-1.01508	1.80690
0.50	0.97942	0.77075	-0.93275	1.70350
0.55	0.97517	0.75008	-0.86026	1.61035
0.60	0.97053	0.72982	-0.79587	1.52569
0.65	0.96552	0.70996	-0.73824	1.44820
0.70	0.96015	0.69050	-0.68634	1.37684
0.75	0.95441	0.67144	-0.63935	1.31079
0.80	0.94833	0.65277	-0.59662	1.24939
0.85	0.94191	0.63450	-0.55759	1.19209
0.90	0.93515	0.61661	-0.52184	1.13844
0.95	0.92807	0.59910	-0.48897	1.08809
1.00	0.92067	0.58198	-0.45868	1.04065
1.05	0.91298	0.56523	-0.43069	0.99592
1.10	0.90499	0.54886	-0.40477	0.95363
1.15	0.89671	0.53285	-0.38073	0.91358
1.20	0.88817	0.51722	-0.35838	0.87560
1.25	0.87937	0.50194	-0.33758	0.83952
1.30	0.87031	0.48702	-0.31818	0.80520
1.35	0.86102	0.47245	-0.30008	0.77253
1.40	0.85151	0.45824	-0.28315	0.74139
1.45	0.84178	0.44436	-0.26732	0.71168
1.50	0.83185	0.43083	-0.25248	0.68331
1.6	0.81143	0.40475	-0.22552	0.63027
1.7	0.79035	0.37998	-0.20173	0.58171
1.8	0.76869	0.35646	-0.18068	0.53714
1.9	0.74657	0.33416	-0.16201	0.49617
2.0	0.72406	0.31304	-0.14541	0.45845
2.1	0.70127	0.29304	-0.13063	0.42367
2.2	0.67827	0.27414	-0.11744	0.39158
2.3	0.65515	0.25629	-0.10565	0.36194
2.4	0.63200	0.23945	-0.09510	0.33455
2.5	0.60889	0.22356	-0.08565	0.30921
2.6	0.58589	0.20861	-0.07718	0.28578
2.7	0.56307	0.19453	-0.06957	0.26410
2.8	0.54049	0.18129	-0.06274	0.24403
2.9	0.51820	0.16886	-0.05659	0.22545
3.0	0.49627	0.15719	-0.05107	0.20826
3.2	0.45363	0.13598	-0.04162	0.17760
3.4	0.41289	0.11739	-0.03394	0.15133
3.6	0.37429	0.10113	-0.02770	0.12883
3.8	0.33799	0.08695	-0.02262	0.10958
4.0	0.30409	0.07463	-0.01849	0.09311
4.2	0.27264	0.06394	-0.01511	0.07905
4.4	0.24363	0.05469	-0.01235	0.06705
4.6	0.21704	0.04671	-0.01010	0.05681
4.8	0.19277	0.03983	-0.00826	0.04809
5.0	0.17074	0.03392	-0.00676	0.04068
5.2	0.15083	0.02885	-0.00553	0.03438
5.4	0.13290	0.02450	-0.00453	0.02903
5.6	0.11683	0.02078	-0.00370	0.02449
5.8	0.10247	0.01761	-0.00303	0.02065
6.0	0.08968	0.01491	-0.00248	0.01739

After *Handbook of mathematical functions*, National Bureau of Standards AMS No. 55. The right-hand column is related to (13) for the entropy.

duces to the Planck distribution on setting the chemical potential equal to zero. Table 1 gives values of functions related to the Planck distribution.

The asymptotic form of the Planck distribution at high temperatures $T \gg \hbar\omega$ is of special interest. We set $y = \hbar\omega/T$; then for $y \ll 1$ we have

$$\langle n \rangle = \frac{1}{1 + y + \frac{1}{2}y^2 + \cdots - 1} \cong \frac{1}{y} \cdot \frac{1}{1 + \frac{1}{2}y} \cong \frac{1}{y}(1 - \tfrac{1}{2}y) , \quad (6a)$$

or

$$\langle n \rangle \cong \frac{T}{\hbar\omega} - \frac{1}{2} . \quad (6b)$$

If we add the effective zero-point occupancy of $\frac{1}{2}$ to both sides, we have

$$\langle n \rangle + \tfrac{1}{2} \cong \frac{T}{\hbar\omega} . \quad (6c)$$

The value $T/\hbar\omega$ is often given as the result in the classical limit for the excitation of the oscillator. Equation (6c) shows that the inclusion of the zero-point occupancy will improve the asymptotic behavior of the occupancy function (6). We recall that the exact quantum solutions for the energy of a harmonic oscillator are $(n + \frac{1}{2})\hbar\omega$, where n is any integer and ω is the classical frequency of the oscillator.

EXAMPLE. *Entropy of a harmonic oscillator.* The thermal average energy of a harmonic oscillator was worked out in Chapter 6. In our present language the result is the product of the occupancy $\langle n(\epsilon) \rangle$ times the energy ϵ of unit occupancy: $U = \epsilon\langle n(\epsilon) \rangle$, where $\epsilon \equiv \hbar\omega$. The occupancy $\langle n(\epsilon) \rangle$ is described by the Planck distribution, so that

$$U = \frac{\epsilon}{e^{\epsilon/T} - 1} , \quad (7)$$

as in (6.72). In the limit $T \gg \epsilon$ we have the classical result $U \cong T$, of which $\frac{1}{2}T$ arises from the kinetic energy and $\frac{1}{2}T$ from the potential energy; this equal division of the energy of the harmonic oscillator is an example of the equipartition of energy. The equality does not hold for an anharmonic oscillator.

We solve (7) for $1/T$ to obtain

$$\frac{1}{T} = \frac{1}{\epsilon} \log\left(1 + \frac{\epsilon}{U}\right) = \frac{1}{\epsilon}\left[\log\left(1 + \frac{U}{\epsilon}\right) - \log\left(\frac{U}{\epsilon}\right)\right] . \quad (8)$$

We integrate $\partial\sigma/\partial U = 1/T$ to obtain the entropy:

$$\sigma = \int_0^T \frac{dU}{T} = \frac{1}{\epsilon}\int_0^U dU\left[\log\left(1 + \frac{U}{\epsilon}\right) - \log\left(\frac{U}{\epsilon}\right)\right] . \quad (9)$$

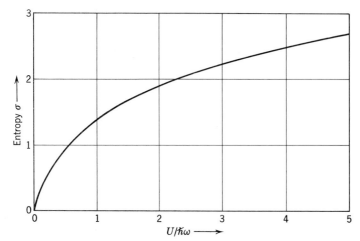

Figure 3 Entropy versus energy for harmonic oscillator of frequency ω.

We use the result that $U = 0$ at $\mathcal{T} = 0$, as follows from the expression (7) for U. We now carry out the integration to obtain

$$\sigma(U) = \left(1 + \frac{U}{\epsilon}\right) \log \left(1 + \frac{U}{\epsilon}\right) - \left(\frac{U}{\epsilon}\right) \log \left(\frac{U}{\epsilon}\right) . \tag{10}$$

This is the form referred to in the historical quotation from Planck at the beginning of this chapter, with $\epsilon = h\nu$.

The result for $\sigma(U)$ versus U is plotted in Fig. 3. The slope is always positive, unlike the two state system treated in Chapter 6. Thus there is no negative temperature regime for a harmonic oscillator. This is because there is no upper bound on the energy, whereas the two state system has an upper bound. Now

$$\frac{U}{\epsilon} = \langle n(\epsilon) \rangle = n ,$$

so that the entropy may be written as

$$\sigma = (1 + n) \log (1 + n) - n \log n , \tag{11}$$

where n denotes $\langle n(\epsilon) \rangle$. The entropy of an assembly of oscillators is the sum of the entropies of the individual oscillators:

$$\sigma = \sum_j [(1 + n_j) \log (1 + n_j) - n_j \log n_j] . \tag{12}$$

We substitute the Planck distribution function (6) for n into (11) to ob-

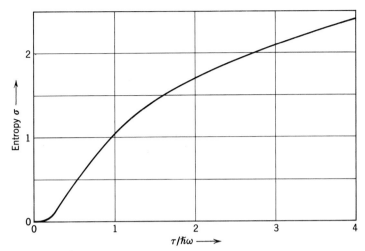

Figure 4 Entropy versus temperature for harmonic oscillator of frequency ω.

tain, after some rearrangement,

$$\sigma(T) = \frac{\hbar\omega/T}{e^{\hbar\omega/T} - 1} - \log\left(1 - e^{-\hbar\omega/T}\right) . \tag{13}$$

The function on the right-hand side is tabulated in Table 1. The variation of entropy with temperature is plotted in Fig. 4.

DENSITY OF PHOTON MODES

The Planck distribution function was originally developed for the study of the energy distribution of electromagnetic radiation which is in thermal equilibrium within a cavity with walls at temperature T. Such radiation is called **thermal radiation** or **black body radiation.**

To solve the problem of the energy distribution we need to find the density of photon modes. In Chapter 14 we introduced the density-of-orbitals function $\mathcal{D}(\epsilon)$, defined as the number of orbitals per unit energy range. We found a general expression for $\mathcal{D}(\epsilon)$ for any kind of particle in terms of the quantum number $n = (n_x^2 + n_y^2 + n_z^2)^{\frac{1}{2}}$ of the plane wave orbital in a cube:

$$\mathcal{D}(\epsilon) = \tfrac{1}{2}\gamma\pi n^2 \frac{dn}{d\epsilon} . \tag{14}$$

For photons there are two independent polarization directions[4] for a given mode, so that $\gamma = 2$.

For photons it is a little more convenient to define

$$\mathfrak{D}(\omega) \equiv \text{number of photon modes per unit frequency range}, \qquad (15)$$

where the frequency is the circular frequency ω. By direct analogy with (14) we have

$$\mathfrak{D}(\omega) = \pi n^2 \frac{dn}{d\omega} . \qquad (16)$$

Here $\mathfrak{D}(\omega)\, d\omega$ is the number of photon modes with frequencies between ω and $\omega + d\omega$. In (16) the symbol n is the magnitude of the quantum number $\mathbf{n}$; it is not to be confused with the occupancy. The factor $dn/d\omega$ in (16) depends on the type of particle (photon, electron, etc.) with which we are concerned. The factor n^2 is independent of the type of particle.

To evaluate (16) we need $dn/d\omega$ for photons. The frequency of photons is related to the quantum numbers n_x, n_y, n_z by the electromagnetic wave equation

$$\left(\frac{\partial^2}{\partial x^2} + \frac{\partial^2}{\partial y^2} + \frac{\partial^2}{\partial z^2} \right)\psi_\mathbf{n} = \frac{1}{c^2} \frac{\partial^2 \psi_\mathbf{n}}{\partial t^2} , \qquad (17)$$

where c is the speed of light and ψ may be any component of the electric or magnetic field intensity.

For example, if ψ is E_z, the z component of the electric field intensity, then

$$\psi_\mathbf{n} = E_z = Ce^{-i\omega t} \sin (n_x \pi x/L) \sin (n_y \pi y/L) \cos (n_z \pi z/L) , \qquad (18)$$

for radiation confined within a metal cavity in the form of a cube of edge L. Here C is a constant. The result (18) satisfies the boundary condition that the tangential component of the electric field intensity vanish at the walls of the cavity. Thus E_z vanishes at the planes $x = 0, L$ and at $y = 0, L$. (We do not need to enter into a complete discussion of the electromagnetic modes of a cavity, for the subject is treated in textbooks on electricity and magnetism.)

By substitution in (17) we see that (18) is a solution of the wave equation if

$$\frac{\omega^2}{c^2} = \frac{\pi^2}{L^2} (n_x^2 + n_y^2 + n_z^2) . \qquad (19)$$

[4] The spin of the photon is one, but because of relativistic considerations a particle of spin one which travels with the speed of light has only two independent polarizations and not three. See E. P. Wigner, "Relativistic invariance and quantum phenomena," Reviews of Modern Physics **29**, 255–268 (1957).

Now $n^2 = n_x{}^2 + n_y{}^2 + n_z{}^2$, so that (19) may be written as

$$n = \frac{L\omega}{\pi c} \ .$$ (20)

Thus

$$\frac{dn}{d\omega} = \frac{L}{\pi c} \ .$$ (21)

On substitution of (20) and (21) into (16) we find for the density of photon modes

$$\mathcal{D}(\omega) = \pi \left(\frac{L}{\pi c}\right)^3 \omega^2 \ .$$ (22)

With the volume $V = L^3$ we have

$$\mathcal{D}(\omega) = \frac{V}{\pi^2 c^3}\, \omega^2$$ (23)

as the number of photon modes per unit frequency range. This result applies in vacuum and will be modified if the medium is dispersive [see Kittel, *Elementary statistical physics*, Wiley, 1958, p. 180].

PLANCK RADIATION LAW

The thermal energy in a single mode of the electromagnetic field is $\langle n(\omega)\rangle \hbar\omega$. The thermal energy of all the modes that lie in a unit frequency range is given by the energy in a single mode times the number of modes in a unit frequency range, or $\mathcal{D}(\omega)$. We denote the thermal energy per unit frequency range by $u(\omega)$ and obtain

$$u(\omega) = \langle n\rangle \hbar\omega \cdot \mathcal{D}(\omega) = \frac{V\hbar}{\pi^2 c^3}\cdot\frac{\omega^3}{e^{\hbar\omega/T}-1} \ ,$$ (24)

from (6) for $\langle n\rangle$ and (23) for $\mathcal{D}(\omega)$. The result is the famous **Planck radiation law** for the distribution in frequency of thermal radiation. A plot of $u(\omega)$ versus ω is given in Fig. 5.

The total electromagnetic energy in the cavity is obtained by integrating $u(\omega)$ over all frequencies. We have

$$U(T) = \int_0^\infty d\omega\, u(\omega) = \frac{VT^4}{\pi^2\hbar^3 c^3}\int_0^\infty dx\,\frac{x^3}{e^x-1} \ ,$$ (25)

after the substitution $x \equiv \hbar\omega/T$. The definite integral[5] on the right is dimen-

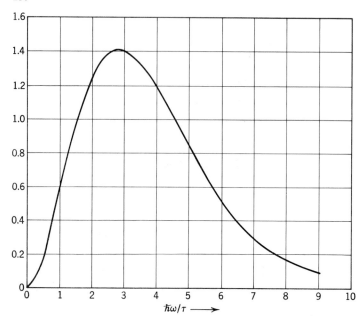

Figure 5 Plot of $x^3/(e^x - 1)$ with $x \equiv \hbar\omega/T$. This function is involved in the Planck radiation law. The temperature of a black body may be found from the frequency ω_{max} at which the radiant energy density is a maximum, per unit frequency range. This frequency is directly proportional to the temperature.

sionless and has the value $\frac{1}{15}\pi^4$. The total energy increases with temperature as T^4:

$$U(T) = \frac{\pi^2 V T^4}{15\hbar^3 c^3} = \left(\frac{\pi^2 V k_B^4}{15\hbar^3 c^3}\right) T^4 \ . \tag{26}$$

The result that the radiant energy density is proportional to T^4 is known as the **Stefan-Boltzmann law of radiation.** It was first derived by the thermodynamic argument in Problem 4 below.

We consider the flux of radiant energy from a small hole in the wall of the cavity. The rate of energy emission through a hole of unit area in a cavity at temperature T is a quantity of the order of the energy density U/V times the velocity of light, or

$$\frac{cU(T)}{V} \ . \tag{27}$$

The exact result is lower than (27) by a geometrical factor equal to $\frac{1}{4}$.

[5] We have

$$\int_0^\infty dx \, \frac{x^3}{e^x - 1} = \int_0^\infty dx \, x^3 e^{-x} \cdot \frac{1}{1 - e^{-x}} = \int_0^\infty dx \, x^3 \sum_{p=1}^\infty e^{-px} = \left(\sum_{p=1}^\infty \frac{1}{p^4}\right) \int_0^\infty dy \, y^3 e^{-y} \ . \tag{26a}$$

The integral is elementary and has the value 6. The sum converges rapidly to the value 1.0823; the exact value is $\pi^4/90$, as shown by Zemansky, p. 637.

We denote the radiant energy flux by J_U, so that

$$J_U = \frac{cU}{4V} = \left(\frac{\pi^2 k_B^4}{60\hbar^3 c^2}\right) T^4 . \tag{28}$$

This is called **Stefan's law** for the rate of radiant energy emission. The coefficient of T^4 is the **Stefan-Boltzmann constant** σ and has the value 5.67×10^{-5} erg cm^{-2} sec^{-1} deg^{-4}. A body that radiates at the rate (28) is said to radiate as a black body.

Problem 1. *Flux of radiant energy.* Derive the geometrical factor $\frac{1}{4}$ that enters the result (28).

Problem 2. *Surface temperature of the Sun.* The value of the total radiant energy flux at the Earth from the Sun is called the **solar constant** of the Earth. The observed value integrated over all emission wavelengths and referred to one astronomical unit (defined as the mean Earth-Sun distance) is

$$\text{solar constant} = 0.136 \text{ J sec}^{-1} \text{ cm}^{-2} , \tag{29}$$

where J denotes joules. (a) Show that the total rate of energy generation of the Sun is 4×10^{26} J sec^{-1}. (b) From this result and Stefan's law, $J_U = 5.67 \times 10^{-12} T^4$ J sec^{-1} cm^{-2}, show that the effective temperature of the surface of the Sun treated as a black body is $T \cong 6000$ K. Take the distance of the Earth from the Sun as 1.5×10^{13} cm and the radius of the Sun as 7×10^{10} cm.

Problem 3. *Pressure of radiation.* Show for a photon gas that

(a) $$p = -\left(\frac{\partial U}{\partial V}\right)_\sigma = -\sum_l n_l \hbar \frac{d\omega_l}{dV} , \tag{30}$$

where n_l is the number of photons in the mode l;

(b) $$\frac{d\omega_l}{dV} = -\frac{\omega_l}{3V} ; \tag{31}$$

(c) $$p = \frac{U}{3V} . \tag{32}$$

Thus the radiation pressure is equal to $\frac{1}{3} \times$ (energy density). Another derivation of this result is given in Chapter 18.

(d) Compare the pressure of radiation with the kinetic pressure of a gas of H atoms at a concentration of 1 mol cm^{-3} as for the Sun. At what temperature (roughly) are the two pressures equal? [30×10^6 K] The average temperature of the Sun is believed to be near 20×10^6 deg. The concentration is highly nonuniform, rising to near 100 moles cm^{-3} at the center, where the kinetic pressure will be very considerably higher than the radiation pressure.

Problem 4. *Stefan-Boltzmann law.* We can employ the result (32) to give a thermodynamic derivation of the temperature dependence of the radiant energy density as given in (26). We cannot use the argument below to find the factor in front of the T^4 in (26), but we can find the T^4. Use the relation

$$\left(\frac{\partial U}{\partial V}\right)_T = T\left(\frac{\partial p}{\partial T}\right)_V - p$$

from (12.39) to show that

$$U(T) = U_0 T^4 \ ,$$

where U_0 is a constant independent of temperature.

Problem 5. *Average temperature of the interior of the Sun.* (a) Estimate as in Problem 14.7a the order of magnitude of the gravitational self-energy of the Sun, with $M_\odot = 2 \times 10^{33}$ gm and $R_\odot = 7 \times 10^{10}$ cm. The gravitational constant G is 6.6×10^{-8} dyne cm^{-2} gm^{-2}. (b) Assume that the total thermal energy of the atoms in the Sun is equal to $-\frac{1}{2}$ times the gravitational energy. This is the result of the virial theorem, Appendix E. Estimate the average temperature of the Sun. Take the average mass of the particles as the average mass of an electron and a proton. This estimate gives too low a temperature, because the density of the Sun is far from uniform.

"The range in central temperature for different stars, excluding only those composed of degenerate matter for which the law of perfect gases does not hold (white dwarfs) and those which have excessively small average densities (giants and supergiants), is between 1.5 and 3.0 $\times$ 10^7 degrees." (O. Struve, B. Lynds, and H. Pillans, *Elementary astronomy*, Oxford University Press, 1959.)

ESTIMATION OF THE SURFACE TEMPERATURE OF A STAR

The total rate of radiation from a star may be estimated as the surface area times the energy flux J_U, as in Problem 2. The energy flux depends on the surface temperature of the star. The surface temperature may also be estimated from the frequency at which the maximum emission of radiant energy takes place from the star (see Fig. 5). What this frequency is depends on what we are using as the abscissa of the graph, frequency or wavelength. If we use frequency the maximum is given from the Planck law, Eq. (24), as

$$\frac{d}{dx}\left(\frac{x^3}{e^x - 1}\right) = 0 \ , \tag{33}$$

or

$$3 - 3e^{-x} = x \ .$$

This equation may be solved numerically. The root is

$$\frac{\hbar\omega_{max}}{k_B T} = x_{max} \simeq 2.82 \ . \tag{34}$$

The wavelength at the maximum of Fig. 5 is $\lambda_{max} = 2\pi c/\omega_{max}$. With λ_{max} expressed in centimeters, we have

$$\lambda_{max}(cm) \simeq \frac{2\pi\hbar c}{2.82 k_B T} \simeq \frac{0.51}{T(\deg)} \ . \tag{35}$$

The maximum of λ is given in Table 2 below for several situations of interest. We emphasize that the λ_{max} values given here apply when the horizontal axis is the frequency. Notice that on such a plot the maximum for the Sun's radiation is in the infrared.

Table 2

Object	Temperature, in deg	λ_{max}, in cm
Primeval radiation of universe[°]	3	0.16
Earth (surface)	300	1.6×10^{-3}
Sun (surface)	6000	0.8×10^{-4}

[°] See P. J. E. Peebles and D. T. Wilkinson, Scientific American, June 1967, pp. 28–37.

Problem 6. *Surface temperature of the Earth.* Calculate the temperature of the surface of the Earth on the assumption that as a black body in thermal equilibrium it reradiates as much thermal radiation as it receives from the Sun. Assume also that the surface of the Earth is at a constant temperature over the day-night cycle. The input values you need are $T_\odot = 5800$ K; $R_\odot = 7 \times 10^{10}$ cm; and the Earth-Sun distance of 1.5×10^{13} cm.

Problem 7. *Neutrino gas.* Neutrinos are fermions and have a spin of $\frac{1}{2}$. They are massless particles and travel with the velocity of light. Their wavefunction satisfies the wave equation (20) just as for photons. (a) Show that the energy ϵ is related to $\mathbf{n}$ by

$$\frac{\hbar^2 \pi^2 n^2}{L^2} = \frac{\epsilon^2}{c^2} . \tag{36}$$

(b) Show that the number of orbitals per unit energy range is

$$\mathfrak{D}(\epsilon) = \frac{V}{\pi^2 \hbar^3 c^3} \epsilon^2 . \tag{37}$$

It has been suggested[6] that the universe is filled everywhere with a degenerate gas of neutrinos that were formed as a byproduct of nuclear transformations. Neutrinos have only a very weak interaction with matter, and the presence of a degenerate neutrino gas is difficult to detect.

PHOTON FLUCTUATIONS

In Problem 11.7 we showed that enormous fluctuations in the number of particles in an orbital may occur for bosons. We now give a qualitative physical interpretation of the large fluctuations for photons.

Suppose that photons are produced with random phases from a large number N of uncorrelated monochromatic sources. We let n be the number of such photons in a small volume, taking $N \gg n$. We work only in the classical limit, $n \gg 1$. The value of n will be proportional to the square of the electric field intensity, because the number of photons is proportional to the field energy and the field energy is proportional to $E^*E = |E|^2$. The asterisk denotes complex conjugate. Thus

$$n \propto E^*E \tag{38}$$

where E is the electric field intensity.

---

[6] See S. Weinberg, "Universal neutrino degeneracy," Physical Review **128**, 1457 (1962).

Let the Greek letter ξ (called xi) denote the electric field intensity for a single source. Then for N sources the total field energy is proportional to

$$E^*E = \xi^2 \left(\sum_k e^{i\varphi_k}\right)^* \left(\sum_j e^{i\varphi_j}\right) = \xi^2 \left(\sum_k e^{-i\varphi_k}\right)\left(\sum_j e^{i\varphi_j}\right), \tag{39}$$

where φ_j is the phase of the jth source. Thus

$$E^*E = \xi^2 \left[N + \sum_{jk}{}' e^{i(\varphi_j - \varphi_k)}\right] = \xi^2 \left[N + 2\sum_{j>k} \cos(\varphi_j - \varphi_k)\right], \tag{40}$$

where the term in N comes from those parts of the double summation in (39) for which $k = j$.

The term $\cos(\varphi_j - \varphi_k)$ when averaged over random values of the phases gives zero. Thus

$$\langle E^*E \rangle = N\xi^2 . \tag{41}$$

Now we know that it is a general property of the mean square fluctuation that

$$\langle(\Delta n)^2\rangle \equiv \langle(n - \langle n\rangle)^2\rangle = \langle n^2\rangle - \langle n\rangle^2 , \tag{42}$$

so that the mean square fractional fluctuation is

$$\frac{\langle(\Delta n)^2\rangle}{\langle n\rangle^2} = \frac{\langle(E^*E)^2\rangle - \langle E^*E\rangle^2}{\langle E^*E\rangle^2} . \tag{43}$$

To evaluate (43) we need $\langle(E^*E)^2\rangle$. We square (40) to obtain

$$(E^*E)^2 = \xi^4 \left[N + \sum_{jk}{}' e^{i(\varphi_j - \varphi_k)}\right]^2$$
$$= \xi^4 \left[N^2 + 2N\sum_{jk}{}' e^{i(\varphi_j - \varphi_k)} + \sum_{lm}{}' e^{i(\varphi_l - \varphi_m)}\sum_{jk}{}' e^{i(\varphi_j - \varphi_k)}\right]. \tag{44}$$

When we average over phases the middle term vanishes; the right-hand term does not vanish only for the $N(N - 1)$ terms for which $l = k$ and $m = j$, when it is equal to unity. Thus in the limit $N \gg 1$ we have

$$\langle(E^*E)^2\rangle \cong 2N^2\xi^4 . \tag{45}$$

Furthermore, from (41), $\langle E^*E\rangle^2 = N^2\xi^4$, so that (43) becomes

$$\frac{\langle(\Delta n)^2\rangle}{\langle n\rangle^2} \cong \frac{2N^2\xi^4 - N^2\xi^4}{N^2\xi^4} = 1 . \tag{46}$$

This result is based on a semiclassical model of electromagnetic waves. It shows that the fractional fluctuations in the number of photons are not smoothed as the average number of photons increases. We may express the sense of (46) by saying that photons like to travel in packs.

Experiments bearing on large photon fluctuations are reported by Brown and Twiss.[7] They find positive correlations between photons in two coherent

[7] R. H. Brown and R. Q. Twiss, Nature **177**, 27 (1956).

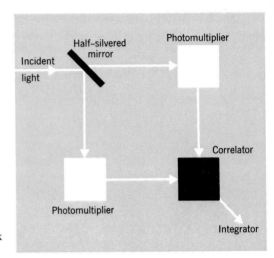

Figure 6 Experimental arrangement in the work of Brown and Twiss on photon correlation.

beams of light.[8] The arrangement is indicated in Fig. 6. They calculate the correlation from classical electromagnetic theory, with results in good agreement with experiment. The correlation naturally depends on the square of the number of quanta per unit time in the beam.

Purcell[9] has given a simple explanation of the extra fluctuations of photons in terms of a wave-packet model. We think of a stream of wave packets, each about $c/\Delta\nu$ long, in a random sequence. Each packet contains one photon. There is a certain probability that two such wave trains accidentally overlap. When the packets overlap they interfere, and the result is a packet with something in between 0 and 4 photons, thus the photon density fluctuations are large. A similar experiment carried out with electrons would show a suppression of the normal fluctuations, instead of an enhancement, because the Pauli principle excludes the accidentally overlapping wave trains.

REFERENCES

D. H. Menzel, P. L. Bhatnagar, and H. K. Sen, *Stellar interiors*, Wiley, 1963.

M. Planck, *Theory of heat radiation*, Dover, New York, 1959, paperback edition.

L. H. Aller, *Astrophysics, nuclear transformations, stellar interiors, and nebulae*, Ronald, New York, 1954

[8] That is, when a photon is registered in one channel, the probability that a photon is also registered in the second channel in the same time interval is *higher* than if the events were uncorrelated.

[9] E. M. Purcell, Nature **178**, 1449 (1956).

CHAPTER **16**

Phonons in Solids: Debye Theory

"So I decided to calculate the spectral distribution of the possible free vibrations for a continuous solid and to consider this distribution as a good enough approximation to the actual distribution. The sonic spectrum of a lattice must, of course, deviate from this as soon as the wavelength becomes comparable to the distances of the atoms. . . . The only thing which had to be done was to adjust to the fact that every solid of finite dimensions contains a finite number of atoms and therefore has a finite number of free vibrations. . . . At low enough temperatures, and in perfect analogy to the radiation law of Stefan-Boltzmann in radiation, the vibrational energy content of a solid will be proportional to T^4." (P. Debye)

The energy of an elastic wave in a solid is quantized just as the energy of an electromagnetic wave in a cavity is quantized. The quantum of energy in an elastic wave is called a **phonon**. The thermal average of the number of phonons in an elastic wave of frequency ω is given by the Planck distribution function of Chapter 15, exactly as for photons:

$$\langle n(\omega) \rangle = \frac{1}{e^{\hbar\omega/\tau} - 1} . \tag{1}$$

We assume that the frequency of an elastic wave is independent of the amplitude of the elastic strain. This result follows from the assumption of linear elastic theory that the strain is directly proportional to the stress, as in Hooke's law.

We want to find the energy and heat capacity of the elastic waves in solids.[1] Several of the results obtained for photons may be carried over to phonons. The results are simple if we assume that the velocity of all elastic waves are equal, that is, independent of frequency, direction of propagation, and direction of polarization. Such an assumption is not very accurate, but it does account well for the general trend of the observed results in many solids. At the expense of some extra computation we may always dispense with the assumption that all waves have the same velocity.

There are two important features of the experimental results: the heat capacity of a nonmetallic solid varies as T^3 at low temperatures, and at high temperatures the heat capacity is independent of the temperature and equal to $3k_B$ for each atom in the specimen.

NUMBER OF PHONON ORBITALS

There is no limit to the number of possible electromagnetic modes in a cavity, but the number of elastic modes in a finite solid is bounded. If the solid consists of N atoms, the total number of elastic modes is $3N$, because each atom has three degrees of freedom.

An elastic wave has three possible polarizations, two transverse and one longitudinal, in contrast to the two possible polarizations of an electromagnetic wave. In a transverse elastic wave the displacement of the atoms is perpendicular to the propagation direction of the wave; in a longitudinal wave the displacement is parallel to the propagation direction. The density of phonon modes is increased by the factor of $\frac{3}{2}$ for an elastic wave as compared with an electro-

---

[1] For further discussion of the thermal properties of elastic waves in solids, see Chapters 4, 5, and 6 of ISSP and references cited there.

magnetic wave. The number of phonon orbitals per unit frequency range is found from (15.23) for photons, with this modification:

$$\mathcal{D}(\omega) = \frac{3V}{2\pi^2 v_s^3}\,\omega^2 \ . \tag{2}$$

Here v_s is a suitable average elastic wave velocity, found from the average of $1/v^3$ over polarization, frequency, and direction. The volume of the solid is V.

This result for $\mathcal{D}(\omega)$ applies up to a maximum frequency which we denote by ω_D. We suppose that there are no phonon modes at frequencies above ω_D:

$$\mathcal{D}(\omega > \omega_D) = 0 \ . \tag{3}$$

Thus the phonon spectrum is assumed to be cut off at ω_D; this frequency is called the **Debye frequency.**[2]

The value of ω_D is determined by the requirement that the total number of modes be equal to the number of degrees of freedom $3N$:

$$3N = \int_0^{\omega_D} d\omega\, \mathcal{D}(\omega) = \frac{V}{2\pi^2 v_s^3}\,\omega_D^3 \ , \tag{4}$$

or

$$\omega_D = \left(\frac{6\pi^2 N v_s^3}{V}\right)^{\frac{1}{3}} \ . \tag{5}$$

The ratio V/N is the volume per atom. We can use (5) to write the density of orbitals in (2) as

$$\mathcal{D}(\omega < \omega_D) = 9N \cdot \frac{\omega^2}{\omega_D^3} \ . \tag{6}$$

DEBYE T^3 LAW

The total elastic energy of a solid at temperature T is the integral over ω of the energy in a mode at frequency ω times the number of phonon modes per unit frequency range:

$$U(T) = \int_0^{\omega_D} d\omega\, \mathcal{D}(\omega)\, \langle n \rangle\, \hbar\omega = \frac{9N\hbar}{\omega_D^3} \int_0^{\omega_D} d\omega\, \frac{\omega^3}{e^{\hbar\omega/T} - 1} \ , \tag{7}$$

where the occupancy $\langle n \rangle$ is given by the Planck distribution function (1). We substitute $x = \hbar\omega/T$, whence

$$U(T) = \frac{9NT^4}{(\hbar\omega_D)^3} \int_0^{x_D} dx\, \frac{x^3}{e^x - 1} \ . \tag{8}$$

[2] A related cutoff wavelength is defined by $\lambda_D \equiv v_s/\omega_D$; with (5) for ω_D we have $\lambda_D = (V/6\pi^2 N)^{\frac{1}{3}}$, which is a length of the order of a lattice constant.

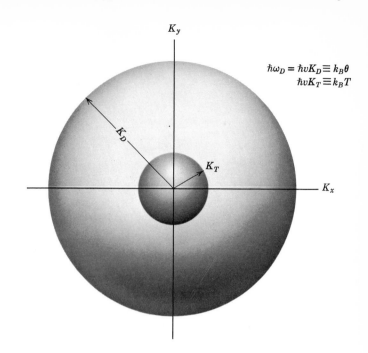

Figure 1 We give a qualitative explanation of the Debye T^3 law in terms of traveling wave modes of the form $e^{i(\mathbf{K} \cdot \mathbf{r} - \omega t)}$. We suppose that all phonon modes of wavevector less than K_T have the classical thermal energy $k_B T$ and that modes between K_T and the Debye cutoff K_D are not excited at all. Of the $3N$ possible modes, the fraction that are excited is $(K_T/K_D)^3 = (T/\theta)^3$, for this is the ratio of the volume of the inner sphere to the outer sphere. Thus the energy is

$$U \approx k_B T \cdot 3N \left(\frac{T}{\theta}\right)^3$$

and the heat capacity is

$$C_V \approx 12Nk_B \left(\frac{T}{\theta}\right)^3 .$$

$\hbar\omega_D = \hbar v K_D \equiv k_B \theta$
$\hbar v K_T \equiv k_B T$

Here

$$x_D \equiv \frac{\hbar\omega_D}{T} \equiv \frac{\theta}{T} , \tag{9}$$

where the **Debye temperature** θ is defined by $\theta \equiv \hbar\omega_D/k_B$.

The result (7) for the energy is of special interest at low temperatures such that $T \ll \theta$. Here the limit x_D on the integral is much larger than unity, and x_D may be replaced by infinity. We note from Fig. 15.5 that there is little contribution to the integrand out beyond $x = 10$. For the definite integral we have

$$\int_0^\infty dx \frac{x^3}{e^x - 1} = \frac{\pi^4}{15} , \tag{10}$$

as in Chapter 15. Thus the energy in the low temperature limit is

$$U(T) \cong \frac{3\pi^4 Nk_B T^4}{5\theta^3} , \tag{11}$$

proportional to T^4.

The heat capacity is, for $T \ll \theta$,

$$C_V = \left(\frac{\partial U}{\partial T}\right)_V = \frac{12\pi^4 Nk_B}{5} \left(\frac{T}{\theta}\right)^3 . \tag{12}$$

This result is known as the **Debye T^3 law**.[3] A qualitative physical argument which leads to the T^3 law is given in Fig. 1. The high temperature limit is the subject of Problem 1. Experimental results for argon are plotted in Fig. 2.

---

[3] P. Debye, Annalen der Physik **39**, 789 (1912); M. Born and T. v. Kármán, Physikalische Zeitschrift **13**, 297 (1912); **14**, 65 (1913).

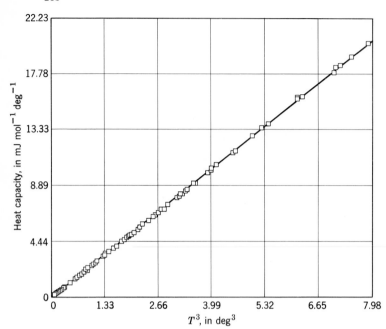

Figure 2 Low temperature heat capacity of solid argon, plotted against T^3 to show the excellent agreement with the Debye T^3 law. The value of θ from these data is 92 deg. (Courtesy of L. Feingold and N. E. Phillips.)

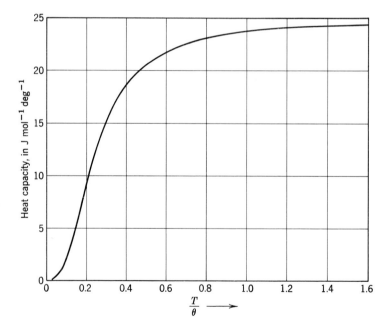

Figure 3 Heat capacity C_V of a solid, according to the Debye approximation. The vertical scale is in J mol^{-1} deg^{-1}. The horizontal scale is the temperature normalized to the Debye temperature θ. The region of the T^3 law is below 0.1θ. The asymptotic value at high values of T/θ is 24.943 J mol^{-1} deg^{-1}.

Table 1   Debye Temperature θ_0 in deg K

(The subscript zero on the θ denotes the low temperature limit of the experimental values.)

Li 344	Be 1440											B	C 2230	N	O	F	Ne 75
Na 158	Mg 400											Al 428	Si 645	P	S	Cl	Ar 92
K 91	Ca 230	Sc 360	Ti 420	V 380	Cr 630	Mn 410	Fe 470	Co 445	Ni 450	Cu 343	Zn 327	Ga 320	Ge 374	As 282	Se 90	Br	Kr 72
Rb 56	Sr 147	Y 280	Zr 291	Nb 275	Mo 450	Tc	Ru 600	Rh 480	Pd 274	Ag 225	Cd 209	In 108	Sn w 200	Sb 211	Te 153	I	Xe 64
Cs 38	Ba 110	Laβ 142	Hf 252	Ta 240	W 400	Re 430	Os 500	Ir 420	Pt 240	Au 165	Hg 71.9	Tl 78.5	Pb 105	Bi 119	Po	At	Rn
Fr	Ra	Ac															

Ce	Pr	Nd	Pm	Sm	Eu	Gd 200	Tb	Dy 210	Ho	Er	Tm	Yb 120	Lu 210
Th 163	Pa	U 207	Np	Pu	Am	Cm	Bk	Cf	Es	Fm	Md	No	Lw

Most of the data were supplied by N. Pearlman; references are given in the *A.I.P. Handbook*, 3rd ed.

Representative experimental values of the Debye temperature are given in Table 1. The calculated variation of C_V versus T/θ is plotted in Fig. 3. Several related thermodynamic functions for a Debye solid are given in Table 2 and are plotted in Fig. 4.

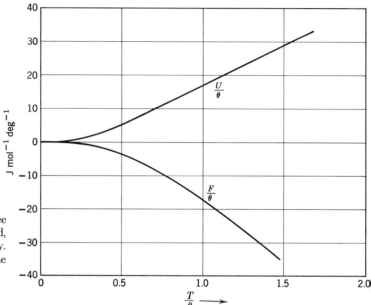

Figure 4 Energy U and free energy $F \equiv U - TS$ of a solid, according to the Debye theory. The Debye temperature of the solid is θ.

Table 2 *Values of C_V, S, U, and F on the Debye Theory*

[The values are based on Eqs. (7) and (9). For more complete tables, see pp. 745–747 of the Landolt-Bornstein tables, 6th ed., Vol. 2, Part 4.]

$\dfrac{\theta}{T}$	C_V, in J mol^{-1} deg^{-1}	U/θ, in J mol^{-1} deg^{-1}	F/θ, in J mol^{-1} deg^{-1}	S, in J mol^{-1} deg^{-1}
0	24.943	∞		∞
0.1	24.93	240.2	-666.8	90.70
0.2	24.89	115.6	-251	73.43
0.3	24.83	74.2	-137	63.34
0.4	24.75	53.5	-87	56.21
0.5	24.63	41.16	-60.3	50.70
0.6	24.50	32.9	-44.1	46.22
0.7	24.34	27.1	-33.5	42.46
0.8	24.16	22.8	-26.2	39.22
0.9	23.96	19.5	-20.9	36.38
1.0	23.74	16.82	-17.05	33.87
1.5	22.35	9.1	-7.23	24.49
2	20.59	5.5	-3.64	18.30
3	16.53	2.36	-1.21	10.71
4	12.55	1.13	-0.49	6.51
5	9.20	0.58	-0.23	4.08
6	6.23	0.323	-0.118	2.64
7	4.76	0.187	-0.066	1.77
8	3.45	0.114	-0.039	1.22
9	2.53	0.073	-0.025	0.874
10	1.891	0.048	-0.016	0.643
15	0.576	0.0096	-0.0032	0.192

Problem 1. *Heat capacity of solids in high temperature limit.* Show that $C_V \cong 3Nk_B$ in the limit $T \gg \theta$. That is, the heat capacity is $3k_B$ for each atom.

Problem 2. *Heat capacity of photons and phonons.* Consider a dielectric solid with a Debye temperature equal to 100 K and with 10^{22} atoms cm^{-3}. Estimate the temperature at which the photon contribution to the heat capacity would be equal to the phonon contribution evaluated at 1 K.

Problem 3. *Energy fluctuations in a solid at low temperatures.* Consider a solid of N atoms in the temperature region in which the Debye T^3 law is valid. The solid is in thermal contact with a heat reservoir. Use the results on energy fluctuations from Chapter 6 to show that the root mean square fractional energy fluctuation is of the order of

$$\mathfrak{F} = \left[\frac{1}{N} \left(\frac{\theta}{T} \right)^3 \right]^{\frac{1}{2}} .$$

Suppose that $T = 10^{-2}$ K; $\theta = 200$ K; and $N \approx 10^{15}$ for a particle 0.01 cm on a side. Then

$$\mathfrak{F} \approx 0.01 ,$$

which is not inappreciable. At 10^{-5} K the fractional fluctuation in energy is of the order of unity for a dielectric particle of volume 1 cm^3.

Problem 4. *Heat capacity of liquid He^4 at low temperatures.* The velocity of longitudinal sound waves in liquid He^4 at temperatures below 0.6 K is 2.383×10^4 cm sec^{-1}. There are no transverse sound waves in the liquid. The density is 0.145 gm cm^{-3}. (a) Calculate the Debye temperature. (b) Calculate the heat capacity per gram on the Debye theory and compare with the experimental value $C_V = 0.0204 \times T^3$ J gm^{-1} deg^{-1}. The T^3 dependence of the experimental value suggests that phonons are the most important excitations in liquid He^4 below 0.6 K. Note that the experimental value has been expressed per gram of liquid. The experiments are due to J. Wiebes, C. G. Niels-Hakkenberg, and H. C. Kramers, Physica **23**, 625 (1957). In liquid He^3, however, the heat capacity below 0.1 K is dominated by a term directly proportional to T, just as for a metal, as appropriate for single particle fermion excitations.

Boson Physics: Einstein Condensation and Liquid He^4

"In his papers on gas degeneracy (1924–1925) Einstein mentioned a peculiar condensation phenomenon of the ideal Bose-Einstein gas. This interesting discovery, however, has been almost entirely forgotten in the meantime . . . ." [F. London, Journal of Physical Chemistry 43, 49 (1939).]

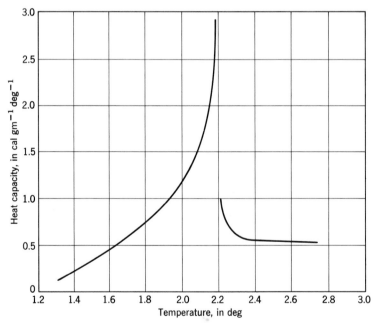

Figure 1 Heat capacity of liquid helium (He⁴). The sharp peak near 2.17 K is evidence of an important transition in the nature of the liquid. The viscosity of the liquid above the transition temperature is typical of normal liquids, whereas the viscosity below the transition as determined by rate of flow through narrow slits is vanishingly small, at least 10⁶ times smaller than the viscosity above the transition. The transition is often called a lambda transition merely because of the shape of the graph. (After Keesom et al.)

BOSON PHYSICS: EINSTEIN CONDENSATION AND LIQUID He^4

A very remarkable effect occurs in a gas of noninteracting bosons at a certain transition temperature, for below this temperature a substantial fraction of the total number of particles in the system will occupy the single orbital of lowest energy, which is called the ground orbital. Any other single orbital, including the orbital of second lowest energy, will at the same temperature be occupied by a relatively negligible number of particles. The total occupancy of all orbitals will always be equal to the specified number of particles in the system. The effect is called the **Einstein condensation.**

There would be nothing surprising to us in this result for the ground state occupancy if it were valid only in the immediate vicinity of absolute zero, say below 10^{-20} K. Such a temperature is mentioned because it is low in comparison with the energy spacing between the lowest and second lowest orbitals of a system of volume 1 cm^3, as you will see if you work out Problem 1. But the condensation temperature for a gas of fictitious noninteracting helium atoms calculated for the observed density of liquid helium is much higher, about 3 K, as we shall show.

The calculated temperature of 3 K is suggestively close to the actual temperature of 2.17 K at which a transition to a new state of matter is observed to take place in liquid helium (Fig. 1). We believe that in liquid He^4 below 2.17 K there is a condensation[1] of a substantial fraction of the atoms of He^4 into the ground orbital of the system. Evidently the interatomic forces that lead to the liquefaction of He^4 at 4.2 K under a pressure of one atmosphere are too weak to destroy the boson condensation or Einstein condensation at 2.17 K. In this respect the liquid behaves as a gas. The condensation into the ground orbital is certainly connected with the properties of bosons. The condensation is not permitted for fermions, and no corresponding transition in the properties of liquid He^3 has ever been observed, at least down to temperatures of about the order of 0.001 K. Atoms of He^3 have spin $\frac{1}{2}$ and are fermions.

We can give several arguments in support of our view of liquid helium as a gas of noninteracting particles. At first sight this is a drastic oversimplification of the problem, but there certainly are some important features of liquid helium for which the view is correct.

(a) The molar volume of liquid He^4 at absolute zero is 3.1 times the volume that we calculate from the known interactions of helium atoms. The

---

[1] This is different from the condensation in coordinate space that occurs in the condensation of a gas to a liquid.

interaction forces between pairs of helium atoms are well-known experimentally and theoretically, and from these forces by standard elementary methods of solid state physics we can calculate the equilibrium volume of a **static** lattice of helium atoms. In a typical calculation we find the molar volume to be $9 \text{ cm}^3 \text{ mol}^{-1}$, as compared with the observed $27.5 \text{ cm}^3 \text{ mol}^{-1}$. Thus the kinetic motion of the helium atoms has a large effect on the liquid state and leads to an expanded structure in which the atoms to a certain extent can move freely over appreciable distances. We can say that the quantum zero-point motion is responsible for the expansion of the molar volume.

(b) The transport properties of liquid helium in the normal state are not very different from those of a normal classical gas. In particular, the ratio of the thermal conductivity K to the product of the viscosity η times the heat capacity $\hat{C}_V$ per unit volume has the values

$$\frac{K}{\eta \hat{C}_V} = \begin{cases} 2.6, & \text{at} \quad 2.8 \text{ K} \\ 3.2, & \text{at} \quad 4.0 \text{ K} \end{cases}$$

These values are quite close to the estimate of Chapter 13 for a classical gas, where the value of the ratio was found to be unity. Improved calculations for the classical gas give values of the ratio even closer to the above values observed for liquid helium. The values of the transport coefficients themselves in the liquid are within an order of magnitude of those calculated for the gas at the same density.

(c) The forces in the liquid are relatively weak, and the liquid does not exist above the critical temperature of 5.2 K, which is the maximum boiling point observed. The binding energy would be perhaps ten times stronger in the equilibrium configuration of a static lattice, but the expansion of the molar volume by the zero-point motion of the atoms is responsible for the reduction in the binding energy to the observed value. The value of the critical temperature is directly proportional to the binding energy.

(d) The liquid is stable at absolute zero at pressures under 25 atm; above 25 atm the solid is more stable.

The new state of matter into which liquid He⁴ enters when cooled below 2.17 K has quite astonishing properties. The viscosity as measured in a flow experiment[2] is essentially zero (Fig. 9.1), and the thermal conductivity is very high. We say that liquid He⁴ below the transition temperature is a superfluid. More precisely, we denote liquid He⁴ below the transition temperature as liquid helium II, and we say that liquid helium II is a mixture of normal fluid and superfluid components. The normal fluid component consists of the

[2] In other arrangements there may be an effective viscosity: this is true of a disk oscillating in liquid He II at any finite temperature below the condensation temperature. For a combination of two fluids of different viscosities, some experiments measure the average viscosity and other experiments measure the average of $1/\eta$, or the average fluidity.

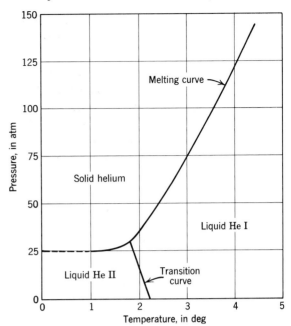

Figure 2   The melting curve of liquid and solid helium (He⁴), and the transition curve between the two forms of liquid helium, He I and He II. The liquid He II form exhibits superflow properties as a consequence of the condensation of atoms into the ground orbital of the system. Note that helium is a liquid at absolute zero at pressures below 25 atm. [After C. A. Swenson, *Physical Review* **79**, 626 (1950).]

helium atoms in (thermally) excited orbitals, and the superfluid component consists of the helium atoms condensed into the ground orbital.

We speak of liquid He⁴ above the transition temperature as liquid He I. There is no superfluid component in liquid He I, for here the ground orbital occupancy is negligible, being of the same order of magnitude as the occupancy of any other low-lying orbital, as we shall see. The regions of pressure and temperature in which liquid helium I and II exist are shown in Fig. 2.

The development of superfluid properties is not an automatic consequence of the Einstein condensation of atoms into the ground orbital. Advanced calculations show that it is the existence of some form (almost any form) of interaction among atoms that leads to the development of superfluid properties in the atoms condensed in the ground orbital.

CHEMICAL POTENTIAL NEAR ABSOLUTE ZERO

The key to the Einstein condensation is the behavior of the chemical potential of a boson system at low temperatures. The chemical potential is responsible for the apparent stabilization of a large population of particles in the ground orbital.

We consider a system composed of a large number N of noninteracting bosons. When the system is at absolute zero all particles occupy the lowest-energy orbital and the system is in the state of minimum energy. It is certainly not surprising that at $T = 0$ all particles should be in the orbital of lowest energy. We proceed to show that a substantial fraction remain in the ground orbital at finite temperatures.

If we choose the energy of the ground orbital to be zero, then from the Bose-Einstein distribution function

$$n(\epsilon, T) = \frac{1}{e^{(\epsilon - \mu)/T} - 1} \tag{1}$$

we obtain the occupancy of the ground orbital at $\epsilon = 0$ as

$$n(0, T) = \frac{1}{e^{-\mu/T} - 1} . \tag{2}$$

When $T = 0$ the occupancy of the ground orbital is equal to the total number of particles in the system, so that

$$n(0, 0) = N = \lim_{T \to 0} \frac{1}{e^{-\mu/T} - 1} = \lim_{T \to 0} \frac{1}{1 - \frac{\mu}{T} - 1} . \tag{3}$$

Here we have made use of the series expansion $e^{-x} = 1 - x + \cdots$. We know that x, which is μ/T, must be small in comparison with unity, for otherwise the total number of particles N could not be large.

From (3) we find

$$N = -\frac{T}{\mu} ; \qquad \mu = -\frac{T}{N} , \tag{4}$$

as $T \to 0$. To confirm this result for μ we form the limit:

$$\lim_{T \to 0} n(0, T) = \frac{1}{e^{1/N} - 1} \cong \frac{1}{\left(1 + \frac{1}{N} \cdots\right) - 1} \cong N ,$$

which is valid to high accuracy when $N \gg 1$. We note further from (4) that

$$\lambda \equiv e^{\mu/T} \cong 1 - \frac{1}{N} , \tag{5}$$

as $T \to 0$. Note that the chemical potential in a boson system must always be lower in energy than the ground state orbital, in order that the occupancy of every orbital be non-negative.

Problem 1. *First excited orbital.* (a) For an atom of He⁴ in a cube of volume 1 cm³, calculate in ergs the difference in energy $\Delta\epsilon$ between the first excited orbital and the ground orbital. (b) For $N = 10^{22}$ atoms, how many atoms are in the first excited orbital at $T = 1$ K? You may take λ as given by (6), but observe that $1/N$ may now be neglected in comparison with unity and in comparison with $\Delta\epsilon/T$. (c) Compare the result of (b) with the naïve and incorrect answer which comes if you use the Boltzmann factor to give the ratio of the population of the first excited orbital to the population of the ground orbital.

GROUND ORBITAL OCCUPANCY VERSUS TEMPERATURE

We saw in Chapter 14 that the number of free particle orbitals per unit energy range is

$$\mathcal{D}(\epsilon) = \frac{V}{4\pi^2}\left(\frac{2M}{\hbar^2}\right)^{\frac{3}{2}}\epsilon^{\frac{1}{2}} , \qquad (6)$$

for a particle of spin zero.

The total number of atoms of He⁴ in the ground and excited orbitals is given by the sum of the occupancies of all orbitals:

$$N = \sum_j n_j = N_0(T) + N_e(T) = N_0(T) + \int_0^\infty d\epsilon\, \mathcal{D}(\epsilon)\, n(\epsilon, T) . \qquad (7)$$

We have separated the sum over j into two parts. Here $N_0(T)$ has been written for $n(0, T)$, the number of atoms in the ground orbital at temperature T. The integral[3] in (7) gives the number of atoms $N_e(T)$ in all excited orbitals, with $n(\epsilon, T)$ as the BE distribution function. The integral gives only the number of atoms in excited orbitals and excludes the atoms in the ground orbital, because the function $\mathcal{D}(\epsilon)$ is zero at $\epsilon = 0$. To count the atoms correctly we must count separately the occupancy N_0 of the orbital with $\epsilon = 0$. Although only a single orbital is involved, the value of N_0 may be very

---

[3] We can use the integral form for the population of the excited orbitals because no one excited orbital has a "macroscopic" occupation.

large in a gas of bosons. The feature is without importance except for boson gases, where it is absolutely crucial. We shall call N_0 the number of atoms in the **superfluid component** and N_e the number of atoms in the **normal fluid component** of liquid helium II. The whole secret of the results which follow is that at low temperatures the chemical potential μ is very much closer in energy to the ground state orbital than the first excited orbital is to the ground state orbital. This closeness of μ to the ground orbital dumps most of the population of the system into the ground orbital.

The BE distribution function when written for the orbital at $\epsilon = 0$ is

$$N_0(T) = \frac{1}{\lambda^{-1} - 1} \, , \tag{8}$$

as in (2), where λ will depend on the temperature T.

The number of particles in all excited orbitals increases as $T^{\frac{3}{2}}$:

$$N_e(T) = \frac{V}{4\pi^2} \left(\frac{2M}{\hbar^2}\right)^{\frac{3}{2}} \int_0^\infty d\epsilon \, \frac{\epsilon^{\frac{1}{2}}}{\lambda^{-1} e^{\epsilon/T} - 1} \, ,$$

or, with $x \equiv \epsilon/T$,

$$N_e(T) = \frac{V}{4\pi^2} \left(\frac{2M}{\hbar^2}\right)^{\frac{3}{2}} T^{\frac{3}{2}} \int_0^\infty dx \, \frac{x^{\frac{1}{2}}}{\lambda^{-1} e^x - 1} \, . \tag{9}$$

Notice here the factor $T^{\frac{3}{2}}$ which gives the temperature dependence of N_e.

At sufficiently low temperatures the number of particles in the ground state will be a very large number. Equation (8) tells us that λ must be very close to unity whenever N_0 is $\gg 1$. Thus λ is very accurately constant throughout the temperature region of the liquid helium II phase, because a macroscopic value of N_0 forces λ to be close to unity.

The value of the integral[4] in (9) is, when $\lambda = 1$,

$$\int_0^\infty dx \, \frac{x^{\frac{1}{2}}}{e^x - 1} = 1.306 \, \pi^{\frac{1}{2}} \, . \tag{11}$$

[4] To evaluate the integral we write

$$\int_0^\infty dx \, \frac{x^{\frac{1}{2}}}{e^x - 1} = \int_0^\infty dx \, \frac{x^{\frac{1}{2}} e^{-x}}{1 - e^{-x}} = \sum_{s=1}^\infty \int_0^\infty dx \, x^{\frac{1}{2}} e^{-sx} = \left(\sum_{s=1}^\infty \frac{1}{s^{\frac{3}{2}}}\right) \int_0^\infty dy \, y^{\frac{1}{2}} e^{-y} \, . \tag{10}$$

The infinite sum is easily evaluated numerically to be 2.612. The integral may be transformed with $y = u^2$ to give

$$2 \int_0^\infty du \, u^2 \, e^{-u^2} = \tfrac{1}{2}\sqrt{\pi} \, ,$$

by (2.51).

Thus the number of atoms in excited states is

$$N_e = \frac{1.306\ V}{4}\left(\frac{2MT}{\pi\hbar^2}\right)^{\frac{3}{2}} = \frac{2.612\ V}{V_Q}\ , \tag{12}$$

where we have used the notation

$$V_Q \equiv \left(\frac{2\pi\hbar^2}{MT}\right)^{\frac{3}{2}} \tag{13}$$

for the quantum volume, as in Chapter 11.

We divide N_e by N to obtain the fraction of atoms in excited orbitals:

$$\frac{N_e}{N} \cong 2.612\,\frac{V}{NV_Q} = \frac{2.612}{cV_Q}\ , \tag{14}$$

where $c \equiv N/V$ is the concentration. The value $\lambda \cong 1$ or $1 - 1/N$ which led to (14) is valid as long as a large number of atoms are in the ground state.

EINSTEIN CONDENSATION TEMPERATURE

We define the **Einstein condensation temperature**[5] T_0 as the temperature for which the number of atoms in excited states is equal to the total number of atoms. That is, $N_e(T_0) = N$. Above T_0 the occupancy of the ground orbital is not a macroscopic number; below T_0 the occupancy is macroscopic. From (12) with N for N_e we find for the condensation temperature

$$\boxed{T_0 \equiv \frac{2\pi\hbar^2}{M}\left(\frac{N}{2.612\ V}\right)^{\frac{2}{3}}.} \tag{15}$$

Now (14) may be written as

$$\frac{N_e}{N} \cong \left(\frac{T}{T_0}\right)^{\frac{3}{2}}\ , \tag{16}$$

where N is the total number of atoms. We see that the number of atoms in excited orbitals varies as $T^{\frac{3}{2}}$ at temperatures below T_0, as shown in Fig. 3.

The Einstein condensation temperature T_0 for bosons is comparable with the Fermi temperature

$$T_F = \left(\frac{\hbar^2}{2M}\right)\left(\frac{3\pi^2 N}{V}\right)^{\frac{2}{3}}$$

[5] A. Einstein, Akademie der Wissenschaften, Berlin, Sitzungsberichte **1924**, 261; **1925**, 3.

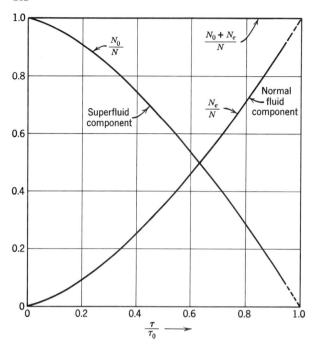

Figure 3 Condensed boson gas: temperature dependence of the proportion N_0/N of atoms in the ground orbital and of the proportion N_e/N of atoms in all excited orbitals.

for fermions. If the particle masses and concentrations are equal, we have

$$\frac{T_F}{T_0} = \frac{T_F}{T_0} = \frac{[3\pi^2(2.612)]^{\frac{2}{3}}}{4\pi} \cong 1.45 \ . \tag{17}$$

The value of T_F for electrons in metals is $\approx 5 \times 10^5$ K while the calculated value of T_0 for atoms of He^4 is ≈ 3 K. The large difference in values here is caused by the large difference in masses. The physical behavior of a system below T_0 or T_F is quite different for bosons and for fermions, because of the differences in occupancies.

The number of particles in the ground orbital is found from (16):

$$N_0 = N - N_e = N\left[1 - \left(\frac{T}{T_0}\right)^{\frac{3}{2}}\right] \ . \tag{18}$$

We note that N may be of the order of 10^{22}. For T even slightly less than T_0 a large number of particles will be in the ground orbital, as we see in Fig. 3. We have said that the particles in the ground orbital below T_0 form the condensed phase or the superfluid phase.

The condensation temperature in deg K is given by the numerical relation

$$T_0 = \frac{115}{V_M^{\frac{2}{3}} M} \text{ deg K} \ , \tag{19}$$

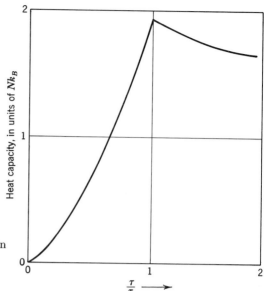

Figure 4   Heat capacity of an ideal Bose-Einstein gas at constant volume.

where V_M is the molar volume in cm³ mol⁻¹ and M is the molecular weight. For liquid helium $V_M = 27.6$ cm³ mol⁻¹ and $M = 4$; thus $T_0 = 3.1$ K. In liquid He⁴ the experimental value of the transition temperature between the low temperature (He II) phase which shows superfluid properties and the high temperature phase (He I) which behaves as a normal liquid is 2.17 K.

Problem 2. *Energy and heat capacity below T_0.* Find expressions as a function of temperature in the region $T < T_0$ for the energy and heat capacity of a gas of N noninteracting bosons of spin zero confined to a volume V. Put the definite integral in dimensionless form; it need not be evaluated. The calculated heat capacity above and below T_c is shown in Fig. 4. The experimental curve was shown in Fig. 1. The difference between the two curves is marked: it is ascribed to the effect of interactions between the atoms.

Problem 3. *Boson gas in one dimension.* Calculate the integral for $N_e(T)$ for a one-dimensional gas of noninteracting bosons, and show that the integral does not converge. Take $\lambda = 1$ for the calculation. (The problem should really be treated by means of a sum over orbitals on a finite line.) *Note.* The problem of the ground state occupancy in a boson gas confined to a finite thin volume has been treated by D. L. Mills, Physical Review **134**, A306 (1964); Fig. 5 is from his work.

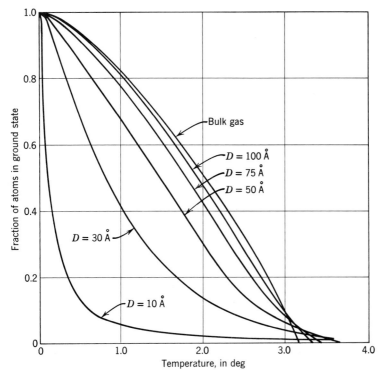

Figure 5 Occupation of the ground orbital by noninteracting He⁴ atoms as calculated for a square slab of side 1000 Å and various thicknesses D. [After D. L. Mills, Physical Review **134**, A306 (1964).]

PHASE RELATIONS OF HELIUM

The phase diagram[6] of He⁴ was shown in Fig. 2. The liquid-vapor curve can be followed from the critical point of 5.2 K down to absolute zero without any appearance of the solid. At the transition temperature the normal liquid, called He I, makes a transition to the form with superfluid properties, called He II. A temperature called the λ point is the triple point at which liquid He I, liquid He II, and vapor coexist. Keesom, who first solidified helium, found that the solid[7] did not exist below a pressure of 25 atm. Another triple point exists at 1.743 K: here the solid is in equilibrium with the two liquid modifications, He I and He II. The two triple points are connected by a line that separates the regions of existence of He II and He I.

---

[6] Phase diagrams are discussed in Chapter 20.

[7] An interesting discussion of solid helium is given by B. Bertram and R. A. Guyer, Scientific American, August 1967, pp. 85–95. Solid He⁴ exists in three crystal structures according to the conditions of temperature and pressure.

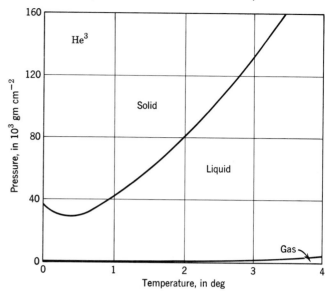

Figure 6   Phase diagram for liquid He³. Unlike that for liquid He⁴, this diagram contains only one liquid form. No superfluid properties are exhibited. The region of negative slope on the phase boundary implies that here the solid is more disordered than the liquid; this is discussed in Chapter 20. In this region we have to heat the liquid to freeze it! There are several crystalline forms in the solid phases of both He³ and He⁴; we have not shown their phase boundaries because they are not involved in the properties of the liquids. (After B. Bertram and R. A. Guyer.)

The phase diagram of He³ is shown in Fig. 6. There is only a single liquid phase, unlike He⁴, and the liquid phase does not show superfluid properties. The phase diagram does show in a remarkable way the importance of the fermion nature of He³, for at low temperatures the negative slope of the curve shows that the entropy of the Fermi liquid is lower than the entropy of the solid, as explained in Chapter 20.

QUASIPARTICLES AND SUPERFLUIDITY

For many purposes the superfluid component of liquid helium II behaves as if it were a vacuum, as if it were not there at all. The N_0 atoms of the superfluid are condensed into the ground orbital and have no excitation energy, for the ground orbital by definition has no excitation energy. The superfluid has energy only when the center of mass of the superfluid is given a velocity relative to the laboratory reference frame—as when the superfluid is set into flow relative to the laboratory.

The condensed component of N_0 atoms will flow with zero viscosity so long as the flow does not create excitations in the superfluid—that is, so long as no atoms make transitions between the ground orbital and the excited orbitals. Such transitions might be caused by collisions of helium atoms with irregularities in the wall of the tube through which the helium atoms are flowing. The transitions if they occur are a cause of energy loss and of momentum loss from the moving fluid, and the flow is not resistanceless if such collisions can occur.

The criterion for superfluidity involves the energy and momentum relationship of the excitations in liquid He II. If the excited orbitals were really like the orbitals of free atoms, with a free particle relation

$$\epsilon = \tfrac{1}{2}Mv^2 = \frac{1}{2M}(\hbar k)^2 \tag{20}$$

between the energy ϵ and the momentum Mv or $\hbar k$ of an atom, then we can show that superfluidity would not be expected. Here $k = 2\pi/\text{wavelength}$. But because of the existence of interactions between the atoms the low energy excitations do not resemble free particle excitations, but are longitudinal sound waves, longitudinal phonons. After all, it is not unreasonable that a longitudinal sound wave should propagate in any liquid, even though we have no previous experience of superliquids.

A language has grown up to describe the low-lying excited states of a system of many atoms. These states are called **elementary excitations** and in their particle aspect the states are called **quasiparticles.** Longitudinal phonons are the elementary excitations of liquid He II. We shall give the clear-cut experimental evidence for this, but first we derive a necessary condition for superfluidity. This condition will show us why the phonon-like nature of the elementary excitations leads to the superfluid behavior of liquid He II.

We consider in Fig. 7 a body, perhaps a ball bearing or perhaps a neutron, of mass M_0 falling with velocity V down a column of liquid helium at rest at absolute zero, so that initially no elementary excitations are excited. If the motion of the body generates elementary excitations, there will be a damping force on the body. In order to generate an elementary excitation of energy ϵ_k and momentum $\hbar k$, we must satisfy the law of conservation of energy:

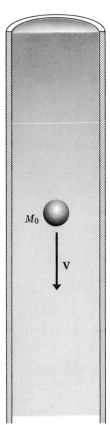

Figure 7   Body of mass M_0 moving with velocity $\mathbf{V}$ down a cylinder that contains liquid He II at absolute zero.

$$\tfrac{1}{2}M_0V^2 = \tfrac{1}{2}M_0V'^2 + \epsilon_\mathbf{k} , \tag{21}$$

where V' is the velocity of the body after creation of the elementary excitation. Furthermore, we must satisfy the law of conservation of momentum

$$M_0\mathbf{V} = M_0\mathbf{V}' + \hbar\mathbf{k} . \tag{22}$$

The two conservation laws cannot always be satisfied at the same time even if the direction of the excitation created in the process is unrestricted. To show this we rewrite (22) as

$$M_0\mathbf{V} - \hbar\mathbf{k} = M_0\mathbf{V}'$$

and take the square of both sides:

$$M_0^2V^2 - 2M_0\hbar\mathbf{V}\cdot\mathbf{k} + \hbar^2k^2 = M_0^2V'^2 .$$

On multiplication by $1/2M_0$ we have

$$\tfrac{1}{2}M_0V^2 - \hbar\mathbf{V}\cdot\mathbf{k} + \frac{1}{2M_0}\hbar^2k^2 = \tfrac{1}{2}M_0V'^2 . \tag{23}$$

We subtract (23) from (21) to obtain

$$\hbar \mathbf{V} \cdot \mathbf{k} - \frac{1}{2M_0} \hbar^2 k^2 \gtrless \epsilon_{\mathbf{k}} . \tag{24}$$

There is a lowest value of the magnitude of the velocity $\mathbf{V}$ for which this equation can be satisfied. The lowest value will occur when the direction of $\mathbf{k}$ is parallel to that of $\mathbf{V}$. This critical velocity is given by

$$V_c = \text{minimum of } \frac{\epsilon_{\mathbf{k}} + \frac{1}{2M_0} \hbar^2 k^2}{\hbar k} . \tag{25}$$

The condition is a little simpler to express if we let the mass M_0 of the body become very large, for then

$$\boxed{V_c = \text{minimum of } \frac{\epsilon_{\mathbf{k}}}{\hbar k} .} \tag{26}$$

A body moving with a lower velocity than V_c will not be able to create excitations in the liquid, so that the motion will be resistanceless. A body moving with higher velocity will encounter resistance because of the generation of excitations.

There is a simple geometrical construction for (26). We make a plot of the energy $\epsilon_{\mathbf{k}}$ of an elementary excitation as a function of the momentum $\hbar k$ of the excitation. We construct the straight line from the origin which just touches the curve from below. The slope of this line is equal to the critical velocity. If $\epsilon_{\mathbf{k}} = \hbar^2 k^2 / 2M$, as for the excitation of a free atom, the straight line has slope and the critical velocity is zero:

Free atoms: $\qquad V_c = \text{minimum of } \dfrac{\hbar k}{2M} = 0 .$ \hfill (27)

For a low energy phonon in liquid He II the energy is

$$\epsilon_{\mathbf{k}} = \hbar \omega_k = \hbar v_s k \tag{28}$$

in the frequency region of sound waves where the product of wavelength and frequency is equal to the velocity of sound v_s, or where the circular frequency ω_k is equal to the product of v_s times the wavevector k. Now the critical velocity is

Phonons: $\qquad V_c = \text{minimum of } \dfrac{\hbar v_s k}{\hbar k} = v_s .$ \hfill (29)

The critical velocity V_c is equal to the velocity of sound if (28) is valid for all wavevectors, which it is not in liquid helium II. The observed critical flow velocities are indeed finite, but are considerably lower than the velocity of

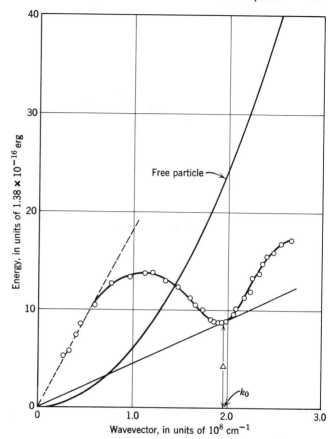

Figure 8 Energy ϵ_k versus wavevector k of elementary excitations in liquid helium at 1.12 K. The parabolic curve rising from the origin represents the theoretically calculated curve for free helium atoms at absolute zero. The open circles correspond to the energy and momentum of the measured excitations. A smooth curve has been drawn through the points. The broken curve rising linearly from the origin is the theoretical phonon branch calculated from a velocity of sound of 237 meters sec⁻¹. The solid straight line gives the critical velocity, in appropriate units: the line gives the minimum of ϵ_k/k over the region of k covered in these experiments. The linear region of the curve between 0 and about 0.6×10^8 cm⁻¹ is called the **phonon region.** The region centered about the minimum at 1.9×10^8 cm⁻¹ is called the **roton region** of the spectrum. [After D. G. Henshaw and A. D. B. Woods, Physical Review **121**, 1266 (1961).]

sound and usually lower than the solid straight line in Fig. 8, presumably because the plot of ϵ_k versus $\hbar k$ turns downward at very high $\hbar k$.

The actual spectrum of elementary excitations in liquid helium II has been determined by the observations on the inelastic scattering of slow neutrons.[8] The experimental results are shown in Fig. 8. The solid straight line is the Landau critical velocity for the range of wavevectors covered by the neutron experiments, and for this line the critical velocity is

$$V_c = \frac{\Delta}{\hbar k_0} \approx 5 \times 10^3 \text{ cm}^{-1} , \qquad (30)$$

where Δ and k_0 are identified on the figure.

---

[8] The method is discussed in ISSP, Chapter 5.

Charged ions of helium in solution in liquid helium II under certain experimental conditions of pressure and temperature have been observed[9] to move almost like free particles and to have a limiting drift velocity near 5×10^3 cm sec^{-1}, closely equal to the calculated value of (30). Under other experimental conditions the motion of the ions is limited at a lower velocity by the creation of vortex rings. Such vortex rings have values of the wavevector above the range covered by Fig. 8 and do not appear there.

Our result (29) for a necessary condition for the critical velocity is more general than the calculation we have given. Our calculation demonstrates that a body will move without resistance through liquid He II at absolute zero if the velocity V of the body is less than the critical velocity V_c. However, at temperatures above absolute zero, but below the Einstein temperature, there will be a normal fluid component of elementary excitations that are thermally excited. The normal fluid component is the source of resistance to the motion of the body. The superflow aspect appears first in experiments in which the liquid flows out through a fine tube in the side of a container. The normal fluid component may remain behind in the container while the superfluid component leaks out without resistance. The derivation we have given of the critical velocity also holds for this situation, with V as the velocity of the superfluid relative to the walls of the tube; M_0 is the mass of the fluid. The excitations would be created above V_c by the interaction between the flow of the liquid and any mechanical irregularity in the walls.

REFERENCES

R. B. Dingle, "Theories of helium II," Advances in Physics **1**, 111 (1952). Excellent review of the two-fluid theory.

J. Wilks, *Properties of liquid and solid helium,* Oxford University Press, 1967. Very thorough interpretative review of the experimental data on He^3 and He^4.

[9] L. Meyer and F. Reif, Physical Review **123**, 727 (1961); G. W. Rayfield, Physical Review Letters **16**, 934 (1966).

Consider a system allowed to expand reversibly at constant temperature. In the expansion the system does work against an external piston: the work performed by the system on the piston in an expansion from volume V_1 to V_2 is given by

$$\int_{V_1}^{V_2} p \, dV \; ,$$

which is the area under the curve in Fig. 1. The work performed by the system is supplied from two sources, the heat flow into the system through the walls as required to maintain the constant temperature and the decrease in the internal energy of the system with the increase of volume. The internal energy change is zero for the special case of an ideal gas, but in general it is not zero.

The contributions from these two sources to the work performed by the system can be expressed in terms of a single quantity, the **free energy** F. One motivation for the introduction of the free energy into the subject of thermal physics is just this property: it tells us how much work can be performed by the system in a process at constant temperature.

The free energy has other important and useful properties:

(a) It is a minimum in equilibrium for a system of constant volume in thermal contact with a heat reservoir.

(b) It is obtained directly from the partition function Z by $F = -\mathcal{T} \log Z$.

(c) The entropy may be calculated directly from the free energy.

Figure 1 A general system, which need not be an ideal gas, is allowed to expand reversibly from V_1 to V_2 while the temperature is kept constant. The work done by the system on the external piston is equal to the area under the p-V curve. This work is equal to the decrease $F_1 - F_2$ in the free energy of the system, as we shall see.

In figure: p; $\mathcal{T} = $ constant; F_1; Work done by system $= \int_{V_1}^{V_2} p \, dV$; F_2; V_1; V_2; V

PRESSURE

We first establish the connection between the pressure and the free energy. When we introduced the pressure in Chapter 7 we showed that

$$p = -\left(\frac{\partial U}{\partial V}\right)_{\sigma, N} . \tag{1}$$

The pressure is related to the rate of change of energy with volume, the derivative being taken at constant entropy. Constant entropy suggests a constant probability that the system remain in any given quantum state l throughout the expansion. We require an expression for the pressure in terms of derivatives at constant temperature, because experiments are often carried out at constant temperature. The result that follows was the subject of a problem in Chapter 7, but it is important enough to rederive here.

To obtain the desired expression for the pressure we start from the thermodynamic identity

$$\tau \, d\sigma = dU - \mu \, dN + p \, dV . \tag{2}$$

We form derivatives with respect to volume, with τ and N constant:

$$\tau \left(\frac{\partial \sigma}{\partial V}\right)_{T, N} = \left(\frac{\partial U}{\partial V}\right)_{T, N} + p , \tag{3}$$

whence

$$\boxed{p = -\left(\frac{\partial U}{\partial V}\right)_{T, N} + \tau \left(\frac{\partial \sigma}{\partial V}\right)_{T, N} .} \tag{4}$$

The difference between this result and (1) arises because the entropy is held constant in (1), whereas the temperature is held constant in (4).

Notice in (4) the two contributions to the pressure. The term $-(\partial U/\partial V)_{T, N}$ is, roughly speaking, of mechanical origin, similar to the elastic forces in a normal crystalline solid or a steel coil spring. This term is zero for an ideal gas. The contribution $\tau(\partial \sigma/\partial V)_{T, N}$ to the pressure arises from the dependence of the entropy on the volume; for example, we know in a gas that the entropy increases with an increase of volume. This term is solely responsible for the pressure of an ideal gas. At absolute zero the term $\tau(\partial \sigma/\partial V)_{T, N}$ is zero: only the mechanical pressure is present at absolute zero.

The form of (4) leads us to introduce the quantity called the **free energy:**[1]

$$\boxed{F \equiv U - \tau \sigma .} \tag{5}$$

---

[1] This is also called the Helmholtz free energy.

The free energy acts as the "effective potential energy" for the work performed in isothermal changes. To prove this property we differentiate both sides of (5) with respect to the volume, holding $\mathcal{T}$ constant:

$$\left(\frac{\partial F}{\partial V}\right)_{T,N} = \left(\frac{\partial U}{\partial V}\right)_{T,N} - \mathcal{T}\left(\frac{\partial \sigma}{\partial V}\right)_{T,N} . \tag{6}$$

On comparison with (4) we see that

$$\boxed{p = -\left(\frac{\partial F}{\partial V}\right)_{T,N} .} \tag{7}$$

The pressure is simply related to the dependence of the free energy on the volume. By comparison with (1) we see that U is the "effective potential energy" at constant entropy and F is the "effective potential energy" at constant temperature.

AVAILABLE WORK AT CONSTANT TEMPERATURE

The work performed by the system on a piston in a reversible isothermal expansion is found on integration of both sides of (7) from the initial volume V_1 to the final volume V_2:

$$\text{Work performed} = \int_{V_1}^{V_2} p\, dV = -\int_{V_1}^{V_2} dV \left(\frac{\partial F}{\partial V}\right)_{T,N} = F(V_1) - F(V_2) . \tag{8}$$

Thus the work delivered in a reversible process by a system in thermal contact with a heat reservoir is equal to the decrease in the free energy of the system. The free energy is a measure of the energy that the system has available to give us in an isothermal process. Part of the energy is supplied by a change in the mechanical deformation of the system and part is supplied by heat that flows into the system from the reservoir that maintains the temperature constant.

Problem 1. *Force on a linear polymer.* Find the analogues of (4) and (7) for the force f on a linear polymer in terms of isothermal derivatives with respect to the length l.

ENTROPY AND ENERGY

Given the free energy of a system, how do we obtain the entropy and the energy? To find an expression for the entropy, we form the differential of the free energy:

$$dF = d(U - \mathcal{T}\sigma) = dU - \mathcal{T}\,d\sigma - \sigma\,d\mathcal{T} \ . \tag{9}$$

By the thermodynamic identity

$$dU - \mathcal{T}\,d\sigma = \mu\,dN - p\,dV \ ; \tag{10}$$

we use this to write dF as

$$dF = \mu\,dN - p\,dV - \sigma\,d\mathcal{T} \ . \tag{11}$$

This result gives dF in terms of dN, dV, and $d\mathcal{T}$ for reversible changes; here μ, p, and σ are equilibrium values.

We say because of (11) that the free energy F has N, V, $\mathcal{T}$ as natural independent variables. The differential dF in terms of these variables is

$$dF = \left(\frac{\partial F}{\partial N}\right)_{V,\mathcal{T}} dN + \left(\frac{\partial F}{\partial V}\right)_{\mathcal{T},N} dV + \left(\frac{\partial F}{\partial \mathcal{T}}\right)_{V,N} d\mathcal{T} \ . \tag{12}$$

On comparison with (11) we find

$$\mu = \left(\frac{\partial F}{\partial N}\right)_{V,\mathcal{T}} \ ; \qquad p = -\left(\frac{\partial F}{\partial V}\right)_{\mathcal{T},N} \ ; \qquad \sigma = -\left(\frac{\partial F}{\partial \mathcal{T}}\right)_{V,N} \ . \tag{13}$$

We obtained earlier [in (7)] the relation for the pressure.

The relation for the entropy is particularly useful. We use it first to obtain the energy of the system from the free energy. By (13) we have

$$F \equiv U - \mathcal{T}\sigma = U + \mathcal{T}\left(\frac{\partial F}{\partial \mathcal{T}}\right)_{V,N} , \tag{14}$$

or

$$U = F - \mathcal{T}\left(\frac{\partial F}{\partial \mathcal{T}}\right)_{V,N} = -\mathcal{T}^2\left(\frac{\partial}{\partial \mathcal{T}}\frac{F}{\mathcal{T}}\right)_{V,N} \ . \tag{15}$$

Problem 2. Maxwell relation and the bulk modulus. (a) Prove that

$$\left(\frac{\partial p}{\partial \mathcal{T}}\right)_{V,N} = \left(\frac{\partial \sigma}{\partial V}\right)_{\mathcal{T},N} \ ; \qquad \left(\frac{\partial \mu}{\partial V}\right)_{\mathcal{T},N} = -\left(\frac{\partial p}{\partial N}\right)_{\mathcal{T},V} \ . \tag{16a}$$

A full discussion of the Maxwell relations is given in Chapter 7 of H. B. Callen, *Thermodynamics*, Wiley, 1960.

(b) The isothermal bulk modulus is defined as

$$B_T \equiv -V \left(\frac{\partial p}{\partial V} \right)_{T,N} .$$ (16b)

Prove using the second Maxwell relation above that this may be written for a gas or a liquid as

$$B = \frac{N^2}{V} \left(\frac{\partial \mu}{\partial N} \right)_{T,V} ,$$ (16c)

a form widely used in the theory of Fermi liquids. *Hint.* Use the fact that the pressure and the chemical potential depend on the number of particles N and the volume V only through the concentration $c = N/V$.

FREE ENERGY AND THE PARTITION FUNCTION

The free energy has a simple and direct relation to the partition function

$$Z = \sum_l e^{-\epsilon_l/T} ,$$ (17)

and for this reason many calculations in thermal physics have the free energy as their starting point.

It was the subject of Problem 7.2 to show that the result

$$\boxed{F = -T \log Z}$$ (18)

follows from the Boltzmann definition of the entropy, (7.46). Here we merely confirm that (18) is consistent[2] with the definition of F as $U - T\sigma$. We form

$$T \left(\frac{\partial F}{\partial T} \right)_{V,N} = -T \log Z - \frac{\sum \epsilon_l e^{-\epsilon_l/T}}{Z} .$$ (19)

The right-hand side is seen to be equal to $F - U$, so that (19) may be written as

$$T \left(\frac{\partial F}{\partial T} \right)_{V,N} = F - U ,$$ (20)

which is exactly the result (14).

It is occasionally helpful to write (18) as

$$e^{-F/T} = Z .$$ (21)

[2] R. Gray has pointed out that other functions satisfy a differential equation of the form (14); for example, $-T \log \alpha Z$, where α is any positive constant. Only the function (18) leads to values of σ consistent with the original definition $\sigma = \log g$.

Problem 3. *Free energy of a two state system.* (a) Find an expression for the free energy of a system with two states, one at energy 0 and one at energy ϵ. (b) From the free energy find expressions for the energy, entropy, and heat capacity of the system.

Problem 4. *Free energy of a harmonic oscillator.* (a) Show that for a harmonic oscillator the free energy is

$$F = T \log\left(1 - e^{-\hbar\omega/T}\right) = -\tfrac{1}{2}\hbar\omega + T \log\left(2 \sinh \frac{\hbar\omega}{2T}\right). \tag{22}$$

Note that at high temperatures such that $T \gg \hbar\omega$ we may expand the argument of the logarithm to obtain

$$F \cong T \log \frac{\hbar\omega}{T} .$$

(b) From the middle form in (22), show that the entropy is

$$\sigma = \frac{\hbar\omega/T}{e^{\hbar\omega/T} - 1} - \log\left(1 - e^{-\hbar\omega/T}\right) . \tag{23}$$

Find the form of the entropy when $T \gg \hbar\omega$.

Problem 5. *Radiation pressure of a photon gas.* (a) Show that the partition function of a photon gas is given by

$$Z = \frac{1}{\displaystyle\prod_l \left(1 - e^{-\hbar\omega_l/T}\right)} , \tag{24}$$

where the product is over the orbitals l.

(b) Show that the free energy is

$$F = T \sum_l \log\left(1 - e^{-\hbar\omega_l/T}\right) . \tag{25}$$

(c) Show that the pressure is

$$p = -\hbar \sum_l \frac{d\omega_l/dV}{e^{\hbar\omega_l/T} - 1} . \tag{26}$$

For photons we showed in Problem 15.3 that

$$\frac{d\omega_l}{dV} = -\frac{\omega_l}{3V} \tag{27}$$

by virtue of the boundary conditions on the photon modes, whence we found that the radiation pressure is

$$p = \tfrac{1}{3}u \, , \tag{28}$$

where u is the energy density per unit volume. This is a famous result. It plays an important part in the theory of the internal constitution of stars.

(d) Show that for a photon gas

$$F = -\frac{\pi^2 V T^4}{45 c^3 \hbar^3} \, . \tag{29}$$

Hint. You may find it convenient to use the relation (15) and the known expression for the total radiant energy in a cavity.

FREE ENERGY A MINIMUM IN EQUILIBRIUM

We now show that in the most probable configuration of a system of constant volume in thermal contact with a reservoir the free energy of the system is a minimum. The minimum is with respect to energy exchange with the reservoir and also with respect to any other quantity or internal parameter of the system. It is understood that along with the volume, the number of particles and the temperature of the system are held constant.

The total energy of the system + reservoir is always constant:

$$dU_s + dU_r = 0 \, . \tag{30}$$

The subscript s refers to the system and r to the reservoir. The total entropy[3] is a maximum in the most probable configuration, as we saw in Chapter 4:

$$d\sigma_s + d\sigma_r = 0 \, . \tag{31}$$

By the definition of temperature

$$\frac{1}{T} = \frac{d\sigma_r}{dU_r} \, ; \qquad dU_r = T \, d\sigma_r \, , \tag{32}$$

where in the differentiation V_r and N_r are constant by our initial assumption. Thus (30) becomes

$$dU_s + T \, d\sigma_r = 0 \, , \tag{33}$$

and by (31)

$$dU_s - T \, d\sigma_s = d(U_s - T\sigma_s) = dF_s = 0 \, , \tag{34}$$

---

[3] Strictly, we mean the generalized entropy in the sense of Chapter 4.

so that at constant temperature and volume the free energy F_s of the system is an extremum in the most probable configuration.

To show that the extremum of F_s is a minimum we must consider finite changes $\Delta\sigma_s$ and $\Delta\sigma_r$. Because the entropy is a maximum in equilibrium we know that

$$\Delta\sigma_s + \Delta\sigma_r \leq 0 \; ; \qquad \Delta\sigma_r \leq -\Delta\sigma_s \; . \tag{35}$$

Equation (33) for finite changes reads

$$\Delta U_s + T\,\Delta\sigma_r = 0 \; , \tag{36}$$

so that on passing to (34) we have for finite changes

$$\Delta F_s = \Delta U_s - T\,\Delta\sigma_s \geq 0 \; . \tag{37}$$

Thus the free energy of the system increases for any departure from the most probable configuration.

This argument is quite general. In thermal equilibrium at constant temperature and volume the free energy is a minimum. With respect to what is the free energy a minimum? We clarify this question by the example which follows. The example concerns a magnetic system. For a magnetic system in thermal equilibrium at constant temperature and magnetic field intensity the free energy is a minimum. This is the standard statement, and it is true with respect to any variation in the system within the restraints prescribed. But it is more instructive to consider a specific variation, for example in the magnetic moment. We introduce a generalized free energy function defined for any value (equilibrium or not) of the magnetization and such that the minimum of this function with respect to the magnetization (at constant T and H) defines the free energy and gives the equilibrium value of the magnetization.

EXAMPLE. *Landau free energy function and the paramagnetic suscepti-bility.* As an example of the application of the minimum property of the free energy we calculate the free energy of our model system of N spins in an external magnetic field H. The spins are independent of one another, as in Chapter 2. We want to find the free energy and the equilibrium value of the magnetization when the system is in thermal contact with a reservoir at temperature T.

We introduce the Landau free energy function

$$\tilde{F}(m; T, H) \equiv U(m; T, H) - \sigma(m; T, H) \; ,$$

where the spin excess $2m$ is defined in Chapter 2 as the difference between the number of up spins and number of down spins. The minimum of the Landau function with respect to m defines the (equilibrium) free energy

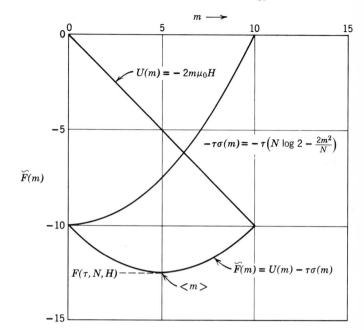

Figure 2 Determination of equilibrium value $\langle m \rangle$ of the spin excess $2m$ for particular values of temperature $\mathcal{T}$, number of particles N, and magnetic field H. The equilibrium is at the minimum of $\widehat{F}(m)$. At low temperatures F is dominated by U, but at high temperatures $-\mathcal{T}\sigma$ may be the dominant term.

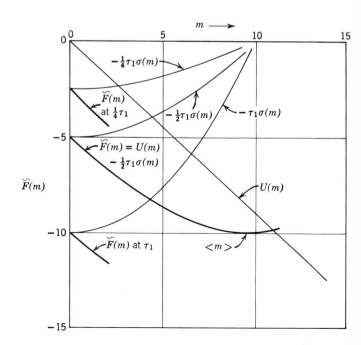

Figure 3 Plot of $U(m)$ and of $-\mathcal{T}\sigma(m)$ at fixed H and at three temperatures, $\frac{1}{4}\mathcal{T}_1$, $\frac{1}{2}\mathcal{T}_1$, and $\mathcal{T}_1$. The minimum of the Landau free energy function occurs at $\langle m \rangle = 20$, 10, and 5, respectively. The free energy is plotted in full only for temperature $\frac{1}{2}\mathcal{T}_1$, for comparison with Fig. 2 at $\mathcal{T}_1$. Note the change in $\langle m \rangle$ from that in Fig. 2.

$F(T, H)$ of the system. Here we are being purists: we now view the term free energy as defined only in thermal equilibrium. The Landau function $\tilde{F}$ is defined for any situation, equilibrium or not. The value of m or of the magnetization at the minimum of F defines the thermal equilibrium value of m or of the magnetization. The explicit introduction of the Landau function avoids much of the confusion and inconsistency that often accompany the exploitation of the extremal property of the free energy. The function is named after the Soviet theoretical physicist who used it in many problems, notably in the study of fluctuations in liquids, and in the analysis of transitions in ferroelectrics, superconductors, and antiferromagnetic crystals. The use of $\tilde{F}$ as a free energy generalized to nonequilibrium conditions is analogous to the use of the generalized entropy σ_G in Chapter 4; in fact, the function $\sigma(m; T, H)$ above and in (39) is an example of a generalized entropy.

Consider the system in a configuration with $\frac{1}{2}N + m$ spins up and $\frac{1}{2}N - m$ spins down. The energy is

$$U(m) = -2m\mu_0 H , \tag{38}$$

where μ_0 is the magnetic moment of a spin. The entropy of the system in this configuration was found in Chapter 4 to be

$$\sigma(m) = N \log 2 - \frac{2m^2}{N} , \tag{39}$$

in the approximation $|m| \ll N$. We do not have to make this approximation, but the argument is more explicit if we do so.

The Landau free energy function as a function of m is

$$\tilde{F}(m; T, H) \equiv U(m) - T\sigma(m) = -2m\mu_0 H - TN \log 2 + \frac{2Tm^2}{N} . \tag{40}$$

This function is plotted in Figs. 2 and 3. We see that the magnetic field tends to reduce the free energy at a given value of m, and the temperature tends to increase it.

In the equilibrium condition of the system $\tilde{F}(m; T, H)$ is a minimum with respect to m, at constant T and H. We have

$$\left(\frac{\partial \tilde{F}}{\partial m}\right)_{T, H} = 0 = -2\mu_0 H + \frac{4Tm}{N} . \tag{41}$$

The value of m that satisfies this equilibrium condition is denoted by $\langle m \rangle$, where from (41)

$$\langle m \rangle = \frac{N\mu_0 H}{2T} . \tag{42}$$

We substitute the value of $\langle m \rangle$ in $\tilde{F}(m; T, H)$ to obtain the free energy of the system in equilibrium

$$F(T, H) = -NT \log 2 - \frac{N\mu_0^2 H^2}{2T} . \qquad (43)$$

In this problem the free energy F depends on the independent variables H, T.

As a check on (42) and (43), we draw on the relation (23.37):

$$\left(\frac{\partial F}{\partial H}\right)_T = -M ,$$

per unit volume; now from (43)

$$\left(\frac{\partial F}{\partial H}\right)_T = -\frac{N\mu_0^2 H}{T} ; \qquad M = 2\mu_0\langle m \rangle = \frac{N\mu_0^2 H}{T} ,$$

in agreement with (42). Here M is the magnetization.

EXAMPLE. *Mean field theory of spin-spin interactions.*[4] In our model system above the spins interacted only with the external magnetic field, but not with each other. We now add an interaction among the spins. We do not inquire here into the physical origin of the interaction. We require it to be ferromagnetic: the energy of the system is lowered as more spins become parallel to each other. We suppose further—this is called the **mean field approximation**—that the interaction energy depends only on the magnitude of the excess of spins that point up over the number of spins that point down; that is, on $|2m|$. The mean field approximation is a brutal, but powerful, assumption. It does not take proper account of the range dependence of the actual interaction: in a real solid the strength of the interaction between a pair of spins will depend on the distance apart of the spins, so that the probability that a pair of spins are parallel will depend on the distance apart.

Formally, the mean field approximation consists in adding to the free energy function a term $-\alpha m^2$, where α is a positive constant. The term provides that the greater the spin excess $2m$, the lower the value of the free energy function. Thus (40) is changed to

$$\tilde{F}(m) = -\alpha m^2 - 2m\mu_0 H - TN \log 2 + \frac{2Tm^2}{N} , \qquad (44)$$

[4] See also ISSP, Chapter 15, where a more physical method of solution is given to the same problem. The present formal solution is included for the exercise it provides in the application of the free energy function. For an application to ferroelectrics, see ISSP, Chapter 13.

provided as before that $|m| \ll N$. In equilibrium

$$\left(\frac{\partial F}{\partial m}\right)_{T, H} = 0 = -2\alpha m - 2\mu_0 H + \frac{4Tm}{N} , \tag{45}$$

whence

$$\langle m \rangle = \frac{\frac{1}{2}N\mu_0 H}{T - \frac{1}{2}\alpha N} , \tag{46}$$

in place of (42).

The result suggests that something drastic happens to $\langle m \rangle$ at the temperature at which the denominator vanishes:

$$T_c = \tfrac{1}{2}\alpha N . \tag{47}$$

This temperature is called the **Curie temperature** and it marks the onset of ferromagnetism. At temperatures below T_c it is possible for $\langle m \rangle$ to be non-zero even for $H = 0$; this is what we mean by ferromagnetism. But we made the approximation $|m| \ll N$ in the expression (39) for $\sigma(m)$, and this approximation is not good enough to allow us to discuss the finer details of the ferromagnetic state.

Problem 6. *Ferromagnetic region.* Add a term βm^4 to the free energy function (44), where β is positive. Set $H = 0$ and find $\langle m \rangle$ as a function of T for $T \leq T_c$. Sketch the result.

FREE ENERGY AND PARTITION FUNCTION OF AN IDEAL GAS

We use the results of Chapter 11 to find an expression for the free energy and the partition function of an ideal monatomic gas. The results have an astonishing feature.

We start from the partition function Z_N of a system of N free particles:

$$Z_N = e^{-F/T} = e^{-(U - T\sigma)/T} , \tag{48}$$

from (21). We use the result $U = \frac{3}{2}NT$ and σ as given by the Sackur-Tetrode equation in Chapter 11 to obtain, for atoms of spin zero,

$$Z_N = e^{-\frac{3}{2}N}\left[\left(\frac{MT}{2\pi\hbar^2}\right)^{\frac{3}{2}}\left(\frac{V}{N}\right)\right]^N e^{\frac{5}{2}N}$$

$$= e^N N^{-N} V^N \left(\frac{MT}{2\pi\hbar^2}\right)^{\frac{3}{2}N} . \tag{49}$$

We combine $e^N N^{-N}$ by use of the Stirling approximation

$$e^{-N} N^N \cong N! \tag{50}$$

We then have

$$Z_N = \frac{V^N}{N!\,(2\pi\hbar^2/MT)^{\frac{3N}{2}}} = \frac{1}{N!}\left(\frac{V}{V_Q}\right)^N, \tag{51}$$

which is the correct partition function in the ideal gas limit. We can appreciate this important result more fully if we attempt to derive it by the rocky route followed historically.

The partition function of an ideal gas composed of N atoms is

$$Z_N = \sum_l \exp\left[-\epsilon_l(N)/T\right], \tag{52}$$

where $\epsilon_l(N)$ is the energy eigenvalue of state l of the N-particle system. We might expect to be able to factor (52) as the product of separate partition functions for each particle:[5]

$$Z_N \overset{?}{=} \left[\sum_n e^{-\epsilon_n/T}\right]^N, \tag{53}$$

The product form is suggested because the particles are independent of each other. Furthermore, (53) ensures that the free energy $F_N = -T \log Z_N$ is additive for each particle. Here ϵ_n is the energy eigenvalue of a single free particle orbital:

$$\epsilon_n = \frac{\hbar^2}{2M}\left(\frac{\pi}{L}\right)^2 (n_x{}^2 + n_y{}^2 + n_z{}^2), \tag{54}$$

as in Chapter 10.

Then (53) becomes

$$Z \overset{?}{=} \left[\sum_{n_x} e^{-\alpha n_x{}^2}\right]^{3N}, \tag{55}$$

with

$$\alpha \equiv \frac{\hbar^2 \pi^2}{2ML^2 T}. \tag{56}$$

The sum is over all positive integers n_x.

If we approximate the sum in (55) by an integral, we find

$$\sum_{n_x} e^{-\alpha n_x{}^2} = \int_0^\infty dn_x\, e^{-\alpha n_x{}^2} = \left(\frac{\pi}{4\alpha}\right)^{\frac{1}{2}} = (2\pi MT)^{\frac{1}{2}}\left(\frac{L}{2\pi\hbar}\right). \tag{57}$$

---

[5] The question mark over the equality sign in (53) means, as we shall discover, that the equation is incorrect.

Thus (55) becomes, with $L^3 = V$,

$$Z \overset{?}{=} \frac{V^N}{(2\pi\hbar)^{3N}} (2\pi MT)^{\frac{3}{2}N} = \left(\frac{V}{V_Q}\right)^N . \tag{58}$$

This differs from the correct result (51) by the presence of the factor $1/N!$ in the correct result. Yet we have used quantum mechanics in deriving (58): the answer even involves $\hbar$. What has gone wrong with our argument?

The expression (58) gives too large a value for the partition function, too large by a factor $N!$. The discrepancy arises from the quantum mechanics of a gas of N identical particles. We have overcounted in (58) the number of states of the N particle system. Even if the particles are completely independent, in quantum mechanics we must account for what is called the indistinguishability of identical particles. This is another aspect of the Pauli principle, but an aspect that affects both fermions and bosons. It has automatically been taken into account correctly in our earlier chapters. The effect for the ideal gas problem is to reduce by $1/N!$ the number of states of the N-particle system and hence the sum over all states in (53). The mistake was made in writing (53). By hindsight it appears that we should instead have written

$$Z = \frac{1}{N!}\left[\sum e^{-\epsilon_n/T}\right]^N \tag{59}$$

in place of (53). An advantage of our correct (and direct) result (49) through (51) is that the whole question of indistinguishability is handled **automatically,** with no apology or confusion. A further discussion of classical statistical mechanics is given in Appendix E.

The free energy of an ideal monatomic gas of N atoms is found from the definition $F \equiv U - T\sigma$ or from the relation $F = -T \log Z$, with (49) for Z. We have

$$F = -NT\left(1 + \log\frac{V}{NV_Q}\right) , \tag{60}$$

with $V_Q \equiv (2\pi\hbar^2/MT)^{\frac{3}{2}}$.

The thermodynamic potential G is defined in the next chapter. It is equal to $U - T\sigma + pV$, or $F + pV$. For an ideal gas $pV = NT$, whence (60) leads us to

$$G = NT \log c V_Q , \tag{61}$$

where $c = N/V$ is the concentration. It is usual to express G as a function of p, T, and N, so that for an ideal gas

$$G = NT \log \frac{pV_Q}{T} . \tag{62}$$

From the result $\mu = (\partial F / \partial N)_{T, V}$ for the chemical potential we have for the ideal gas

$$\boxed{\mu = T \log c V_Q ,} \qquad (63)$$

as found in Chapter 11. On comparison with (62) we see that $G = N\mu$; this is actually a general result which is derived in the next chapter.

EXAMPLE. *Free energy of excitation.* Consider a complex molecule or system containing a subunit that exists in either of two conditions, 1 and 2. We denote the excitation energy of the subunit when isolated from the rest of the molecule by

$$\Delta \epsilon = \epsilon_2 - \epsilon_1 .$$

The excitation energy will change to

$$\Delta \epsilon' = \epsilon_2' - \epsilon_1'$$

when the subunit is attached to the molecule because the molecule is distorted elastically near the point of attachment. The energy of distortion may depend on whether the subunit is in the condition 1 or 2. Furthermore, the vibrational energy spectrum of the molecule may depend on the condition of the subunit. Thus when the subunit changes from one condition to the other, the energy of each vibrational state of the system changes, along with the thermal occupancy of the vibrational states. The effect of the condition of the subunit on the energy of the system will therefore depend on the temperature.

This is a complicated situation in which neither the energy difference $\Delta \epsilon$ nor $\Delta \epsilon'$ is the relevant quantity to use in the Boltzmann factor for the probability the subunit is in condition 1 or 2. The correct energy difference appears to depend on the temperature, but our entire computational apparatus was set up with all energy states independent of the temperature, because our states by definition are states of the entire system. Our scheme is to feed into the theory the energies ϵ_l, and we then calculate all thermal effects. We can, however, find a quantity analogous to $\Delta \epsilon$ or $\Delta \epsilon'$ which gives the desired correct results for the subunit.

The temperature-independent states are those of the system as a whole. For the system the partition function is

$$Z = \sum_l e^{-\epsilon_l / T} = e^{-F/T} , \qquad (64)$$

where F is the free energy, as in (21). We now assume that the set of states ϵ_l can be divided into two parts, one for states for which the subunit is in the condition 1 and one for states for which the subunit is in the condition 2. Then the partition function is the sum of two parts:

$$Z = Z_1 + Z_2 = e^{-F_1/T} + e^{-F_2/T} \ , \tag{65}$$

where the free energies F_1 and F_2 are defined by

$$F_1 \equiv -T \log Z_1 \ ; \qquad F_2 = -T \log Z_2 \ . \tag{66}$$

The probability that the system will be found with the subunit in the condition 1 is

$$P_1 = \frac{Z_1}{Z} = \frac{e^{-F_1/T}}{Z} \ , \tag{67}$$

and similarly for P_2.

The ratio of the probability the subunit is in condition 2 to the probability the subunit is in condition 1 is

$$\frac{P_2}{P_1} = \frac{e^{-F_2/T}}{e^{-F_1/T}} = e^{-(F_2 - F_1)/T} \ . \tag{68}$$

Thus the change in free energy

$$\Delta F = F_2 - F_1 \tag{69}$$

between the two conditions of the system acts as the effective energy in the Boltzmann factor for a process in which many energy states of the system are affected by a specific change in the condition of a part of the system. This result is widely used in chemistry, but it sometimes takes physicists by surprise. The quantity ΔF may be called the free energy of the excitation; it will in general be a function of the temperature.

This result was derived for a system at constant volume, but it may be extended to a system at constant pressure. The analogue to (68) becomes

$$\frac{P_2}{P_1} = \frac{e^{-(pV_2 + F_2)/T}}{e^{-(pV_1 + F_1)/T}} = \frac{e^{-G_2/T}}{e^{-G_1/T}} \ , \tag{70}$$

where the thermodynamic potential $G \equiv U - T\sigma + pV$ is treated in Chapter 19. The result (70) is readily derived by the methods of Chapter 6 extended to a system in thermal and mechanical contact with a reservoir. The entropy change which accompanies an "exchange" of volume between the system and the reservoir is responsible for the appearance of $-pV/T$ in the exponent.

REFERENCE

E. A. Guggenheim, "Grand partition functions and the so-called 'thermodynamic probability.'" Journal of Chemical Physics **7**, 103 (1939).

CHAPTER 19

Thermodynamic Potential, Grand Potential, and Heat Function

CHAPTER 19 THERMODYNAMIC POTENTIAL, GRAND POTENTIAL, AND HEAT FUNCTION

The free energy is useful for the discussion of the equilibrium configuration of a system at constant volume and temperature. But many experiments, and in particular many chemical reactions, are performed at constant pressure, which may often be one atmosphere. It is appropriate to introduce another function to treat the equilibrium configuration at constant pressure and temperature. The new function is closely related to the free energy F.

We define the **thermodynamic potential** G as

$$G \equiv U - T\sigma + pV .$$

(1)

This is usually called the **Gibbs free energy**; chemists often call it the free energy. Exactly analogous to Chapter 18, we can also define a generalized or Landau thermodynamic potential function $\tilde{G}$ to describe nonequilibrium situations. We would then reserve the quantity G to denote the value of the thermodynamic potential in thermal equilibrium.

PROPERTIES OF THE THERMODYNAMIC POTENTIAL

The central properties of the thermodynamic potential are:

(a) The thermodynamic potential is a minimum in equilibrium for a system at constant pressure in thermal contact with a heat reservoir.

(b) For a system of a single chemical component the thermodynamic potential divided by the number of particles is equal to the chemical potential.

Chapters 20 and 21 treat the applications of the thermodynamic potential to phase equilibria and to chemical equilibria; both are problems of wide interest.

ENTROPY AND CHEMICAL POTENTIAL

The differential dG of the thermodynamic potential is

$$dG = dU - T\,d\sigma - \sigma\,dT + p\,dV + V\,dp .$$

(2)

In a reversible change the thermodynamic identity holds:

$$T\,d\sigma = dU - \mu\,dN + p\,dV ,$$

(3)

whence (2) becomes

$$dG = \mu \, dN - \sigma \, dT + V \, dp \ . \tag{4}$$

We see that G appears as a function of the variables N, T, p, so that the differential may be written as

$$dG = \left(\frac{\partial G}{\partial N}\right)_{T, p} dN + \left(\frac{\partial G}{\partial T}\right)_{N, p} dT + \left(\frac{\partial G}{\partial p}\right)_{N, T} dp \ . \tag{5}$$

By comparison of (4) and (5) we have the relations

$$\left(\frac{\partial G}{\partial N}\right)_{T, p} = \mu \ ; \qquad \left(\frac{\partial G}{\partial T}\right)_{N, p} = -\sigma \ ; \qquad \left(\frac{\partial G}{\partial p}\right)_{N, T} = V \ . \tag{6}$$

Let us consider the thermodynamic potential as a function of N, T, and p. Here T and p are **intensive quantities:** they do not change value when two identical systems are put together. But G is linear in the number of particles N: the value of G doubles when two identical systems are put together. This result follows because U, σ, and V are linear in N, if we neglect possible surface effects. Thus the functional dependence of G on N, T, and p must be such that

$$G = N\Phi(p, T) \ , \tag{7}$$

where $\Phi(p, T)$ is a function only of p and T, and independent of N. From (7) we have

$$\left(\frac{\partial G}{\partial N}\right)_{p, T} = \Phi(p, T) \ . \tag{8a}$$

We saw in (6) that

$$\left(\frac{\partial G}{\partial N}\right)_{p, T} = \mu \ , \tag{8b}$$

so that Φ must be identical with μ, and we have

$$G(N, p, T) = N \mu(p, T) \ . \tag{9}$$

Thus for a single-component system the chemical potential is equal to the thermodynamic potential per particle, G/N. If more than one chemical species is present, (9) is replaced by a sum over all species:

$$G = \sum_i N_i \mu_i \ . \tag{10}$$

In Chapter 21 we shall develop the theory of chemical equilibria by exploiting the property that $G = \Sigma N_i \mu_i$ is a minimum with respect to changes in the numbers of reacting molecules under constant temperature and pressure: see especially (21.18)–(21.22).

EXAMPLE. *Thermal expansion as* $\mathcal{T} \to 0$. We easily find another Maxwell relation:

$$\left(\frac{\partial V}{\partial \mathcal{T}}\right)_p = -\left(\frac{\partial \sigma}{\partial p}\right)_\mathcal{T} , \tag{11}$$

if we equate $\partial^2 G / \partial \mathcal{T} \, \partial p$ and $\partial^2 G / \partial p \, \partial \mathcal{T}$. If, as suggested by the third law of thermodynamics, the entropy approaches a constant limiting value as $\mathcal{T} \to 0$, then $(\partial \sigma / \partial p)_\mathcal{T} \to 0$ as $\mathcal{T} \to 0$. It then follows from (11) that $(\partial V / \partial \mathcal{T})_p \to 0$ as $\mathcal{T} \to 0$.

The volume coefficient of thermal expansion is defined as

$$\alpha = \frac{1}{V}\left(\frac{\partial V}{\partial \mathcal{T}}\right)_p , \tag{12}$$

which will $\to 0$ as $\mathcal{T} \to 0$.

EXAMPLE. *Thermodynamic potential of a monatomic ideal gas.* In (18.62) we found

$$G = N\mathcal{T} \log V_Q + N\mathcal{T} \log \frac{p}{\mathcal{T}} . \tag{13a}$$

We can obtain the chemical potential by use of (8):

$$\mu(p, \mathcal{T}) = \left(\frac{\partial G}{\partial N}\right)_{p, \mathcal{T}} = \mathcal{T} \log V_Q + \mathcal{T} \log \frac{p}{\mathcal{T}} , \tag{13b}$$

which agrees with the result of Chapter 11.

EXAMPLE.[1] *Effective force in superfluid helium.* Suppose that we add one atom of He⁴ to the superfluid component of liquid He II, while keeping the volume constant. The expression for the change in internal energy of the system is

$$dU = \mathcal{T} \, d\sigma + \mu \, dN - p \, dV . \tag{14a}$$

---

[1] This example may be deferred on the first reading of the text.

But $dV = 0$ and the entropy of the superfluid component is zero, so that $d\sigma = 0$. Thus for one atom

$$\Delta U = \mu \ . \tag{14b}$$

This result tells us that the chemical potential μ of liquid helium is the effective potential energy for the motion of the superfluid component. The acceleration equation of mechanics therefore assumes the form

$$M\frac{dv_s}{dt} = -\operatorname{grad}\mu \ , \tag{14c}$$

where v_s is the velocity of the superfluid and M is the mass of an atom of He[4]. This equation is used in the two-fluid model of liquid He II.

THERMODYNAMIC POTENTIAL A MINIMUM IN EQUILIBRIUM

We now consider a system that is in thermal contact with a heat reservoir (1) at temperature T and in mechanical contact with a pressure reservoir (2) which maintains the pressure p, as in Fig. 1. The total energy of the system plus the two reservoirs is assumed to be constant:

$$dU_s + dU_{r1} + dU_{r2} = 0 \ . \tag{15}$$

The total entropy of the system plus the heat reservoir in the most probable configuration is a maximum:

$$d\sigma_s + d\sigma_{r1} = 0 \ . \tag{16}$$

We do not include the pressure reservoir in the entropy condition because the pressure reservoir is not in thermal contact either with the system or with the heat reservoir. From the thermodynamic identity and (16) we have

$$dU_{r1} = T\,d\sigma_{r1} = -T\,d\sigma_s \ , \tag{17}$$

in a reversible change.

The pressure reservoir maintains a constant pressure on the system by displacement of the plunger that separates the system from the pressure reservoir. A decrease of volume of the pressure reservoir causes an equal increase in volume of the system:

$$dV_s = -dV_{r2} \ . \tag{18}$$

The pressure reservoir is insulated, so that its entropy does not change when the volume is changed. Thus

$$dU_{r2} = -p\,dV_{r2} = p\,dV_s \ . \tag{19}$$

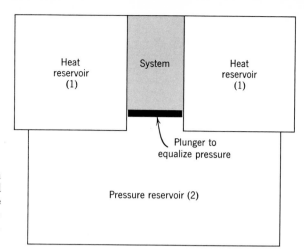

Figure 1 A system in thermal equilibrium with a heat reservoir and in mechanical equilibrium with a barystat or pressure reservoir which maintains a constant pressure on the system. The barystat is thermally insulated.

We substitute (17) and (19) in (15) to obtain

$$0 = dU_s - T\, d\sigma_s + p\, dV_s = d(U_s - T\sigma_s + pV_s) = dG_s \ . \qquad (20)$$

Thus at constant temperature and pressure the thermodynamic potential is an extremum in the most probable configuration of the system.

To show that the extremum of G_s is a minimum, we must consider finite changes $\Delta\sigma_s$ and $\Delta\sigma_{r1}$. Because the entropy is a maximum in equilibrium, we have

$$\Delta\sigma_s + \Delta\sigma_{r1} \le 0\ ; \qquad \Delta\sigma_{r1} \le -\Delta\sigma_s\ , \qquad (21)$$

in place of (16). Equation (20) for finite changes reads

$$\Delta U_s - T\, \Delta\sigma_s + p\, \Delta V_s \ge 0\ , \qquad (22)$$

so that

$$\Delta G_s \ge 0\ . \qquad (23)$$

Thus at constant temperature and pressure the thermodynamic potential increases for any departure from the most probable condition of the system.

GRAND POTENTIAL

What thermodynamic function has the same relation to the grand sum $\mathfrak{Z}$ as the free energy has to the partition function Z? We found in Chapter 18 that

$$F = -T \log Z\ . \qquad (24)$$

By analogy we introduce a function Ω defined by

$$\mathfrak{Z} \equiv e^{-\Omega/T}\ ; \qquad \text{or} \qquad \Omega \equiv -T \log \mathfrak{Z}\ , \qquad (25)$$

where $\mathfrak{Z}$ is the grand sum. The symbol Ω is the Greek letter capital omega. We may call Ω as defined by (25) the **grand potential**.

The grand potential has interesting properties. We first prove that the definition (25) is consistent with

$$\Omega = U - T\sigma - N\mu \ . \tag{26}$$

We form the differential

$$d\Omega = dU - T\,d\sigma - \sigma\,dT - \mu\,dN - N\,d\mu \ , \tag{27}$$

and use the thermodynamic identity $T\,d\sigma = dU + p\,dV - \mu\,dN$ to obtain

$$d\Omega = -\sigma\,dT - p\,dV - N\,d\mu \ , \tag{28}$$

for a reversible process. Thus Ω is a function of T, V, μ. From (28) we have

$$\left(\frac{\partial\Omega}{\partial T}\right)_{V,\,\mu} = -\sigma \ ; \qquad \left(\frac{\partial\Omega}{\partial V}\right)_{T,\,\mu} = -p \ ; \qquad \left(\frac{\partial\Omega}{\partial\mu}\right)_{T,\,V} = -N \ . \tag{29}$$

To prove that $\Omega = U - T\sigma - N\mu$, we use the definition of the grand sum

$$\mathcal{Z} = \sum_N \sum_l e^{(N\mu - \epsilon_l)/T}$$

and differentiate $\Omega = -T \log \mathcal{Z}$ to obtain

$$\frac{\partial\Omega}{\partial T} = \frac{\partial}{\partial T}(-T \log \mathcal{Z}) = -\log \mathcal{Z} + \frac{1}{T}(N\mu - U) = \frac{\Omega + N\mu - U}{T} \ . \tag{30}$$

The left-hand side is equal to $-\sigma$, by (29), while by (26) the right-hand side is also equal to $-\sigma$. Thus (25) and (26) are consistent; that they are equal follows directly on substitution of the Gibbs factor into the form (7.50) of the Boltzmann definition of the entropy, with use of (9).

Because

$$G = U - T\sigma + pV = N\mu \tag{31}$$

by (1) and (9), we may write the grand potential (26) as

$$\boxed{\Omega = -pV = -T \log \mathcal{Z} \ .} \tag{32}$$

This simple relation is useful in the theory of real gases. A real or imperfect gas is a nonideal gas; it may be degenerate and the atoms may interact with each other.

The generalized potential Ξ, Greek capital xi, is

$$\Xi(T, p, N) \equiv \sum_s \sum_l e^{-(\epsilon_l + pV_s)/T} \ ; \tag{33}$$

it has the interesting property that $G = -T \log \Xi$. To prove this, form $d(\log \Xi)$ with T, p, N as independent variables; use $\mu = \partial U/\partial N$; and compare with the differential of $-G/T$, using the thermodynamic identity.

Problem 1. Equation of state of a Fermi gas. (a) Show that for a Fermi gas

$$pV = T \int_0^\infty d\epsilon \, \mathcal{D}(\epsilon) \log \left(1 + e^{(\mu - \epsilon)/T}\right) , \tag{34}$$

where $\mathcal{D}(\epsilon)$ is the number of orbitals per unit energy range. (b) Use the explicit form of $\mathcal{D}(\epsilon)$ for free particles and integrate (34) by parts to obtain

$$pV = \tfrac{2}{3} \int_0^\infty d\epsilon \, \epsilon \, \mathcal{D}(\epsilon) \frac{1}{e^{(\epsilon - \mu)/T} + 1} = \tfrac{2}{3} U , \tag{35}$$

in agreement with the general result given in Problem 11.4.

Problem 2. Equation of state of a noninteracting lattice gas. (a) For the lattice gas of Appendix B, show that

$$\mathfrak{Z} = (1 + \lambda)^{N_0} , \tag{36}$$

where N_0 is the number of sites, and λ is determined in terms of the number of atoms N by the relation

$$\frac{N}{N_0} = \frac{\lambda}{1 + \lambda} . \tag{37}$$

(b) Show that

$$pV = N_0 T \log \frac{N_0}{N_0 - N} . \tag{38}$$

(c) Show that for $N \ll N_0$ this reduces to the form of the ideal gas law,

$$pV \cong NT , \tag{39}$$

in agreement with Appendix B.

HEAT FUNCTION

The **heat function** H is defined as

$$\boxed{H \equiv U + pV .} \tag{40}$$

This function is also called the **enthalpy** or the **heat content**. The use of H as a symbol for the heat function should not be confused with the magnetic field or with the hamiltonian.

The differential dH is

$$dH = dU + p \, dV + V \, dp . \tag{41}$$

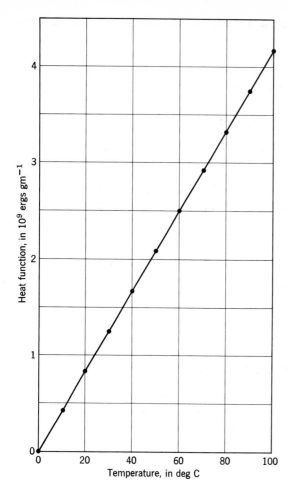

Figure 2   Heat function of water (air-saturated) at a pressure of one atmosphere, referred to 0 C. The dependence on T is approximately linear because C_p is independent of temperature over this range.

With the thermodynamic identity

$$T \, d\sigma = dU - \mu \, dN + p \, dV \,, \tag{42}$$

we have

$$dH = T \, d\sigma + V \, dp + \mu \, dN \,. \tag{43}$$

If heat is added reversibly to a system at constant pressure $(dp = 0)$ and constant composition $(dN = 0)$, then

$$dH = T \, d\sigma = DQ \,. \tag{44}$$

The change in H under these conditions is equal to the heat added. This is the origin of the name heat function for the function $H = U + pV$.

From (43) we see that

$$\left(\frac{\partial H}{\partial \sigma} \right)_{p, N} = T \,; \qquad \left(\frac{\partial H}{\partial N} \right)_{\sigma, p} = \mu \,; \qquad \left(\frac{\partial H}{\partial p} \right)_{\sigma, N} = V \,. \tag{45}$$

Observe that, from (43),

$$\left(\frac{\partial H}{\partial T}\right)_{p,\,N} = T\left(\frac{\partial \sigma}{\partial T}\right)_{p,\,N} ,\qquad (46)$$

so that the heat capacity at constant pressure is given by

$$C_p \equiv T\left(\frac{\partial S}{\partial T}\right)_{p,\,N} = \left(\frac{\partial H}{\partial T}\right)_{p,\,N} . \qquad (47)$$

Experimentally it is usually simpler to measure C_p than C_V. Thus chemists often provide us with tables of the heat function as obtained from the integration of C_p in (47):

$$\boxed{\; H(T) = H(0) + \int_0^T dT\, C_p(T) \; .} \qquad (48)$$

Results for liquid water are plotted in Fig. 2.

EXAMPLE. *General relation between C_V and C_p.* It is convenient to express $C_p - C_V$ in terms of $(\partial V/\partial T)_p$ and $(\partial V/\partial p)_T$, for these derivatives are easily accessible to direct measurement. We found earlier in Chapter 11 the result $C_p - C_V = R$ for one mole of an ideal gas. The general result below is a triumph of the manipulation of thermodynamic relations.

We consider the entropy S as a function of the temperature T and the pressure p. We take the differential

$$dS = \left(\frac{\partial S}{\partial T}\right)_p dT + \left(\frac{\partial S}{\partial p}\right)_T dp \;. \qquad (49)$$

We form $(\partial S/\partial T)_V$:

$$\left(\frac{\partial S}{\partial T}\right)_V = \left(\frac{\partial S}{\partial T}\right)_p + \left(\frac{\partial S}{\partial p}\right)_T \left(\frac{\partial p}{\partial T}\right)_V , \qquad (50)$$

or, after multiplication by T,

$$C_V = C_p + T \left(\frac{\partial S}{\partial p}\right)_T \left(\frac{\partial p}{\partial T}\right)_V . \qquad (51)$$

By the Maxwell relation $(\partial S/\partial p)_T = -(\partial V/\partial T)_p$ as derived in (11) we have

$$C_V = C_p - T \left(\frac{\partial V}{\partial T}\right)_p \left(\frac{\partial p}{\partial T}\right)_V . \qquad (52)$$

To proceed further we form

$$dV = \left(\frac{\partial V}{\partial T}\right)_p dT + \left(\frac{\partial V}{\partial p}\right)_T dp , \qquad (53)$$

whence for a process at constant volume

$$0 = \left(\frac{\partial V}{\partial T}\right)_p + \left(\frac{\partial V}{\partial p}\right)_T \left(\frac{\partial p}{\partial T}\right)_V . \qquad (54)$$

We rearrange (54) to obtain

$$\left(\frac{\partial p}{\partial T}\right)_V = - \frac{\left(\dfrac{\partial V}{\partial T}\right)_p}{\left(\dfrac{\partial V}{\partial p}\right)_T} . \qquad (55)$$

Now combine (52) and (55):

$$C_V = C_p + T\frac{\left(\dfrac{\partial V}{\partial T}\right)_p^2}{\left(\dfrac{\partial V}{\partial p}\right)_T} . \qquad (56)$$

This important relation can be expressed in terms of the **thermal expansivity**

$$\alpha \equiv \frac{1}{V}\left(\frac{\partial V}{\partial T}\right)_p \qquad (57)$$

and the **isothermal compressibility**

$$K_T \equiv -\frac{1}{V}\left(\frac{\partial V}{\partial p}\right)_T . \qquad (58)$$

Thus the difference of the heat capacities is

$$\boxed{C_p - C_V = \frac{TV\alpha^2}{K_T} ,} \qquad (59)$$

where C_p and C_V refer to a volume V of matter.

We know from (12) that $\alpha \to 0$ as $T \to 0$, whereas the compressibility of a real substance approaches a finite limit. Therefore $C_p \to C_V$ as $T \to 0$.

The ratio C_p/C_V of the heat capacity at constant pressure to the heat capacity at constant volume is denoted by γ, the Greek letter gamma. From (59) we have

$$\gamma \equiv \frac{C_p}{C_V} = 1 + \frac{TV\alpha^2}{C_V K_T} . \qquad (60)$$

Summary of Useful Thermodynamic Relations

(a) Given the entropy $\sigma(U, N, V)$:

$$\frac{1}{T} = \left(\frac{\partial \sigma}{\partial U}\right)_{N, V} \;; \qquad -\frac{\mu}{T} = \left(\frac{\partial \sigma}{\partial N}\right)_{U, V} \;; \qquad \frac{p}{T} = \left(\frac{\partial \sigma}{\partial V}\right)_{U, N} .$$

(b) Given the energy $U(\sigma, V, N)$:

$$T = \left(\frac{\partial U}{\partial \sigma}\right)_{V, N} \;; \qquad -p = \left(\frac{\partial U}{\partial V}\right)_{\sigma, N} \;; \qquad \mu = \left(\frac{\partial U}{\partial N}\right)_{\sigma, V} .$$

(c) Given the energy $U(T, V, N)$:

$$\sigma = \int_0^U \frac{dU}{T} \;; \qquad C_V = \left(\frac{\partial U}{\partial T}\right)_{V, N} = k_B \left(\frac{\partial U}{\partial T}\right)_{V, N} .$$

(d) Given the chemical potential $\mu(T, V, N)$:

$$\sigma = -\int_0^N dN \cdot \frac{\mu}{T} \;.$$

(e) Given the partition function Z or the free energy $F(T, V, N) = U - T\sigma = -T \log Z$:

$$\sigma = -\left(\frac{\partial F}{\partial T}\right)_{V, N} = T\frac{\partial}{\partial T} \log Z + \log Z \;;$$

$$p = -\frac{\partial F}{\partial V} = T\frac{\partial}{\partial V} \log Z \;;$$

$$U = -T^2 \frac{\partial}{\partial T} \frac{F}{T} = T^2 \frac{\partial}{\partial T} \log Z \;;$$

$$\mu = \left(\frac{\partial F}{\partial N}\right)_{T, V} = -T\frac{\partial}{\partial N} \log Z \;.$$

(f) Given the grand sum $\mathfrak{Z}(T, \mu, V)$:

$$\sigma = \frac{\partial}{\partial T}(T \log \mathfrak{Z}) \;; \qquad \text{[From (19.30)]}$$

$$pV = T \log \mathfrak{Z} \;;$$

$$N = T\frac{\partial}{\partial \mu} \log \mathfrak{Z} = \lambda\frac{\partial}{\partial \lambda} \log \mathfrak{Z} \;.$$

(g) Given the thermodynamic potential $G(T, p, N) = U - T\sigma + pV$:

$$\sigma = -\left(\frac{\partial G}{\partial T}\right)_{p,N} \quad ; \qquad V = \left(\frac{\partial G}{\partial p}\right)_{T,N} \quad ; \qquad \mu = \left(\frac{\partial G}{\partial N}\right)_{T,p} \quad ; \qquad G = N\mu \ .$$

(h) Given the heat function $H(\sigma, p, N)$:

$$T = \left(\frac{\partial H}{\partial \sigma}\right)_{p,N} \quad ; \qquad V = \left(\frac{\partial H}{\partial p}\right)_{\sigma,N} \quad ; \qquad \mu = \left(\frac{\partial H}{\partial N}\right)_{\sigma,p} \quad ;$$

$$C_p = \left(\frac{\partial H}{\partial T}\right)_{p,N} = k_B \left(\frac{\partial H}{\partial T}\right)_{p,N} \ .$$

CHAPTER **20**

Vapor Pressure Equation

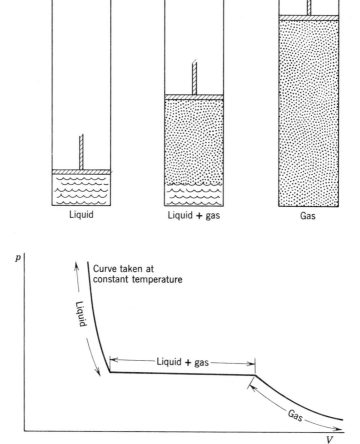

Figure 1 Pressure-volume isotherm of a real gas at a temperature such that liquid and gas phases may coexist, that is, $T < T_c$. In the two-phase region of liquid + gas the pressure is constant, but the volume may change. At a given temperature there is only a single value of the pressure for which a liquid and its vapor are in equilibrium. If at this pressure we move the piston down, some of the gas is condensed to liquid, but the pressure remains unchanged as long as any gas remains.

ISOTHERMS

The curve of pressure versus volume for a quantity of matter held at constant temperature is determined by the thermodynamic properties of the substance. Such a curve is called an **isotherm.** We are concerned in this chapter with the isotherms of a real gas in which the atoms or molecules interact with one another and under appropriate conditions can be bound together in a liquid or solid phase. A **phase** is a portion of a system that is uniform and has a definite boundary.

An isotherm of a real gas may show a region in the p-V plane in which liquid and gas coexist in equilibrium with each other. As in Fig. 1, a part of the volume contains atoms in the gas or vapor phase. Vapor is a term used for a gas when the gas is in equilibrium with its liquid or solid form.

There are isotherms at low temperatures for which solid and gas coexist. Everything we say for the liquid-gas equilibrium holds also for the solid-gas equilibrium and the liquid-solid equilibrium.

Liquid and vapor may coexist on a section of an isotherm only if the temperature of the isotherm lies below some **critical temperature,** T_c. On an isotherm above the critical temperature only a single phase, the fluid phase, exists, no matter how great the pressure. (There is no more reason to call it a gas than a liquid, so we call it a fluid.) Values of the critical temperature for representative gases are given in Table 1.

Liquid and gas will never coexist along the entire extent of an isotherm from zero volume to infinite volume, but at most only along a section. For a fixed temperature and number of atoms, there will be a volume above which all atoms present are in the gas phase. A small drop of water placed in an evacuated sealed bell jar at room temperature will evaporate entirely, leaving the bell jar filled with H_2O gas at some low pressure. A drop of water exposed to normal air in a room will evaporate entirely. There will be a volume, however, below which the atoms from the vapor will all be squeezed into the liquid state. The volume relations are suggested by Fig. 1.

Table 1   Critical Temperatures of Gases

	T_c, in deg K		T_c, in deg K
He	5.2	H_2	33.2
Ne	44.4	N_2	126.0
Ar	151	O_2	154.3
Kr	210	H_2O	647.1
Xe	289.7	CO_2	304.2

PHASE EQUILIBRIA

The thermodynamic conditions for the coexistence of two phases are the conditions for the equilibrium of two systems that are in thermal, diffusive, and mechanical contact. These conditions are that $T_1 = T_2$; $\mu_1 = \mu_2$; $p_1 = p_2$, or

$$T_l = T_g \; ; \qquad \mu_l = \mu_g \; ; \qquad p_l = p_g \; , \tag{1}$$

where the subscripts l and g denote respectively the liquid and vapor phases.

The most interesting of these conditions is that the chemical potentials of the two phases be equal if the phases coexist. The values of the chemical potentials are taken at the common pressure and common temperature of the liquid and gas, so that

$$\boxed{\mu_l(p, T) = \mu_g(p, T) \; .} \tag{2}$$

At a general point in the p-T plane at which the two phases do not coexist we do not have the equality (2): if $\mu_l < \mu_g$ the liquid phase alone is stable, and if $\mu_g < \mu_l$ the gas phase alone is stable. We should bear in mind that metastable phases may occur, as in supercooling and superheating.

DERIVATION OF THE COEXISTENCE CURVE, p VERSUS T

Let p_0 be the pressure for which two phases, liquid and gas, coexist at the temperature T_0. Suppose that the two phases also coexist at the nearby point $p_0 + dp$; $T_0 + dT$. The curve in the p, T plane along which the two phases coexist divides the p, T plane into a phase diagram, as given in Fig. 2 for H_2O. It is a condition of coexistence that

$$\mu_g(p_0, T_0) = \mu_l(p_0, T_0) \; , \tag{3}$$

and also that

$$\mu_g(p_0 + dp, T_0 + dT) = \mu_l(p_0 + dp, T_0 + dT) \; . \tag{4}$$

Equations (3) and (4) give a relationship between dp and dT.

We make a series expansion of each side of (4) to obtain

$$\mu_g(p_0, T_0) + \left(\frac{\partial \mu_g}{\partial p}\right)_T dp + \left(\frac{\partial \mu_g}{\partial T}\right)_p dT + \cdots = \mu_l(p_0, T_0) + \left(\frac{\partial \mu_l}{\partial p}\right)_T dp$$

$$+ \left(\frac{\partial \mu_l}{\partial T}\right)_p dT + \cdots \; . \tag{5}$$

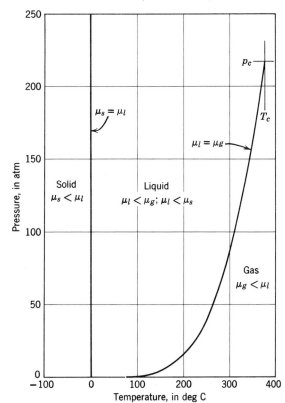

Figure 2   Phase diagram of H_2O. The relationships of the chemical potentials μ_s, μ_l, and μ_g in the solid, liquid, and gas phases are shown. The phase boundary here between ice and water is not exactly vertical; the slope is actually negative, although very large. [After *International Critical Tables*, Vol. 3, and P. W. Bridgman, Proceedings of the American Academy of Sciences **47**, 441 (1912); for the several forms of ice, see Zemansky, p. 375.]

We use (3) and (5) to obtain, in the limit as dp and dT approach zero,

$$\left(\frac{\partial\mu_g}{\partial p}\right)_T dp + \left(\frac{\partial\mu_g}{\partial T}\right)_p dT = \left(\frac{\partial\mu_l}{\partial p}\right)_T dp + \left(\frac{\partial\mu_l}{\partial T}\right)_p dT . \qquad (6)$$

This result may be rearranged to give

$$\frac{dp}{dT} = \frac{\left(\frac{\partial\mu_l}{\partial T}\right)_p - \left(\frac{\partial\mu_g}{\partial T}\right)_p}{\left(\frac{\partial\mu_g}{\partial p}\right)_T - \left(\frac{\partial\mu_l}{\partial p}\right)_T} , \qquad (7)$$

which is the differential equation of the coexistence curve or **vapor pressure curve.**

The derivatives of the chemical potential which occur in (7) may be expressed in terms of quantities readily accessible to measurement. In the treatment of the thermodynamic potential in Chapter 19 we found the relations

$$G = N\mu(p, T) ; \qquad \left(\frac{\partial G}{\partial p}\right)_{N,T} = V ; \qquad \left(\frac{\partial G}{\partial T}\right)_{N,p} = -\sigma . \qquad (8)$$

With the definitions

$$v \equiv \frac{V}{N} \; ; \qquad s \equiv \frac{S}{N} \; , \tag{9}$$

we have

$$\frac{1}{N}\left(\frac{\partial G}{\partial p}\right)_{N,\,T} = \frac{V}{N} = v = \left(\frac{\partial \mu}{\partial p}\right)_{T} \; ; \qquad \frac{1}{N}\left(\frac{\partial G}{\partial T}\right)_{N,\,p} = -\frac{S}{N} = -s = \left(\frac{\partial \mu}{\partial T}\right)_{p} \; . \tag{10}$$

We recall that $\mathcal{T} = k_B T$ and $S = k_B \sigma$. Then (7) for dp/dT becomes

$$\boxed{\frac{dp}{dT} = \frac{s_g - s_l}{v_g - v_l} \; .} \tag{11}$$

Here $s_g - s_l$ is the increase of entropy of the system when we transfer one molecule from the liquid to the gas, and $v_g - v_l$ is the increase of volume of the system when we transfer one molecule from the liquid to the gas.

It is essential to understand that the derivative dp/dT in (11) is not simply taken from the equation of state of the gas. The derivative is not to be calculated for $pV = Nk_B T$ or for any modification of this equation of state. Instead, the derivative refers to the very special change of p and T in which the gas and liquid continue to coexist. The number of atoms in each phase will vary as the volume is varied, subject only to $N_l + N_g = N$, a constant. Here N_l and N_g are the numbers of atoms in the liquid and gas phases, respectively. ɪ

The quantity $s_g - s_l$ is related directly to the quantity of heat that must be added to the system to transfer one molecule quasistatically from the liquid to the gas, while keeping the temperature of the system constant. (If heat is not added to the system from outside in the process, the temperature will decrease when the molecule is transferred to the gas.) The quantity of heat added in the quasistatic transfer is

$$DQ = T(s_g - s_l) \; , \tag{12}$$

by virtue of the connection between heat and the change of entropy in a quasistatic process. The quantity

$$L \equiv T(s_g - s_l) \tag{13}$$

defines the **latent heat of vaporization,** a quantity is easily measured by elementary calorimetry.

We let

$$\Delta v = v_g - v_l \tag{14}$$

denote the change of volume when one molecule is transferred from the liquid to the gas. We combine (11), (13), and (14) to obtain

$$\boxed{\frac{dp}{dT} = \frac{L}{T\,\Delta v}}$$

(15)

This is known as the **Clausius-Clapeyron equation** or the **vapor pressure equation.** The derivation of this equation was a remarkable early accomplishment of the science of thermodynamics. Both sides of (15) are easily determined experimentally, and the equation has been verified to high precision.

We obtain a particularly useful form of (15) if we make two approximations:

(a) We assume that $v_g \gg v_l$: the volume occupied by an atom in the gas phase is very much larger than in the liquid (or solid) phases, so that we may replace Δv by v_g:

$$\Delta v \cong v_g = \frac{V_g}{N_g}\,.$$

(16)

At atmospheric pressure $v_g/v_l \approx 10^3$, so that the approximation is very good.

(b) We assume that the ideal gas law $pV_g = N_g k_B T$ applies to the gas phase, so that (16) may be written as

$$\Delta v \cong \frac{k_B T}{p}\,.$$

(17)

With these modest approximations the vapor pressure equation becomes

$$\frac{dp}{dT} = \frac{L}{k_B T^2}\,p\ ;\qquad \frac{d}{dT}\log p = \frac{L}{k_B T^2}\,,$$

(18)

where L is the latent heat per molecule. Given L as a function of temperature, this equation may be integrated to find the coexistence curve.

If, in addition, the latent heat L is independent of temperature over the temperature range of interest, we may take $L = L_0$ outside the integral. Thus when we integrate (18) we obtain

$$\int \frac{dp}{p} = \frac{L_0}{k_B}\int \frac{dT}{T^2}\,,$$

(19)

whence

$$\log p = -\frac{L_0}{k_B T} + \text{constant}\ ;\qquad p(T) = p_0 e^{-L_0/k_B T}\,,$$

(20)

where p_0 is a constant. We recall that we defined L_0 as the latent heat of

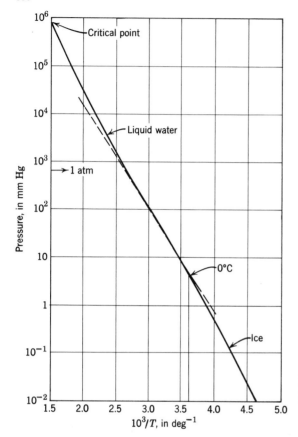

Figure 3 Vapor pressure of water and of ice plotted versus $1/T$. The vertical scale is logarithmic. The dashed line is a straight line.

evaporation of one molecule. If L_0 refers instead to one mole, then (20) becomes

$$p(T) = p_0 e^{-L_0/RT} ,\qquad (21)$$

where R is the gas constant, $R \equiv N_0 k_B$, as in (11.39).

The vapor pressure of water and of ice is plotted in Fig. 3 as $\log p$ versus $1/T$. The curve is linear over substantial regions, as predicted by the approximate result (20).

The vapor pressure of He^4 is plotted in Fig. 4. This vapor pressure curve is widely used in the measurement of temperatures between 1 and 5 K.

In Fig. 17.5 we showed the phase diagram of He^4 at low temperatures. Notice that the liquid-solid curve is closely horizontal below 1.4 K. From (11) we may then infer that the entropy of the liquid is very nearly equal to the entropy of the solid in this region, and there is scarcely any latent heat of fusion. It is rather remarkable that the entropies of the two phases should be so similar, because a normal liquid is significantly more disordered than

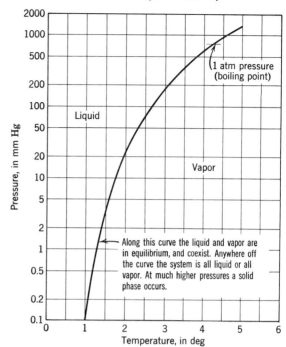

Figure 4 Vapor pressure versus temperature for He⁴. [After H. van Dijk et al., Journal of Research of the National Bureau of Standards **63A**, 12 (1959).]

Within the figure:
- Pressure, in mm Hg (vertical axis)
- Temperature, in deg (horizontal axis)
- Liquid
- Vapor
- 1 atm pressure (boiling point)
- Along this curve the liquid and vapor are in equilibrium, and coexist. Anywhere off the curve the system is all liquid or all vapor. At much higher pressures a solid phase occurs.

a solid. For He³ at low temperatures (Fig. 17.6) the slope of the liquid-solid curve is negative; because the volume of the solid is less than the volume of the liquid, it follows from (11) that in this region the entropy of the liquid is less than the entropy of the solid. The solid is more disordered than the liquid!

EXAMPLE. *Model system for gas-solid equilibrium.* We construct a simple model to describe a solid in equilibrium with a gas, as in Fig. 5. We can easily derive the vapor pressure curve for this model. We treat a solid rather than a liquid just because the model is simpler.

Imagine the solid to consist of N atoms, each bound as a harmonic oscillator of frequency ω to a fixed center of force. The binding energy of each atom in the ground state is ϵ_0; that is, the energy of an atom in its ground state is $-\epsilon_0$ referred to a free atom at rest. The energy states of a single oscillator are $n\hbar\omega - \epsilon_0$, where n is a positive integer or zero (Fig. 6). For the sake of simplicity we suppose that each atom can oscillate only in one dimension. The result for oscillators in three dimensions is left as a problem.

The partition function of a single oscillator in the solid is

$$Z_s = \sum_n e^{-(n\hbar\omega - \epsilon_0)/T} = e^{\epsilon_0/T} \sum_n e^{-n\hbar\omega/T} = \frac{e^{\epsilon_0/T}}{1 - e^{-\hbar\omega/T}} . \tag{22}$$

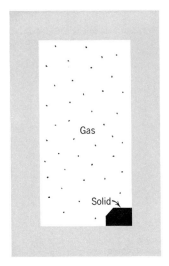

Figure 5 Atoms in a solid in equilibrium with atoms in the gas phase. The equilibrium pressure is a function of temperature. The energy of the atoms in the solid phase is lower than in the gas phase, but the entropy of the atoms tends to be higher in the gas phase. The equilibrium configuration is determined by the counterplay of the two effects. At low temperature most of the atoms are in the solid; at high temperature all or most of the atoms may be in the gas.

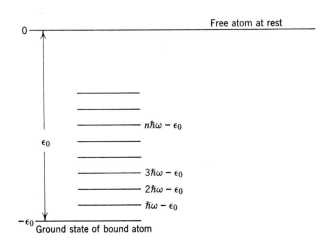

Figure 6 States of an atom bound as a harmonic oscillator of frequency ω. The ground state is assumed to be ϵ_0 below that of a free atom at rest in the gas phase.

The free energy F_s of a single oscillator in the solid is

$$F_s = U_s - T\sigma_s = -T \log Z_s . \tag{23}$$

The thermodynamic potential in the solid is, per atom,

$$G_s = U_s - T\sigma_s + pv_s = F_s + pv_s = \mu_s . \tag{24}$$

The pressure in the solid is equal to that of the gas with which it is in contact, but the volume v_s per atom in the solid phase is much smaller than the volume v_g per atom in the gas phase: $v_s \ll v_g$.

If we neglect the term $p v_s$ we have for the chemical potential of the solid $\mu_s \cong F_s$, whence the absolute activity is

$$\lambda_s \equiv e^{\mu_s/T} \cong e^{F_s/T} = e^{-\log Z_s} = \frac{1}{Z_s} = e^{-\epsilon_0/T}\left(1 - e^{-\hbar\omega/T}\right) . \tag{25}$$

We make the ideal gas approximation to describe the gas phase, and we take the spin of the atom to be zero. From the result of Chapter 11 we have

$$\lambda_g = \frac{N_g V_Q}{V} = \frac{p V_Q}{T} = \frac{p}{T}\left(\frac{2\pi\hbar^2}{MT}\right)^{\frac{3}{2}} . \tag{26}$$

The gas is in equilibrium with the solid when

$$\lambda_g = \lambda_s ,$$

or

$$\frac{p}{T}\left(\frac{2\pi\hbar^2}{MT}\right)^{\frac{3}{2}} = e^{-\epsilon_0/T}\left(1 - e^{-\hbar\omega/T}\right) . \tag{27}$$

We solve this for the vapor pressure as a function of the temperature:

$$p = \left(\frac{M}{2\pi\hbar^2}\right)^{\frac{3}{2}} T^{\frac{5}{2}} e^{-\epsilon_0/T}\left(1 - e^{-\hbar\omega/T}\right) . \tag{28}$$

Now consider two limits, low and high temperatures:

(a) For $T \ll \hbar\omega$ we neglect the term in $e^{-\hbar\omega/T}$. The vapor pressure law involves dp/dT. We form the derivative:

$$\frac{dp}{dT} = \left(\frac{5}{2T} + \frac{\epsilon_0}{T^2}\right)p = \frac{(\epsilon_0 + \frac{5}{2}T)}{T} \cdot \frac{p}{T} = \frac{(\epsilon_0 + \frac{5}{2}T)}{T} \cdot \frac{N_g}{V} , \tag{29}$$

so that in this limit the latent heat per atom is $\epsilon_0 + \frac{5}{2}T$.

(b) For $T \gg \hbar\omega$ we expand

$$\left(1 - e^{-\hbar\omega/T}\right) = \left[1 - \left(1 - \frac{\hbar\omega}{T} + \cdots\right)\right] \cong \frac{\hbar\omega}{T} . \tag{30}$$

Thus

$$p = \left(\frac{M}{2\pi}\right)^{\frac{3}{2}} \frac{T^{\frac{3}{2}}\omega}{\hbar^2} e^{-\epsilon_0/T} , \tag{31}$$

and

$$\frac{dp}{dT} = \frac{(\epsilon_0 + \frac{3}{2}T)}{T} \cdot \frac{N_g}{V} . \tag{32}$$

Problem 1. *Gas-solid equilibrium.* Consider now a more realistic version of the preceding example: we let the oscillators in the solid move in three dimensions. (a) Show that in the high temperature regime ($T \gg \hbar\omega$) the vapor pressure is

$$p \cong \left(\frac{M}{2\pi}\right)^{\frac{3}{2}} \frac{\omega^3}{T^{\frac{1}{2}}} e^{-\epsilon_0/T} . \tag{33}$$

(b) Show that the latent heat per atom is $\epsilon_0 - \frac{1}{2}T$.

Problem 2. *Calculation of dp/dT for water.* Calculate from the vapor pressure equation the value of dT/dp near $p = 1$ atm for the liquid-vapor equilibrium of water. The heat of vaporization at $100\,C$ is given in handbooks as 539.5 cal/gm. Express the result in deg/atm.

Problem 3. *Heat of vaporization of ice.* The pressure of water vapor over ice is 3.88 mm Hg at $-2\,C$ and 4.58 mm Hg at $0\,C$. Estimate in J mol^{-1} the heat of vaporization of ice at $-1\,C$.

Problem 4. *Critical point of the van der Waals equation.* The van der Waals equation of state was introduced in Problem 12.2. When written for n moles of molecules the equation is

$$\left[p + a\left(\frac{n}{V}\right)^2\right](V - nb) = nRT . \tag{34}$$

We define the quantities

$$p_c = \frac{a}{27b^2} ; \qquad V_c = 3nb ; \qquad RT_c = \frac{8a}{27b} . \tag{35}$$

(a) Show that the van der Waals equation becomes

$$\left(\frac{p}{p_c} + \frac{3}{(V/V_c)^2}\right)\left(\frac{V}{V_c} - \frac{1}{3}\right) = \frac{8T}{3T_c} . \tag{36}$$

This equation is plotted in Fig. 7 for several temperatures near the critical temperature. The equation may be written in terms of the dimensionless variables

$$\hat{p} \equiv \frac{p}{p_c} ; \qquad \hat{V} \equiv \frac{V}{V_c} ; \qquad \hat{T} \equiv \frac{T}{T_c} , \tag{37}$$

as

$$\left(\hat{p} + \frac{3}{\hat{V}^2}\right)(\hat{V} - \frac{1}{3}) = \frac{8}{3}\hat{T} . \tag{38}$$

This result is known as the law of corresponding states.

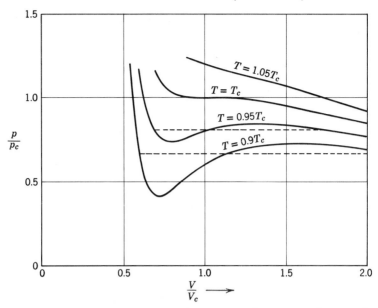

Figure 7   The van der Waals equation of state near the critical temperature. (Courtesy of R. Cahn.)

(b) At the critical point the curve of $\hat{p}$ versus $\hat{V}$ at constant $\hat{T}$ has a point of inflection, for at this point the local maximum and minimum coincide. At a point of inflection

$$\left(\frac{\partial \hat{p}}{\partial \hat{V}}\right)_{\hat{T}} = 0 \; ; \qquad \left(\frac{\partial^2 \hat{p}}{\partial \hat{V}^2}\right)_{\hat{T}} = 0 \; . \tag{39}$$

Show that these conditions are satisfied if

$$\hat{p} = 1 \; ; \qquad \hat{V} = 1 \; ; \qquad \hat{T} = 1 \; . \tag{40}$$

[(c) follows on page 337]

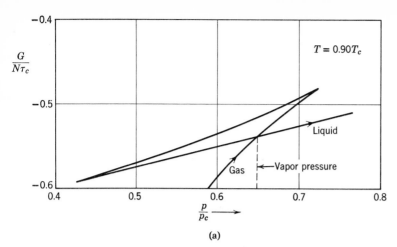

Figure 8a Thermodynamic potential versus pressure for van der Waals equation of state: $T = 0.90T_c$. (Courtesy of R. Cahn.)

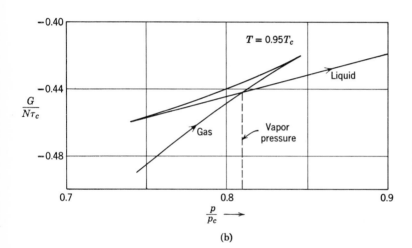

Figure 8b Thermodynamic potential versus pressure for van der Waals equation of state: $T = 0.95T_c$.

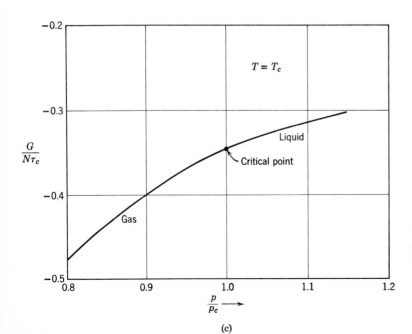

Figure 8c Thermodynamic potential versus pressure for van der Waals equation of state: $T = T_c$.

(c) Show that the thermodynamic potential of a van der Waals gas may be written as (*Hint:* first find F):

$$G = \frac{nRTV}{V - nb} - \frac{2n^2a}{V} - nRT \log (V - nb) + f(T) , \qquad (41)$$

where $f(T)$ is a function of temperature alone. This result cannot conveniently be put into analytic form as a function of pressure and temperature. By numerical methods we obtain the curves of Fig. 8. For $T \geq T_c$ there is only a single value of the thermodynamic potential at each value of the pressure. For $T < T_c$ there exist three values of the thermodynamic potential over a certain range of the pressure. The lowest value represents the stable state. The other branches represent unstable states. The pressure at which the curves cross gives the transition between the gas and the liquid at this temperature. This pressure is the vapor pressure.

REFERENCES

P. W. Bridgman, "Water, in the liquid and five solid forms, under pressure," Proceedings of the American Academy of Arts and Sciences **47**, 441–558 (1912).

P. W. Bridgman, "The phase diagram of water to 45,000 kg/cm²," Journal of Chemical Physics **5**, 964–986 (1937).

B. M. Abraham, D. W. Osborne, and B. Weinstock, "The vapor pressure, critical point, heat of vaporization, and entropy of liquid He³," Physical Review **80**, 366–371 (1950).

S. G. Sydorvak and T. R. Roberts, "Thermodynamic properties of liquid helium three. Vapor pressures below 1°K," Physical Review **l06**, 175–182 (1957).

H. N. V. Temperly, *Changes of state*, Cleaver-Hume Press, London, 1956, Chapter 2. A good general discussion of phase transformations, although there have been many important developments since 1956.

CHAPTER 21

Equilibrium in Reactions

Men who work with thermodynamics mostly use it to predict the equilibrium concentrations among particles which react with one another, or they may use the concentration data to determine the energies of formation of the reactants. The reactants may be atoms, molecules, electrons, ions, or nuclei. The fields in which this type of scientific activity is important include low temperature physics, astrophysics, geophysics, molecular biology, biochemistry, and chemistry. In this chapter we first discuss a simple class of reactions, and then we develop a general theory of the condition for equilibrium among reacting species.

ADSORPTION OF ATOMS ON SITES: LANGMUIR ISOTHERM

We consider an ideal gas in contact with a surface that contains independent sites on each of which a single atom of the gas may be adsorbed or bound. If you think about it, you will see that the adsorption of an atom by a site is a primitive form of a chemical reaction. The problem is to find the fraction of surface sites occupied by atoms, as a function of the concentration of atoms in the gas phase. We neglect any interaction between sites. The problem is relevant to some important biochemical reactions.

Because the surface sites are independent, to calculate the average occupancy it suffices to consider a single site. The grand partition function of a single surface site (Fig. 1) is

$$\mathfrak{Z} = 1 + \lambda e^{-\epsilon/\tau} , \tag{1}$$

where ϵ is the energy of an adsorbed atom relative to an atom at infinite separation from the site. If energy must be added to remove the atom from the site, ϵ will be negative. The first term in (1) arises from occupancy zero; the second term arises from single occupancy of the site. We assume that these are the only possibilities.

Vacant lattice site

Figure 1 Adsorption of an atom by a lattice site, where ϵ is the energy of an adsorbed atom relative to an atom at infinite separation from the site. If energy must be supplied to detach the atom from the site, then ϵ will be a negative number.

Lattice site with
one adsorbed atom

The atoms on the surface are in equilibrium with the atoms in the gas, so that the chemical potentials are equal for the surface and the gas:

$$\mu(\text{surface}) = \mu(\text{gas}) \; ; \qquad \lambda(\text{surface}) = \lambda(\text{gas}) \qquad (2)$$

where $\lambda \equiv e^{\mu/T}$. From Chapter 11 we find the value of λ for the gas in terms of the gas pressure by the relation

$$\lambda = \frac{N V_Q}{V} = \frac{p V_Q}{T} \; , \qquad (3)$$

for an ideal monatomic gas of atoms of zero spin.[1] Here V_Q is the quantum volume. At constant temperature $\lambda(\text{gas})$ is directly proportional to the pressure p, for an ideal gas.

The fraction f of occupied surface sites is found from (1) to be

$$f = \frac{\lambda e^{-\epsilon/T}}{1 + \lambda e^{-\epsilon/T}} = \frac{1}{\lambda^{-1} e^{\epsilon/T} + 1} \; , \qquad (4)$$

which is the same as the Fermi-Dirac distribution function. We substitute (3) in (4) to obtain

$$f = \frac{1}{\left(\dfrac{T e^{\epsilon/T}}{p V_Q}\right) + 1} = \frac{p}{\left(\dfrac{T e^{\epsilon/T}}{V_Q}\right) + p} \; , \qquad (5)$$

or, with $p_0 \equiv (T/V_Q)e^{\epsilon/T}$,

$$f = \frac{p}{p_0 + p} \; , \qquad (6)$$

where p_0 is a constant with respect to pressure, but depends on the temperature. The result (6) is known to physicists as the **Langmuir adsorption isotherm,** Fig. 2. It was developed to describe the adsorption of gases on the surfaces of solids. At low pressures the adsorption is directly proportional to the pressure,

---

[1] If the molecules have spin and also have what we called internal degrees of freedom, then by (11.96a) we replace (3) with

$$\lambda = \frac{N V_Q}{V} \cdot e^{F(\text{int})/T} \; , \qquad (3a)$$

where the free energy of the internal degrees of freedom is related to the internal partition function by

$$F(\text{int}) = -T \log Z(\text{int}) \; . \qquad (3b)$$

For purposes of this chapter we redefine what we mean by internal degrees of freedom to include the spin degeneracy $2I + 1$ in the internal partition function.

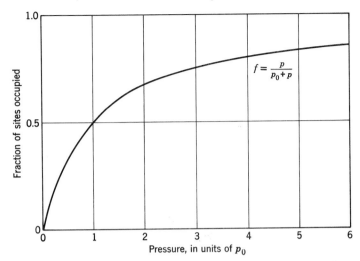

Figure 2 Reaction of atoms with adsorption sites, according to the Langmuir adsorption isotherm.

but the adsorption saturates at high pressures such that $p \gg p_0$, for here most of the sites are occupied.

We may use (6) to express the fraction of occupied sites in terms of the concentration $c \equiv N/V$ of atoms in the gas or liquid phase.

$$f = \frac{c}{c_0 + c} \; ; \qquad c_0 \equiv \frac{e^{\epsilon/\tau}}{V_Q} \; , \tag{7}$$

where c_0 is a constant independent of concentration. Equation (7) is closely related to the Michaelis-Menten equation[2] used in biochemical kinetics.

ADSORPTION OF OXYGEN

The result (6) or (7) describes important chemical and biological processes. A good biochemical example is the reaction:

$$\text{myoglobin} + \text{oxygen} \rightleftharpoons \text{oxymyoglobin} \; ;$$

this may be written as

$$\text{Mb} + \text{O}_2 \rightleftharpoons \text{MbO}_2 \; . \tag{8}$$

[2] L. Michaelis and M. L. Menten, "Die Kinetik der Invertinwirkung", Biochemische Zeitschrift **49**, 333–369 (1913); see also C. Tanford, *Physical chemistry of macromolecules*, Wiley, 1961, pp. 641–645.

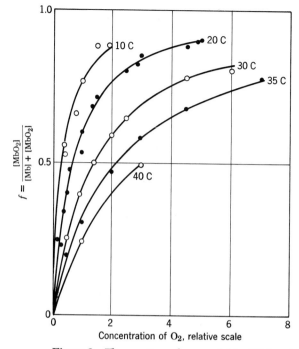

Figure 3 The reaction of a myoglobin (Mb) molecule with oxygen may be viewed as an example of the adsorption of a molecule of O_2 at a site on the large myoglobin molecule. The results follow a Langmuir isotherm quite accurately. Each myoglobin molecule can adsorb one O_2 molecule. These curves show the fraction of myoglobin with adsorbed O_2 as a function of the partial pressure of O_2. The curves are for human myoglobin in solution. Myoglobin is found in muscles; it is responsible for the color of steak. The temperatures are in degrees centigrade. [After A. Rossi-Fanelli and E. Antonini, Archives of Biochemistry and Biophysics **77**, 478 (1958).]

Myoglobin is an important protein of molecular weight about 17,000. The structure of the molecule is known. For our purposes all we need to know is that each molecule of Mb can bind one molecule of oxygen in the form of molecular oxygen. The reaction (8) is studied in water solution.

We are interested in the fraction of myoglobin molecules that have bound an oxygen molecule. Let

$$[Mb] = \text{concentration of myoglobin} \ ;$$
$$[O_2] = \text{concentration of oxygen} \ ;$$
$$[MbO_2] = \text{concentration of oxymyoglobin} \ .$$

Then the desired fraction is

$$f = \frac{[MbO_2]}{[MbO_2] + [Mb]} \ ,$$

or

$$f = \frac{1}{\dfrac{[Mb]}{[MbO_2]} + 1} = \frac{[O_2]}{\dfrac{[Mb][O_2]}{[MbO_2]} + [O_2]} \ . \tag{9}$$

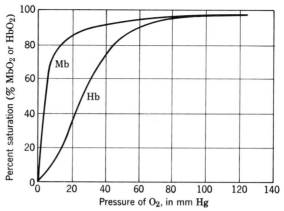

Figure 4 Saturation curves of O_2 bound to myoglobin (Mb) and hemoglobin (Hb) molecules in solution in water. The partial pressure of O_2 is plotted as the horizontal axis. The vertical axis gives the fraction of the molecules of Mb which have one bound O_2 molecule, or the fraction of the strands of Hb which have one bound O_2 molecule. Hemoglobin has a much larger change in oxygen content in the pressure range between the arteries and the veins. This circumstance is of obvious physiological importance. The curve for myoglobin has the predicted form for the reaction $Mb + O_2 \rightleftharpoons MbO_2$. The curve for hemoglobin has a different form, quite possibly because of interactions between O_2 molecules bound to the four strands of the Hb molecule. An alternative explanation has been proposed by J. Monod, J. Wyman, and J.-P. Changeux, "On the nature of allosteric transitions," Journal of Molecular Biology **12**, 88 (1965). The drawing is from Fruton and Simmons. *General biochemistry*, Wiley, 1961.

This expression is of the form $f = c/(c_0 + c)$ as in (7), where now c is the concentration of oxygen molecules in solution in water. Notice that the term

$$\frac{[Mb][O_2]}{[MbO_2]}$$

that appears in the role of c_0 is made up of factors that are themselves variable. It must be, therefore, that this particular combination is in fact independent of the pressure and is a function only of the temperature. We shall see later that this result is a special case of the law of mass action.

Experimental results for the fractional occupancy versus the concentration of oxygen at various temperatures are shown in Fig. 3. In Fig. 4 we compare the observed oxygen saturation curves of myoglobin and hemoglobin. (Hemoglobin is the oxygen-carrying component of blood. It is made up of four molecular strands, each strand nearly identical with the single strand of myoglobin, and each capable of binding a single oxygen molecule.) Historically, the classic work on the adsorption of oxygen by hemoglobin was done by Christian Bohr,[3] the father of Niels Bohr. The oxygen saturation curve for hemoglobin (Hb) has a slower rise at low pressures, so that the binding energy of a single O_2 to a molecule of Hb is lower than for Mb. At higher pressures of oxygen the Hb curve has a region that is concave upwards, a feature never found with Mb, but of physiological importance to man.

[3] See, for example, C. Bohr, Zentralblatt für Physiologie **17**, 682 (1903).

Problem 1. *Adsorption of O_2 on Mb and Hb.* Show that if the four sites on a molecule of Hb which adsorb O_2 do not interact with each other, then the saturation curve must be of the same form as for Mb.

Problem 2. *Thermal ionization of hydrogen.* We may consider the formation of atomic hydrogen in the reaction $e + H^+ \rightleftharpoons H$, where e is an electron, as the adsorption of an electron on a proton H^+. Show that the equilibrium concentrations of the reactants satisfy the relation

$$\frac{[e][H^+]}{[H]} = \frac{e^{-\vartheta/\tau}}{V_Q} \ , \tag{10}$$

where ϑ is the energy required to ionize atomic hydrogen, and the quantum volume V_Q refers to the electron:

$$V_Q = \left(\frac{2\pi\hbar^2}{m\tau}\right)^{\frac{3}{2}} \ ; \tag{11}$$

here m is the mass of the electron. In (10) the brackets denote concentrations. We neglect the spins of the particles; this assumption does not affect the final result.

   If all the electrons and protons arise from the ionization of hydrogen atoms, then the concentration of protons is equal to that of the electrons, and the electron concentration is given by

$$[e] = [H]^{\frac{1}{2}} V_Q^{-\frac{1}{2}} e^{-\vartheta/2\tau} \ . \tag{12}$$

(A similar problem arises in semiconductor physics in connection with the thermal ionization of impurity atoms that are donors of electrons.) Notice that:

   (a) The exponent involves $\frac{1}{2}\vartheta$ and not ϑ, which shows that this is not a simple "Boltzmann factor" problem. (Here ϑ is the ionization energy.)

   (b) The electron concentration is proportional to the square root of the hydrogen atom concentration.

   (c) If we add excess electrons to the system, then (10) tells us that the concentration of protons will decrease.

GENERAL THEORY OF REACTION EQUILIBRIA

   We have successfully treated reactions that may be written as a chemical equation of the form

$$B + C \rightleftharpoons BC \ ; \qquad \text{or} \qquad B + C - BC = 0 \ , \tag{13}$$

where molecules of species B and C are in equilibrium with BC. We need to treat complex reactions with more reacting species. To do this we must develop the theory in a more general form.

We may write a general chemical reaction equation as

$$\nu_1 A_1 + \nu_2 A_2 + \cdots + \nu_l A_l = 0 , \tag{14}$$

or

$$\sum_j \nu_j A_j = 0 , \tag{15}$$

where the A_j denote the chemical species, and the ν_j are the coefficients of the species in the reaction equation. For the reaction (13) we have

$$A_1 = B ; \quad A_2 = C ; \quad A_3 = BC ; \quad \nu_1 = 1 ; \quad \nu_2 = 1 ; \quad \nu_3 = -1 . \tag{16}$$

The discussion of chemical equilibria is usually given for reactions under conditions of constant pressure and temperature. Under these conditions we know from Chapter 19 that in equilibrium the thermodynamic potential

$$G = U - T\sigma + pV \tag{17}$$

is a minimum with respect to changes in the proportions of the reactants. The increment in the thermodynamic potential is

$$dG = -\sigma \, dT + V \, dP + \sum_j \mu_j \, dN_j , \tag{18}$$

from (19.4). Here μ_j is the chemical potential of species j, as defined by (5.13).

At constant pressure $dp = 0$ and at constant temperature $dT = 0$; then (18) reduces to

$$dG = \sum_j \mu_j \, dN_j . \tag{19}$$

The change in the thermodynamic potential in a reaction is seen to be closely related to the chemical potentials of the reactants. In equilibrium this change must be zero.

The change dN_j in the number of molecules of species j is proportional to the coefficient ν_j in the chemical equation $\Sigma \nu_j A_j = 0$. We may write dN_j in the form

$$dN_j = \nu_j \, d\hat{N} , \tag{20}$$

where $d\hat{N}$ is the increment in the number of times the reaction takes place. Thus (19) becomes

$$dG = \left(\sum_j \nu_j \mu_j \right) d\hat{N} . \tag{21}$$

In equilibrium $dG = 0$ at constant temperature and pressure, so that

$$\boxed{\sum_j \nu_j \mu_j = 0 .} \tag{22}$$

This is the general condition for equilibrium in a transformation of matter at constant pressure and temperature.

EQUILIBRIUM FOR IDEAL GASES: LAW OF MASS ACTION

We obtain a simple and very useful form of the general equilibrium condition $\Sigma \nu_j \mu_j = 0$ when we assume that each of the constituents acts as an ideal gas. The gas need not be monatomic. The chemical potential μ_j of species j is then given by

$$e^{\mu_j/T} = \frac{N_j}{V}\left(\frac{2\pi\hbar^2}{M_j T}\right)^{\frac{3}{2}} e^{F_j(\text{int})/T} \ , \tag{23}$$

from (3a) with $F_j(\text{int})$ as the internal free energy of a molecule of species j. The internal excitations include vibration, rotation, and electronic excitations, and all the nuclear orientations.

We let

$$c_j \equiv \frac{N_j}{V} \tag{24}$$

denote the concentration of molecules of species j. We take the logarithm of both sides of (23) and multiply by T to obtain the chemical potential of species j:

$$\mu_j = T \log c_j + \tfrac{3}{2}T \log\left(\frac{2\pi\hbar^2}{M_j T}\right) + F_j(\text{int}) \ . \tag{25}$$

This result for the chemical potential is the sum of a term in the logarithm of the concentration and a term that is a function only of the temperature:

$$\mu_j = T \log c_j + T\chi_j(T) \ , \tag{26}$$

with the definition

$$\chi_j(T) \equiv \tfrac{3}{2}\log\left(\frac{2\pi\hbar^2}{M_j T}\right) + \frac{F_j(\text{int})}{T} \ . \tag{27}$$

Here χ is the Greek letter chi. Note that the internal free energy $F_j(\text{int})$ is an additive term in the chemical potential of the jth chemical component.

The equilibrium condition $\Sigma \nu_j \mu_j = 0$ now becomes

$$\Sigma \nu_j \mu_j = T\Sigma(\nu_j \log c_j + \nu_j \chi_j) = 0 \ , \tag{28}$$

or

$$\Sigma \log c_j^{\nu_j} = -\Sigma \nu_j \chi_j \ . \tag{29}$$

We raise e to powers of both sides of (29) to obtain

$$\prod_j c_j^{\nu_j} = \exp\left(-\Sigma \nu_j \chi_j\right) .$$ (30)

We define the **equilibrium constant** $K_c(T)$ by

$$K_c(T) \equiv \exp\left(-\Sigma \nu_j \chi_j\right) .$$ (31)

The subscript c in K_c refers to the presence of the concentrations c_j on the left-hand side of (30); later we shall introduce a related equilibrium constant K_p in terms of the partial pressures of the chemical components. We now combine (30) and (31) to obtain

$$\boxed{\prod_j c_j^{\nu_j} = K_c(T) ,}$$ (32)

an important result known as the **law of mass action.** The result says that the indicated product of the cencentrations is a function of the temperature alone. We can see that a change in the concentration of any one reactant will force a change in the equilibrium concentration of one or more of the other reactants.

EXAMPLE. *Equilibrium of atomic and molecular hydrogen.* For the reaction $H_2 - 2H = 0$ for the dissociation of molecular hydrogen into atomic hydrogen the statement of the law of mass action (32) is

$$[H_2][H]^{-2} = \frac{[H_2]}{[H]^2} = K_c(T) ,$$ (33)

where $[H_2]$ denotes the concentration of molecular hydrogen, and $[H]$ denotes the concentration of atomic hydrogen. It follows that

$$[H] = \frac{[H_2]^{\frac{1}{2}}}{K_c^{\frac{1}{2}}} ;$$ (34)

that is, the concentration of atomic hydrogen at a given temperature is proportional to the square root of the concentration of molecular hydrogen. This result is of the same form of (12) for the ionization problem.

STANDARD FREE ENERGY CHANGES

Chemists have found a very useful form of expression for the equilibrium constants of reactions in terms of what are called standard free energy changes. We develop the concept in this section as an assistance to the understanding of the chemical and biochemical literature.

In (26) we wrote the chemical potential of species j in the form

$$\mu_j = T \log c_j + T\chi_j(T) \, , \tag{35}$$

where χ_j was given by (27). We may also write (35) in the form

$$\mu_j = \mu_j{}^0 + T \log c_j \, , \tag{36}$$

where

$$\mu_j{}^0 \equiv T\chi_j(T) \, . \tag{37}$$

The definition (36) is such that μ_j is equal to $\mu_j{}^0$ when the concentration c_j is unity. We can think of $\mu_j{}^0$ as containing the concentration-independent parts of the chemical potential. These parts relate to the structure of a molecule of species j.

We may call $\mu_j{}^0$ the **standard chemical potential,** standardized to be the chemical potential of species j at unit concentration. Values given in standard tables often refer to a concentration of one mole per liter.

The equilibrium constant in (31) may be expressed in terms of the standard chemical potentials: this is the use of the $\mu_j{}^0$. We have, with use of (40),

$$K_c(T) \equiv \exp\left(-\Sigma\nu_j\chi_j\right) = \exp\left(-\Sigma\nu_j\mu_j{}^0/T\right) = \exp\left(-\Delta\mu^0/T\right) \, . \tag{38}$$

Here

$$\Delta\mu^0 \equiv \Sigma\nu_j\mu_j{}^0 \tag{39}$$

is called the change in the standard chemical potential in the reaction.[4]

Problem 3. *Heat of reaction and the van't Hoff relation.* We defined $d\hat{N}$ by (20) such that $dN_j = \nu_j \, d\hat{N}$. (a) Show that

$$\left(\frac{\partial H}{\partial \hat{N}}\right)_{T,\,p} = \frac{\partial G}{\partial \hat{N}} - T\frac{\partial}{\partial \hat{N}}\frac{\partial G}{\partial T} = -T\Sigma\nu_j\left(\frac{\partial \mu_j}{\partial T}\right)_p \, , \tag{40}$$

where $H \equiv U + pV$ is the heat function of Chapter 19.

The left-hand side is known as the **heat of reaction.** It is the heat added to the system when $\Delta\hat{N} = 1$ in a reversible change at constant temperature

---

[4] Values in chemical tables may be called standard free energy changes.

and pressure. (When $\Delta \hat{N} = 1$ the chemical reaction is executed once; we defined $\hat{N}$ in (20).)

(b) Show that the law of mass action for ideal gases may be written as

$$\prod_j p_j^{r_j} = K_p(T) \ , \tag{41}$$

where p_j is the partial pressure of chemical component j, and $K_p(T)$ is a function of temperature alone, but a different function than $K_c(T)$ in (31).

(c) Show that the heat of reaction satisfies

$$\left(\frac{\partial H}{\partial \hat{N}}\right)_{T,\,p} = T^2 \left(\frac{\partial}{\partial T} \log K_p(T)\right)_p \ . \tag{42}$$

This is the **law of van't Hoff**; it relates the heat of reaction to the equilibrium constant K_p. The integral form of (42) is often used to obtain $K_p(T)$ experimentally from determinations and extrapolations of the heat of reaction over a suitable temperature range.

REFERENCES

Biological reactions

H. R. Mahler and E. H. Cordes, *Biological chemistry*, Harper and Row, 1966; see particularly Chapter 5, "Equilibria and thermodynamics in biochemical transformations," and Chapter 6, "Enzyme kinetics."

I. M. Klotz, *Energy changes in biochemical reactions*, Academic Press, 1967, pp. 108. A readable short introduction to selected major topics.

Systems in Electric Fields: Work and Energy

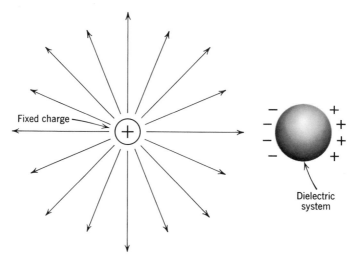

Figure 1 Polarization of a dielectric system by a fixed electric charge. The system is moved from infinity into the field of the fixed charge.

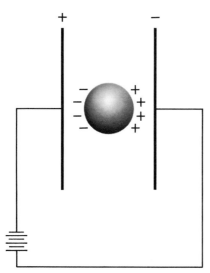

Figure 2 Polarization of a dielectric system by charging the plates of a capacitor. The charges are moved by the battery onto the plates of the capacitor.

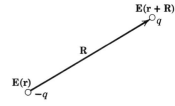

Figure 3 Construction for finding the force on a dipole.

CHAPTER 22   SYSTEMS IN ELECTRIC FIELDS:
WORK AND ENERGY

Some of the most interesting physical applications of thermal physics concern changes in systems when placed in electric or magnetic fields. There is no particular difficulty in the problem, but we need a suitable expression for the work done on the system by an applied field. There are, however, two different ways in which the field can be applied, and the two ways give different results for the work done on the system. Both ways are useful. We shall call one **scheme A** and the other **scheme B.** The difference arises from the value of what we include in the energy of the system and is a matter of what portion of the field energy is included as a part of the system. The central problem of this chapter is to develop the two results for applied electric fields. In Chapter 23 we translate the results to experiments in magnetic fields.

We consider two different methods of polarizing a dielectric system[1] by an electric field: In scheme A we polarize the system by bringing it into the field of a fixed charge (Fig. 1); in scheme B we polarize the system by applying a dc voltage across the plates of a capacitor enclosing the system (Fig. 2).

FORCE ON AN ELECTRIC DIPOLE

We first develop an expression for the force on an electric dipole in an inhomogeneous electric field. Let $E(\mathbf{r})$ denote the electric field produced by a fixed charge external to the polarizable system; thus $\mathbf{E}$ is called the external electric field or applied electric field. We consider a neutral molecule or assembly of molecules at $\mathbf{r}$, and we represent the dipole moment of the molecule by charges $\pm q$ separated by a distance $\mathbf{R}$, as in Fig. 3. The dipole moment may be permanent, or it may be induced by the electric field, or a combination of both. The net force exerted on the molecule by the applied electric field is the difference between the forces exerted on opposite ends of the molecule:

$$\mathbf{F}(\mathbf{r}) = q\{\mathbf{E}(\mathbf{r} + \mathbf{R}) - \mathbf{E}(\mathbf{r})\} \ . \tag{1}$$

The dipole moment $\mathbf{p}$ of the molecule is defined as

$$\mathbf{p} = q\mathbf{R} \ . \tag{2}$$

---

[1] For discussions of the properties of dielectrics, see E. M. Purcell, *Electricity and magnetism,* McGraw-Hill, 1965, Chapter 9, and C. Kittel, *Introduction to solid state physics,* 3rd ed., Wiley, 1966, Chapter 12.

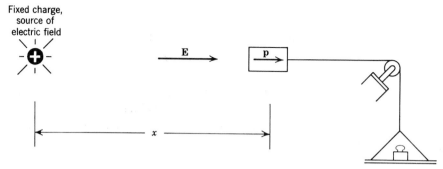

Figure 4 The dipole is attracted by the fixed charge and moved toward it; the work W_A is done on the dipole by the weights in the pan:

$$W_A = -\int_0^{\mathbf{E}_1} \mathbf{p} \cdot d\mathbf{E} \ .$$

We carry out a series expansion of (1). The leading term in the series expansion of the x component of the force is

$$F_x(\mathbf{r}) = q\{E_x(\mathbf{r} + \mathbf{R}) - E_x(\mathbf{r})\} = q\left\{R_x \frac{\partial E_x}{\partial x} + R_y \frac{\partial E_x}{\partial y} + R_z \frac{\partial E_x}{\partial z}\right\}. \quad (3)$$

We know from a Maxwell equation that curl $\mathbf{E} = 0$ for static fields, so that

$$\frac{\partial E_x}{\partial y} = \frac{\partial E_y}{\partial x} \ ; \qquad \frac{\partial E_x}{\partial z} = \frac{\partial E_z}{\partial x} \ . \quad (4)$$

Thus the x component of the force on the molecule may be written as

$$F_x(\mathbf{r}) = q\left\{R_x \frac{\partial E_x}{\partial x} + R_y \frac{\partial E_y}{\partial x} + R_z \frac{\partial E_z}{\partial x}\right\}, \quad (5)$$

whence

$$F_x(\mathbf{r}) = q\mathbf{R} \cdot \frac{\partial \mathbf{E}}{\partial x} = \mathbf{p} \cdot \frac{\partial \mathbf{E}}{\partial x} \ . \quad (6)$$

The force involves the dipole moment and the gradient of the electric field.

This is the force exerted on the molecule by the applied electric field. Now in this imaginary experiment there is an equal but opposite force on the dipole, namely the force that we exert on the dipole in order to hold it at rest at the point $\mathbf{r}$ or to allow it to creep in quasistatically from infinity to $\mathbf{r}$. It is by means of this force

$$\mathbf{F}' = -\mathbf{F} \ ; \quad (7)$$

$$F_x'(\mathbf{r}) = -\mathbf{p} \cdot \frac{\partial \mathbf{E}}{\partial x} \ , \quad (8)$$

that an external mechanical agency, such as the weights in the pan in Fig. 4, does work on the dipole or has work done on it by the dipole.

WORK DONE ON DIPOLE IN DISPLACEMENT
IN AN ELECTRIC FIELD

The work done by our external agency when the dipole is displaced from infinity to the position $\mathbf{r}_1$ in the external electric field defines the work W_A in scheme A. The work is the integral of the force $\mathbf{F}'$ times the element of displacement $d\mathbf{r}$:

$$W_A = \int_{\infty}^{\mathbf{r}_1} \mathbf{F}' \cdot d\mathbf{r} = \int_{\infty}^{\mathbf{r}_1} (F_x' \, dx + F_y' \, dy + F_z' \, dz) \ . \tag{9}$$

From (8) for F_x' and the analogous equations for F_y' and F_z' we have for the work done on the dipole

$$\boxed{W_A = - \int_{\infty}^{\mathbf{r}_1} \mathbf{p} \cdot \left(\frac{\partial \mathbf{E}}{\partial x} \, dx + \frac{\partial \mathbf{E}}{\partial y} \, dy + \frac{\partial \mathbf{E}}{\partial z} \, dz \right) = - \int_{0}^{\mathbf{E}_1} \mathbf{p} \cdot d\mathbf{E}} \tag{10}$$

where 0 is the value of the electric field at infinity and $\mathbf{E}_1$ is the value of the electric field at the position $\mathbf{r}_1$. This is the work done in **scheme A.**

WORK DONE TO POLARIZE THE DIPOLE IN ZERO FIELD

We now ask a different question: how much work must we do to polarize the dipole in zero external electric field? This is the work in scheme B. We can find an indirect process by which such polarization is conceivable. In the process the work we do on the dipole goes exclusively into the internal energy of the dipole; there is no energy of interaction with an external field because there is no external field. We calculate the work in an indirect way by consideration of a reversible process in which:

(a) The molecule is brought from infinity to $\mathbf{r}_1$ in the field of an external charge.

(b) The dipole moment $\mathbf{p}_1$ which exists when the molecule is at $\mathbf{r}_1$ is locked at the value $\mathbf{p}_1$.

(c) The locked dipole moment $\mathbf{p}_1$ is moved from $\mathbf{r}_1$ to infinity, at which position the electric field is zero.

We start off in zero electric field with a permanent dipole moment $\mathbf{p}_0$, which may be zero, and we end up with a dipole moment $\mathbf{p}_1$, also in zero electric field. What is the net work done on the dipole in the process?

In step (a) the work done is W_A, as calculated by (10). In step (b) no work is done, at least in principle. In step (c) we do work on the dipole in bringing it

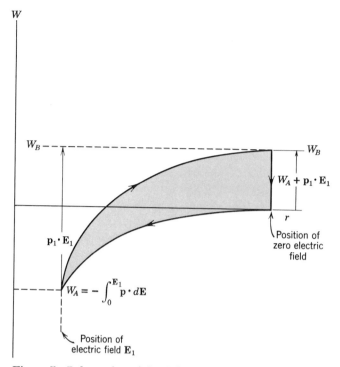

Figure 5   Relationship of the definitions W_A and W_B for a polarizable molecule.

from the field $\mathbf{E}_1$ to zero field; this work is calculated from (10) with $\mathbf{p}$ set equal
to the fixed dipole moment $\mathbf{p}_1$:

$$- \int_{\mathbf{E}_1}^{0} \mathbf{p}_1 \cdot d\mathbf{E} = -\mathbf{p}_1 \cdot \int_{\mathbf{E}_1}^{0} d\mathbf{E} = \mathbf{p}_1 \cdot \mathbf{E}_1 \ . \tag{11}$$

The lower and upper limits of integration now correspond to the electric field
changes in the displacement from $\mathbf{r}_1$ to infinity.

We let W_B denote the total work done on the dipole in the sequence of
steps (a), (b), (c):

$$W_B = W_A + 0 + \mathbf{p}_1 \cdot \mathbf{E}_1 = - \int_{0}^{\mathbf{E}_1} \mathbf{p} \cdot d\mathbf{E} + \mathbf{p}_1 \cdot \mathbf{E}_1 \ . \tag{12}$$

We may simplify this result with the help of two identities:

$$d(\mathbf{p} \cdot \mathbf{E}) \equiv \mathbf{p} \cdot d\mathbf{E} + \mathbf{E} \cdot d\mathbf{p} \ , \tag{13}$$

and

$$\mathbf{p}_1 \cdot \mathbf{E}_1 \equiv \int_{0}^{\mathbf{p}_1 \cdot \mathbf{E}_1} d(\mathbf{p} \cdot \mathbf{E}) = \int_{0}^{\mathbf{E}_1} \mathbf{p} \cdot d\mathbf{E} + \int_{0}^{\mathbf{p}_1} \mathbf{E} \cdot d\mathbf{p} \ . \tag{14}$$

Then (12) becomes

$$W_B = -\int_0^{\mathbf{E}_1} \mathbf{p} \cdot d\mathbf{E} + \int_0^{\mathbf{E}_1} \mathbf{p} \cdot d\mathbf{E} + \int_{\mathbf{p}_0}^{\mathbf{p}_1} \mathbf{E} \cdot d\mathbf{p} \ ,$$

whence

$$\boxed{W_B = \int_{\mathbf{p}_0}^{\mathbf{p}_1} \mathbf{E} \cdot d\mathbf{p} \ .} \tag{15}$$

This is the work done in changing the dipole moment from $\mathbf{p}_0$ to $\mathbf{p}_1$ in zero electric field. It is the work done on the dipole in **scheme B**. Notice that although the scheme is defined in terms of a polarization process in zero electric field, the expression for the work does involve the electric field that would achieve the desired polarization.

The relationship of W_A, W_B and $\mathbf{p}_1 \cdot \mathbf{E}_1$ is given by (12) and is illustrated in Fig. 5. We see by their definitions that W_A and W_B measure the work done in two quite different processes.

EXAMPLE. *Permanent electric dipole moment.* Suppose that the molecule has no induced moment, but only the permanent electric moment $\mathbf{p}_0$. Then the work done in bringing the molecule from infinity to $\mathbf{r}_1$ is

$$W_A = -\int_0^{\mathbf{E}_1} \mathbf{p}_0 \cdot d\mathbf{E} = -\mathbf{p}_0 \cdot \mathbf{E}_1 \ . \tag{16}$$

The work done in polarizing the molecule in zero electric field is

$$W_B = \int_{\mathbf{p}_0}^{\mathbf{p}_1} \mathbf{E} \cdot d\mathbf{p} = 0 \ , \tag{17}$$

because $\mathbf{p}_1 = \mathbf{p}_0$.

We see that no "internal mechanical work" W_B is done on the permanent moment, because we were given a permanent moment that does not change in the sequence of steps (a), (b), (c). The work $W_A = -\mathbf{p}_0 \cdot \mathbf{E}_1$ is simply the energy of interaction of the permanent moment with the field.

EXAMPLE. *Induced electric dipole moment.* Suppose the molecule has zero permanent moment, but has an induced electric dipole moment that is related to the external electric field by

$$\mathbf{p} = \alpha\mathbf{E} \ . \tag{18}$$

Here α is called the **polarizability.** We have assumed in writing (18) that the molecule is isotropic, so that the angle between $\mathbf{E}$ and the axes of the molecule does not affect the value or direction of the induced dipole moment. If we assume the charges $\pm q$ in the dipole are bound by forces that satisfy Hooke's law, then α will be independent of $\mathbf{E}$.

The work we do in bringing the molecule from infinity to $\mathbf{r}_1$ is, from (10),

$$W_A = - \int_0^{\mathbf{E}_1} \mathbf{p} \cdot d\mathbf{E} = - \int_0^{E_1} \alpha E \, dE = -\tfrac{1}{2}\alpha E_1{}^2 \ . \tag{19}$$

The work done in polarizing the molecule in zero electric field is, from (15),

$$W_B = \int_0^{\mathbf{p}_1} \mathbf{E} \cdot d\mathbf{p} = \frac{1}{\alpha} \int_0^{\mathbf{p}_1} p \, dp = \frac{1}{2\alpha} p_1{}^2 \ . \tag{20}$$

Because $p_1 = \alpha E_1$, we may write W_A as

$$W_A = -\tfrac{1}{2} p_1 E_1 = - \frac{1}{2\alpha} p_1{}^2 \ ,$$

which should be contrasted with the result for W_B. We may think of W_A as the sum of the polarization work $W_B = \dfrac{1}{2\alpha} p_1{}^2$ and the energy of interaction $-p_1 E_1$ of the induced dipole moment p_1 with the field E_1.

RELATION OF W_A AND W_B TO THE ENERGY

In a reversible process the thermodynamic identity of Chapter 7 may assume two forms, according to the means by which the work is performed in the electric field. If the work is done (**scheme A**) by moving the dipole in the field of a fixed charge, then with W_A from (10) we have

$$dU_A = \mathcal{T} \, d\sigma + \mu \, dN - \mathbf{p} \cdot d\mathbf{E} \ . \tag{21}$$

If the work is done (**scheme B**) by polarizing the dipole in zero electric field, then with W_B from (15) we have

$$dU_B = \mathcal{T} \, d\sigma + \mu \, dN + \mathbf{E} \cdot d\mathbf{p} \ . \tag{22}$$

With two different expressions for the change of energy, there must be two different meanings of the energy of the system. The energy naturally depends on what is included in the definition of the system, and this is the origin of the different meanings. From (12) we see that in scheme B the interaction $-\mathbf{p}_1 \cdot \mathbf{E}_1$ is counted as a part of the energy of the system. In process B the interaction energy is not counted in the system, but belongs to

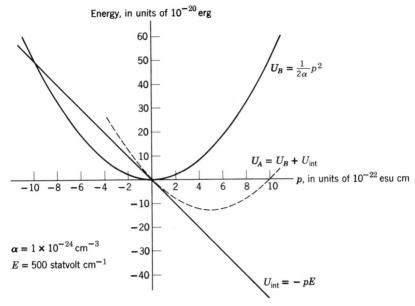

Figure 6   Polarization energies U_A, U_B and interaction energy $U_{int} = -\mathbf{p} \cdot \mathbf{E}$ are plotted as functions of the induced dipole moment p, for a system at constant entropy. The interaction energy is plotted for $E = 500$ statvolt cm^{-1}, with p viewed as an adjustable parameter. The minimum of the curve $U_A = U_B + U_{int}$ gives the equilibrium value of the dipole moment as $p = 5 \times 10^{22}$ esu cm, for the chosen value of E.

the external apparatus that does work on the system. In Appendix F we show that it is convenient in theoretical calculations to work with U_A.

If the entropy is constant and the number of particles is constant, then

$$dU_A = -\mathbf{p} \cdot d\mathbf{E} \; ; \tag{23}$$
$$dU_B = \mathbf{E} \cdot d\mathbf{p} \; . \tag{24}$$

The polarization energies U_A and U_B of an induced electric dipole moment are shown in Fig. 6.

In Chapter 18 we saw that the change in the free energy $F = U - T\sigma$ measures the work done on the system in a reversible change at constant temperature and constant number of particles. The function

$$F_A \equiv U_A - T\sigma \tag{25}$$

has the differential

$$dF_A = dU_A - T\,d\sigma - \sigma\,dT = \mu\,dN - \mathbf{p} \cdot d\mathbf{E} - \sigma\,dT \;, \tag{26}$$

from (21) for dU_A. Thus at constant T and N we have

$$dF_A = -\mathbf{p} \cdot d\mathbf{E} \; . \tag{27}$$

The work done on a dipole in an electric field in process A at constant T, N is equal to the change in F_A.

The free energy F_B is defined as

$$F_B \equiv U_B - T\sigma . \tag{28}$$

The differential is

$$dF_B = dU_B - T\,d\sigma - \sigma\,dT = \mu\,dN + \mathbf{E}\cdot d\mathbf{p} - \sigma\,dT , \tag{29}$$

with (22) for dU_B. At constant T, N we have

$$dF_B = \mathbf{E}\cdot d\mathbf{p} ; \tag{30}$$

the change in F_B is the work done in process B.

From (27) we have the useful relation

$$\left(\frac{\partial F_A}{\partial E}\right)_{T,N} = -p , \tag{31}$$

when $\mathbf{p}$ is parallel to $\mathbf{E}$. We now let F_A refer to unit volume, and we differentiate once again:

$$\left(\frac{\partial^2 F_A}{\partial E^2}\right)_{T,N} = -\left(\frac{\partial P}{\partial E}\right)_{T,N} = -\chi ; \tag{32}$$

here P is the dipole moment per unit volume or the **polarization,** and χ is the **dielectric susceptibility.**

MEASUREMENT OF W_A

We can measure the frequency of a photon emitted by a system in zero electric field in a spectral transition from state l to state l':

$$\hbar\omega(0) = \epsilon_l(0) - \epsilon_{l'}(0) , \tag{33}$$

where the argument of the energy of ϵ and of the frequency ω denotes zero applied electric field. We may use the results of spectroscopic experiments to determine the values of the energies of the individual states with reference to a standard zero of energy.

The energies of the states l, l' are functions of the electric field. The photon energy is also a function of the field, for

$$\hbar\omega(E) = \epsilon_l(E) - \epsilon_{l'}(E) \tag{34}$$

will be shifted with respect to $\hbar\omega(0)$ by the application of an electric field. The dependence of the frequency of a spectral line on the static electric field intensity is known as the Stark effect. We can find the energies $\epsilon_l(E)$ spectroscopically just as we found the $\epsilon_l(0)$.

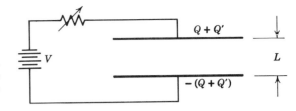

Figure 7 Capacitor. The charge Q' is induced by the dipole moment $\mathcal{P}$ of the system; the charge Q is that induced by the battery when the capacitor is empty.

Suppose that a system in the state l is displaced reversibly from infinity to a point where the electric field intensity is E. The work done on the system in the displacement is

$$W_A = \epsilon_l(E) - \epsilon_l(0) = \Delta U_A . \tag{35}$$

This displacement defines W_A. There is no change in entropy in the process because the system remains in the same state l throughout, so that the work done on the system is equal to the change of energy ΔU_A. In Appendix F we give the quantum theory formulation of the connection between U_A and U_B.

MEASUREMENT OF W_B

How can we measure W_B, defined as the polarization energy in the absence of an electric field? We can measure W_B by an experiment with a capacitor.

Let us consider the energy change in a process in which we hold the dielectric system between the plates of a vacuum capacitor. We charge the plates reversibly from an external voltage source V to produce an electric field $E = V/L$ across the capacitor. Here L is the plate separation in a parallel plate capacitor, Fig. 7.

The work done by the power supply in charging the plates to a given voltage V will be greater when the dielectric specimen is present between the plates than when the specimen is absent. The excess work done is equal to W_B, as we now show.

Let Q be the charge on the positive plate of the capacitor when the voltage is V. When the dielectric system is between the plates, the charge on the positive plate is increased from Q to $Q + Q'$; the charge on the negative plate is $-(Q + Q')$. The extra charge Q', which is induced on the plates by the dipole moment $\mathcal{P}_1$ of the specimen, is given by[2]

$$|Q'| = \frac{|\mathcal{P}_1|}{L} , \tag{36}$$

where we take the direction of $\mathcal{P}_1$ to be normal to the plates. This value of

[2] Proofs of this result are discussed by C.-Y. Fong and C. Kittel, American Journal of Physics **35**, 1091 (1967).

the charge simulates a dipole moment on the plates which is equal but opposite to $\mathcal{P}_1$: because of the induced dipole moment the capacitor screens the field of the original dipole as observed at points outside the capacitor.

The work done by the external power supply in charging the empty capacitor reversibly is

$$W(\text{empty}) = \int_0^Q V \, dQ = L \int_0^Q E(Q) \, dQ = \int_0^{QL} E \, d(QL) \; . \tag{37}$$

We now consider the capacitor with the dielectric included between the plates. If there is a permanent part of the dipole moment, say $\mathcal{P}_0$, then the lower limit in the integral below for the work done in charging the capacitor is $\mathcal{P}_0$. The work done in charging the capacitor to $Q + Q' = Q + \mathcal{P}_1/L$ is

$$W = L \int_{\mathcal{P}_0/L}^{Q + \mathcal{P}_1/L} E(Q) \, dQ = \int_{\mathcal{P}_0}^{QL + \mathcal{P}_1} E \, d(QL) \; . \tag{38}$$

The difference between (38) and (37) is the work done on the dielectric in charging the capacitor:

$$\int_{\mathcal{P}_0}^{\mathcal{P}_1} E \, d(QL) = \int_{\mathcal{P}_0}^{\mathcal{P}_1} E \, d\mathcal{P} \; . \tag{39}$$

This is exactly the result for W_B defined in (15) above. Thus the capacitor experiment actually measures W_B, defined as the work done in polarizing the system in zero electric field.

Problem 1. *Temperature dependence of the polarization at absolute zero.*
(a) Prove that for unit volume of matter

$$\left(\frac{\partial P}{\partial T} \right)_E = \left(\frac{\partial \sigma}{\partial E} \right)_T \; . \tag{40}$$

(b) Show as a consequence of one statement of the third law that

$$\left(\frac{\partial P}{\partial T} \right)_E = 0 \tag{41}$$

at $T = 0$. (This result applies, for example, to the temperature dependence of the spontaneous polarization of a ferroelectric crystal.)

REFERENCES

E. A. Guggenheim, "On magnetic and electrostatic energy," Proceedings of the Royal Society of London **A155**, 49 (1936).

E. A. Guggenheim, "The thermodynamics of magnetization," Proceedings of the Royal Society of London **A155**, 70 (1936).

Systems in Magnetic Fields: Work and Energy

The expressions for the work done by external agencies on systems in magnetic fields can be transcribed directly from those for electric systems as treated in Chapter 22. We denote the total magnetic moment of the specimen by $\mathcal{m}$. We need only substitute in the results of Chapter 22 the magnetic field $\mathbf{H}$ for the electric field $\mathbf{E}$, and the magnetic moment $\mathcal{m}$ for the electric moment p. The magnetization $\mathbf{M}$ replaces the polarization $\mathbf{P}$, both defined as the moment per unit volume. (The apparent nonexistence of free magnetic poles in the physical world requires no change in our approach in calculating the work W_A.) Thus, by analogy with (22.10), in **scheme A**

$$W_A = -\int_0^{\mathbf{H_1}} \mathcal{m} \cdot d\mathbf{H} \tag{1}$$

is the work we must do to take the specimen from the field 0 to the field $\mathbf{H}$, where $\mathbf{H}$ is the field of a permanent magnet external to the specimen.

The field $\mathbf{H}$ is understood to be the field in vacuum due to fixed external permanent sources. We do not need here to distinguish $\mathbf{B}$ and $\mathbf{H}$, for in vacuum $\mathbf{B} = \mathbf{H}$. Workers in magnetism and in solid state physics customarily denote the magnetic field in vacuum by $\mathbf{H}$, and we shall follow their practice. We define W_B as the work needed to magnetize the specimen in zero applied magnetic field. By analogy with (22.15) we write, in **scheme B**,

$$W_B = \int_{\mathcal{m}_0}^{\mathcal{m}_1} \mathbf{H} \cdot d\mathcal{m} \; . \tag{2}$$

The thermodynamic identity for a magnetic system is

Scheme A: $dU_A = T \, d\sigma + \mu \, dN - \mathcal{m} \cdot d\mathbf{H} \; ;$ \hfill (3)

Scheme B: $dU_B = T \, d\sigma + \mu \, dN + \mathbf{H} \cdot d\mathcal{m} \; .$ \hfill (4)

These relations depend on (1) and (2) and are similar to (22.21) and (22.22). Here μ is the chemical potential and not the magnetic moment.

The analogy of magnetic dipoles with electric dipoles reveals the physical significance of W_A. We saw also in Chapter 22 that we can calculate W_B from the work done in charging a capacitor. We need a magnetic analogue to this process, and we shall show that W_B is the work done in magnetizing a specimen in a solenoid.

MEASUREMENT OF THE WORK W_B

A current i flowing in a long empty solenoid produces a magnetic field

$$\mathbf{H} = \frac{4\pi}{c} ni ,$$
(5)

where n is the number of turns per unit length. We now fill the solenoid uniformly with magnetic material. If A is the cross-sectional area of the solenoid, a magnetization change ΔM in the material produces a flux change $4\pi A \, \Delta M$. This flux change causes a voltage

$$V = -\frac{1}{c} 4\pi nlA \frac{dM}{dt}$$
(6)

to be developed across the ends of the solenoid. If the length of the solenoid is l, the total number of turns is nl. The current i that flows against this voltage does work at the rate[1] $-iV$. It follows from (5) and (6) that the work done by the battery is

$$-i\int V \, dt = \frac{1}{c} 4\pi ni(lA) \, \Delta M = \Omega H \, \Delta M ,$$
(7)

where $\Omega = lA$ is the volume of the solenoid. Thus the work done in magnetizing a unit volume of material is $\int H \, dM$, which is identical with W_B as defined by (2).

*RELATION OF W_A TO THE HAMILTONIAN

The energy eigenvalues of the Schrödinger equation with the hamiltonian

$$\mathcal{H}(\mathbf{H}) = \frac{1}{2m} \left(\mathbf{p} - \frac{q}{c} \mathbf{A} \right)^2 - \boldsymbol{\mu}_s \cdot \mathbf{H}$$
(8)

are the energy levels of a system of one particle of charge q and spin magnetic moment μ_s. Here $\mathbf{A}$ is the vector potential of the magnetic field $\mathbf{H}$, and $\mathbf{p}$ is the momentum of the particle. The magnetic work W_A is shown in Appendix F to be the ensemble average of $\mathcal{H}(\mathbf{H}) - \mathcal{H}(0)$:

$$W_A = \langle \mathcal{H}(\mathbf{H}) - \mathcal{H}(0) \rangle .$$
(9)

---

[1] A charge Q which moves against the voltage V does work $-QV$ on the external apparatus; the rate at which work is done is $-(dQ/dt)V$ or $-iV$, by the definition of the current i.

* The material of (8) through (13) is at the advanced level of Appendix F and may be omitted by undergraduate students.

This relation is similar to (F. 12) for the electric field problem. **The importance of process A for the definition of the work is the direct connection with the change in the energy of the system as given by the usual hamiltonian for an atomic system.**

We define the free energy in scheme A as

$$F_A \equiv U_A - T\sigma , \tag{10}$$

whence

$$dF_A = dU_A - T\,d\sigma - \sigma\,dT = \mu\,dN - \sigma\,dT - \mathfrak{m}\cdot d\mathbf{H} , \tag{11}$$

with (3) for dU_A. We may calculate F_A from the partition function

$$Z(H) = \sum_l e^{-\epsilon_l(H)/T} , \tag{12}$$

with the usual relation $F = -T \log Z$, where the $\epsilon_l(H)$ are the energy eigenvalues of $\mathcal{H}(H)$.

From (11) we have, with scalar quantities and referred to unit volume,

$$\left(\frac{\partial F_A}{\partial H}\right)_{T,N} = -M ; \qquad \left(\frac{\partial^2 F_A}{\partial H^2}\right)_{T,N} = -\chi , \tag{13}$$

where M is the magnetization and χ is the magnetic susceptibility.

Problem 1. *Field dependence of the entropy.* Show that

$$\left(\frac{\partial \sigma}{\partial H}\right)_{T,N} = \left(\frac{\partial M}{\partial T}\right)_{H,N} . \tag{14}$$

This relation allows us to find the field dependence of the entropy from the temperature dependence of the magnetization. *Hint.* We know that

$$dF_A = \mu\,dN - \sigma\,dT - M\,dH , \tag{15}$$

for unit volume of material. Thus

$$\left(\frac{\partial F_A}{\partial T}\right)_{H,N} = -\sigma ; \qquad \left(\frac{\partial F_A}{\partial H}\right)_{T,N} = -M . \tag{16}$$

Problem 2. *Ideal paramagnet.* Show that $(\partial U_B/\partial H)_{T,\,N} = 0$ for an ideal para-magnet, defined as one for which the magnetization M is a function of $H/\mathcal{T}$ alone. *Hint.* The thermodynamic identity is

$$dU_B = \mathcal{T}\,d\sigma + \mu\,dN + H\,dM \;, \tag{17}$$

whence

$$\left(\frac{\partial U_B}{\partial H}\right)_{T,\,N} = \mathcal{T}\left(\frac{\partial \sigma}{\partial H}\right)_{T,\,N} + H\left(\frac{\partial M}{\partial H}\right)_{T,\,N}. \tag{18}$$

In Problem 1 we found that $(\partial \sigma/\partial H)_{T,\,N} = (\partial M/\partial \mathcal{T})_{H,\,N}$, whence

$$\left(\frac{\partial U_B}{\partial H}\right)_{T,\,N} = \mathcal{T}\left(\frac{\partial M}{\partial \mathcal{T}}\right)_{H,\,N} + H\left(\frac{\partial M}{\partial H}\right)_{T,\,N}. \tag{19}$$

EXAMPLE. *Stabilization energy of a superconductor.* At low temperatures many metallic elements exhibit a transition from a normal state with finite electrical conductivity to a superconducting state with infinite conductivity.[2] For example, in lead the superconducting state is stable below 7.19 K, and the normal state is stable at temperatures above that. The conductivity in the super-conducting state is infinite as far as we can determine: in one experiment a current in a superconducting ring was estimated to have a decay time not less than 100,000 years.

It is a good thermodynamic problem to determine the stabilization energy of the superconducting state with respect to the normal state. This energy can be determined by direct measurements of the heat capacity of both states over the temperature range from zero to T_c, the transition temperature. Such measurements are possible in both states because we can restore the speci-men to the normal state by application of a sufficiently large magnetic field. For lead a field of 800 gauss will always destroy the superconducting state; thus the heat capacity below T_c is measured for the normal metal in a mag-netic field and for the superconductor in zero magnetic field. We integrate the heat capacity determinations to obtain the difference $U_N - U_S$ in the en-ergy and the difference $\sigma_N - \sigma_S$ in the entropy, and finally we obtain the dif-ference $F_N - F_S$ in the free energy of the two states. The energy difference at absolute zero we call the **stabilization energy** of the superconducting state; the free energy difference at a temperature $\mathcal{T}$ is the stabilization free energy.[3]

---

[2] A survey of the properties of superconductors is given in Chapter 11 of ISSP. The present discus-sion of the magnetic field dependence of the energy of the superconducting state is clearer physically than that in ISSP, 3rd ed., because scheme A for the work done is better adapted to the problem than is scheme B. Both sets of results are equivalent and both derivations are correct.

[3] We assume in using the heat capacity measurements that the thermodynamic properties of the normal state are approximately independent of the field, so that $F_N(H) \cong F_N(0)$. We also use the fact that at the critical temperature the free energies of normal and superconducting phases are equal.

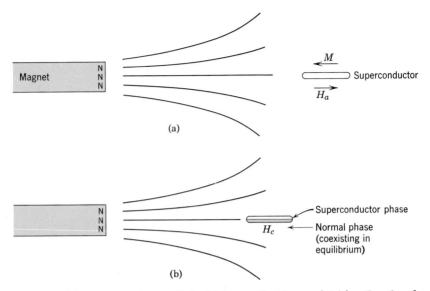

Figure 1 (a) A superconductor (if the Meissner effect is complete) has $B = 0$ and acts as if it had a magnetization $M = -H_a/4\pi$. The work done on the superconductor in a displacement from infinity to the position where the field of the permanent magnet is H_a is given by

$$W = \int \mathbf{M} \cdot d\mathbf{H}_a = \frac{1}{8\pi} H_a{}^2 \ ,$$

per unit volume. (b) When the applied field reaches the value H_c, the normal state can coexist in equilibrium with the superconducting state. In coexistence the free energy densities are equal: $F_N(T, H_c) = F_S(T, H_c)$.

It is also possible to obtain the stabilization energy and the free energy simply from the value H_c of the applied magnetic field which suffices to destroy the superconducting state and carry the specimen into the normal state. The argument depends on the important property of nearly perfect diamagnetism which is exhibited by superconductors (or at least by the common soft superconductors we consider here). According to the Meissner effect, the magnetic induction $\mathbf{B}$ in a bulk superconductor is zero, so that a superconductor simulates a perfect diamagnetic material with a magnetization $\mathbf{M}$ determined by

$$\mathbf{B} \equiv \mathbf{H} + 4\pi\mathbf{M} = 0 \ ; \qquad \mathbf{M} = -\mathbf{H}/4\pi \ . \tag{20}$$

The simplest way to understand the effect of an applied magnetic field H_a on the transition from superconductor to normal is to consider the work done on a superconductor when it is brought to infinity (where the applied field is zero) to a position $\mathbf{r}$ in the field of a permanent magnet. The scheme A for the work done is well-suited to our problem. The work done in the process (Fig. 1) is

$$W = -\int_0^{H_a} \mathbf{M} \cdot d\mathbf{H}_a \ , \tag{21}$$

per unit volume of specimen. If the specimen is in the form of a long needle with axis parallel to the applied field, then $\mathbf{H}_a$ is the field to use in (20), and we have

$$\mathbf{M} = -\frac{\mathbf{H}_a}{4\pi} \tag{22}$$

or

$$W_S = \frac{1}{4\pi} \int_0^{H_a} H_a \, dH_a = \frac{1}{8\pi} H_a^2 \ , \tag{23}$$

per unit volume of specimen. Notice that the work done in the displacement of the superconductor is positive. Supercurrents are induced in a surface layer of the needle to oppose any increase in the flux through the needle as it is moved from field 0 to field H_a.

The thermodynamic identity for this process (a scheme A process) is

$$dU = T \, d\sigma - \mathbf{M} \cdot d\mathbf{H}_a \ , \tag{24}$$

or, for the superconductor,

$$dU_S = T \, d\sigma + \frac{1}{4\pi} H_a \, dH_a \ . \tag{25}$$

Thus at absolute zero the increase in the energy density of the superconductor is

$$\boxed{U_S(H_a) - U_S(0) = \frac{1}{8\pi} H_a^2} \tag{26}$$

on being brought from a position where the applied field is zero to a position where the applied field is H_a. (At $T = 0$ we do not need to consider the term $T \, d\sigma$ in the thermodynamic identity because $d\sigma = 0$.)

We contrast the results (23) and (26) for the superconductor with the corresponding results for a normal nonmagnetic metal. If we neglect the small paramagnetic susceptibility[4] of the conduction electrons of a metal in the normal state, the work done is

$$W_N = -\int_0^{H_a} \mathbf{M} \cdot d\mathbf{H}_a = 0 \ , \tag{27}$$

because the magnetization M is zero. It follows that the change of energy of the specimen is

$$U_N(H_a) - U_N(0) = 0 \tag{28}$$

[4] This is often, but not always, an adequate assumption. The theory is easily handled without the assumption.

on being brought from a position where the applied field is zero to a position where the applied field is H_a.

These results are all we need to determine the stabilization energy of the superconducting state at absolute zero if we are given from experiment the value of the critical magnetic field at absolute zero, $H_c(T = 0)$. At the critical value of the magnetic field the energies are equal in the normal and superconducting states:

$$U_N(H_c) = U_S(H_c) \ . \tag{29}$$

The specimen is equally stable in either state when the applied field is equal to H_c. Now by (28) we have

$$U_N(H_c) = U_N(0) \ , \tag{30}$$

and by (26)

$$U_S(H_c) = U_S(0) + \frac{1}{8\pi} H_c^2 \ ; \tag{31}$$

thus the equilibrium condition (29) becomes

$$U_N(0) = U_S(0) + \frac{1}{8\pi} H_c^2 \ . \tag{32}$$

The stabilization energy of the superconducting state at absolute zero is

$$\boxed{\Delta U \equiv U_N(0) - U_S(0) = \frac{1}{8\pi} H_c^2 \ ,} \tag{33}$$

per unit volume of specimen. For example, the experimental value of H_c for aluminum at absolute zero is 105 gauss, so that

$$\Delta U = \frac{(105)^2}{8\pi} \cong 440 \text{ ergs cm}^{-3} \ , \tag{34}$$

in good agreement with the result of thermal measurements.

At a finite temperature the two phases, normal and superconducting, are in equilibrium not when their energies are equal, but when their free energies are equal, as we now show. We consider the free energy

$$F(T, H) = U - T\sigma \ , \tag{35}$$

which has the differential

$$dF = dU - T \ d\sigma - \sigma \ dT \ . \tag{36}$$

We substitute $dU = \mathcal{T}\,d\sigma - M\,dH_a$ from (24) to obtain

$$dF = -\sigma\,d\mathcal{T} - M\,dH_a .\tag{37}$$

In a change at constant temperature and in a constant applied magnetic field as provided by the permanent magnet, $d\mathcal{T} = 0$ and $dH_a = 0$, whence

$$dF = 0\tag{38}$$

is the condition for thermodynamic equilibrium. In the critical field $H_c(\mathcal{T})$ the two phases coexist. If V_S is the volume in the superconducting phase and V_N the volume in the normal phase, the total free energy is

$$F = F_N\frac{V_N}{V} + F_S\frac{V_S}{V} ; \qquad V_N + V_S = V .\tag{39}$$

To have, as required by (38), $dF = 0$ with respect to changes in the relative volumes of the two phases, we must have

$$\boxed{F_N(\mathcal{T}, H_c) = F_S(\mathcal{T}, H_c) .}\tag{40}$$

The free energies of the two phases are equal in equilibrium. It readily follows by our earlier arguments that the stabilization free energy density is

$$\Delta F \equiv F_N(\mathcal{T}, 0) - F_S(\mathcal{T}, 0) = \frac{1}{8\pi}H_c^2 ,\tag{41}$$

where the critical field H_c is now taken at the temperature $\mathcal{T}$. We have made the approximation that the free energy of the normal phase is independent of the magnetic field.

Scheme A (the permanent magnet) is simpler than scheme B (the solenoid + battery) for this problem, because with the magnet when the superconductor and normal phases coexist in thermal equilibrium at fixed H_c their relative volumes may be changed without any work being done by the magnet. With the solenoid it would be necessary for the battery to do work to maintain fixed $H_a = H_c$, because the flux through the solenoid is changed as the relative volumes of the two phases is changed.

REFERENCE

R. Becker and W. Döring, *Ferromagnetismus*, Springer, Berlin, 1939.

Appendix A
States of a Linear Polymer

We consider a model system closely related to the system of magnetic moments treated in Chapter 2. A linear polymeric molecule[1] consists of a long unbranched chain of small identical chemical units, as in

$$-R-R-R-R-R-\ \cdots\ -R-R\ ,$$

where R denotes the repeating chemical unit. The simplest linear polymer is polymethylene (usually called polyethylene), where R denotes the unit

$$\begin{array}{c} H \\ | \\ -C- \\ | \\ H \end{array}$$

The single bonds connect adjacent repeating units in the chain. Long chain molecules may be prepared which contain up to 10^6 or more units per chain.

Let us represent a repeat unit R of the polymer by an arrow $\rightarrow$ of length ρ which runs from the tail to the head of the unit. No polymer is known with the following property, but it is enlightening to consider as a model system a polymer in which the bond angles connecting adjacent units are equally likely to be either $0°$ or $180°$. Only these angles well be allowed. That is, we may connect two units R as

$$\begin{array}{cc} \rightarrow & \rightarrow \\ 1 & 2 \end{array}$$

or as

$$\begin{array}{c} 1 \\ \rightleftarrows \ , \\ 2 \end{array}$$

where we have slightly displaced the arrows vertically on the page to avoid overprinting.

---

[1] Good discussions of the statistics of linear polymers are given by P. J. Flory, *Principles of polymer chemistry*, 6th ed., Cornell University Press, 1966; C. Tanford, *Physical chemistry of macromolecules*, Wiley, 1963, Chapter 3; T. M. Birshtein and D. B. Ptitsyn, *Conformations of macromolecules*, Interscience, 1966.

The state of a polymer of N units is specified by the sequence of right-directed and left-directed arrows, as in:

$$\begin{array}{ccccccc} \rightarrow & \rightarrow & \leftarrow & \rightarrow & \rightarrow & \leftarrow & \rightarrow \\ 1 & 2 & 3 & 4 & 5 & 6 & \cdots N \end{array} . \tag{1}$$

This line is to be understood symbolically. The straight line distance from the tail of molecule 1 to the head of molecule N is not $N\rho$, as one might guess from the length of the typed line (1). The straight line distance from tail to head is given by the magnitude of

$$\mathbf{r} \equiv \sum_{s=1}^{N} \boldsymbol{\rho}_s \; ; \tag{2}$$

this is called the length.

Exactly as in the model system of N elementary magnets, there are 2^N states of the polymer. The states differ in the relative orientations of the sub-units. The classification of states according to length is conveniently given by the generating function

$$(\rightarrow + \leftarrow)^N , \tag{3}$$

by analogy with (2.11). The arrows here are horizontal because we have arbitrarily selected the horizontal axis as the axis of the chemical units R, whereas in the magnetic problem we took the axis of the spins as the vertical axis.

Problem 1. *Mean square length of model polymer.* Show that the mean square length of the model polymer introduced above is

$$\langle r^2 \rangle = N\rho^2 , \tag{4}$$

on the assumption that all states are equally likely. *Hint.* Evaluate

$$\langle \mathbf{r} \cdot \mathbf{r} \rangle = \left\langle \left(\sum_{s=1}^{N} \boldsymbol{\rho}_s \right) \cdot \left(\sum_{t=1}^{N} \boldsymbol{\rho}_t \right) \right\rangle = \sum_s \rho_s^2 + \left\langle \sum_{s \neq t} \boldsymbol{\rho}_s \cdot \boldsymbol{\rho}_t \right\rangle . \tag{5}$$

What is the value of $\langle \boldsymbol{\rho}_s \cdot \boldsymbol{\rho}_t \rangle$ if $s \neq t$ and if the directions of $\boldsymbol{\rho}_s$ and $\boldsymbol{\rho}_t$ are uncorrelated, as assumed for the model? To say that $\boldsymbol{\rho}_s$ and $\boldsymbol{\rho}_t$ are **uncorrelated** means there is no relation between the orientation $\leftarrow$ or $\rightarrow$ of $\boldsymbol{\rho}_s$ and the orientation $\leftarrow$ or $\rightarrow$ of $\boldsymbol{\rho}_t$.

The value of $\langle r^2 \rangle$ is of importance for the physical properties, such as viscosity and light scattering, of macromolecules in solution. (A polymer is often

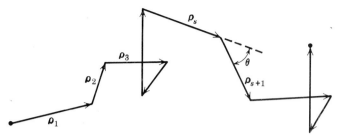

Figure 1   The unrestricted polymer chain. The bond lengths ρ are equal, but the vector ρ_s may terminate with equal probability anywhere on the surface of a sphere drawn about the head of ρ_{s-1} as center.

called a **macromolecule** if $N \gtrsim 100$.) If $N = 10^6$ and $\rho = 3$ Å, then the root mean square length $\sqrt{\langle r^2 \rangle} \approx 3000$ Å, whereas the curvilinear length is $N\rho = 3 \times 10^6$ Å.

Problem 2. *Free association of polymeric units.* Suppose now that the angles between successive groups R are entirely free and no longer restricted to $0°$ and $180°$. The direction of ρ_2 is entirely independent of ρ_1 and may be directed with equal probability in any element of solid angle. This model is a little more realistic than that given above. Show that the mean square length is

$$\langle r^2 \rangle = N\rho^2 , \tag{6}$$

where r is the straight line distance between the tail of the first group and the head of the Nth group. This model is called the unrestricted polymer chain (Fig. 1).

Problem 3. *Hindered bond angles. Suppose that successive chemical units in the chain make a fixed angle θ with each other, but otherwise the bonds are free to rotate. As shown in Fig. 2, the units ρ_s and ρ_{s+1} make an angle θ with each other, and the units ρ_{s+1} and ρ_{s+2} make the same angle θ with each other. In a polymethylene chain θ might be expected to be about $70°$, that is, $180°$ minus the tetrahedral angle $109°28'$ characteristic of the single bonds of the carbon atom. The plane formed by ρ_s and ρ_{s+1} makes a random

---

° This problem is interesting, but it may seem somewhat more difficult than Problem 2.

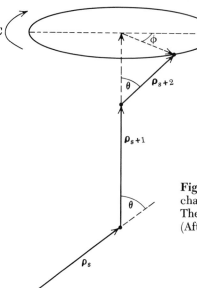

Figure 2 Three successive units of a polymethylene chain. The first two units are in the plane of the figure. The terminus ρ_{s+2} may lie anywhere on the circle C. (After Tanford.)

angle φ with the plane formed by ρ_{s+1} and ρ_{s+2}. On this model ρ_s and ρ_{s+1} are not uncorrelated, for

$$\langle \rho_s \cdot \rho_{s+1} \rangle = \rho^2 \cos \theta \ . \tag{7}$$

With some contemplation you will see that no two units are uncorrelated, for

$$\langle \rho_s \cdot \rho_{s+n} \rangle = (\cos \theta)^n \rho^2 \ . \tag{8}$$

This result is the heart of the problem. Show by summing the appropriate series that

$$\langle r^2 \rangle = N\rho^2 \frac{1 + \cos \theta}{1 - \cos \theta} \ , \tag{9}$$

for $N \gg 1$. By use of the appropriate θ for a tetrahedral bond, show for a polymethylene chain that $\langle r^2 \rangle = 2N\rho^2$.

This model still allows successive bonds more freedom to rotate than the actual polymers possess. In actual polymers the constraints against rotation act to increase the values of the ratio $\langle r^2 \rangle / N\rho^2$ above the value 2 for a tetrahedral bond. Flory gives the following experimental values:

Polymer	Unit	$\langle r^2 \rangle / N\rho^2$
Polymethylene	$-CH_2-$	6.7
Polyoxymethylene	$-CH_2-O-$	8 to 10
Polypeptides	$-NH-CH-CO-$ $\quad\quad\;\; \vert$ $\quad\quad CH_2R'$	8.5 to 9.5

Appendix B

Noninteracting Lattice Gas

The equation of state of the ideal gas of free atoms is

$$pV = N\tau .$$ (1)

This is derived in Chapter 11. From one point of view the physical origin of the pressure of the ideal gas is entirely in the kinetic energy of the atoms striking the walls of the container. It is remarkable that we can find a simple model having no kinetic energy but exhibiting the same equation of state, $pV = N\tau$. The mathematical model is called the **noninteracting lattice gas.** (The model is somewhat artificial, for if there is no energy at all there is really no way to define the temperature.)

The noninteracting lattice gas consists of N noninteracting atoms distributed among N_0 sites, with each site occupied by 0 or 1 atom. Every atom is on a site, but not every site is occupied. For given N_0 and N the number of independent arrangements is exactly the same as the number of states having N spins ↓ and $N_0 - N$ spins ↑, in a system with a total of N_0 spins. The spin problem was considered in Chapter 2; the solution for the number of states or arrangements is

$$g(N_0, N) = \frac{N_0!}{(N_0 - N)!\, N!} ,$$ (2)

when written in the present notation with the substitutions:

$$\tfrac{1}{2}N - m \rightarrow N ; \qquad \tfrac{1}{2}N + m \rightarrow N_0 - N .$$

The entropy is

$$\sigma(N_0, N) = \log g(N_0, N) .$$ (3)

We use the Stirling approximation $\log x! \cong x \log x - x$ to obtain

$$\sigma(N_0, N) \cong N_0 \log N_0 - (N_0 - N) \log (N_0 - N) - N \log N .$$ (4)

The Stirling approximation to the order used here is good if $x \gg 1$.

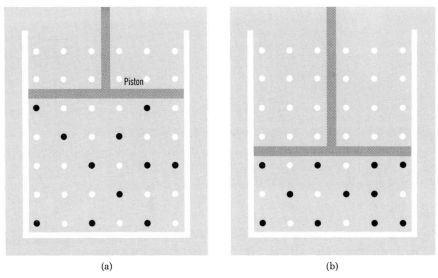

(a)   (b)

Figure 1   Compression of a lattice gas. The hollow circles represent the lattice sites. The solid circles represent atoms on lattice sites. We imagine that the piston compresses only the atoms and leaves the lattice sites unaffected.

The pressure is related to the entropy by

$$\frac{p}{T} = \left(\frac{\partial \sigma}{\partial V}\right)_{N,\,U} = \left(\frac{\partial \sigma}{\partial N_0}\right)_{N,\,U} \frac{dN_0}{dV} \;, \tag{5}$$

as in Chapter 7. Notice that in this problem the number of atoms N is kept constant as the volume is changed. The arrangement is shown in Fig. 1.

From (4) we have

$$\left(\frac{\partial \sigma}{\partial N_0}\right)_{N,\,U} = \log \frac{N_0}{N_0 - N} = -\log\left(1 - \frac{N}{N_0}\right). \tag{6}$$

We have all we need to solve for the pressure at any value of the fractional occupancy N/N_0. The result reduces to the ideal gas law when the fractional occupancy is small. For $N \ll N_0$ the logarithm may be expanded to give

$$\left(\frac{\partial \sigma}{\partial N_0}\right)_{N,\,U} = \frac{N}{N_0}. \tag{7}$$

The number of sites N_0 is related to the concentration of sites n and the volume V by

$$N_0 = nV \;; \qquad \frac{dN_0}{dV} = n. \tag{8}$$

We assume that the number of sites per unit volume is constant, independent of the volume.

Thus (5) becomes, from (7) and (8),

$$\frac{p}{T} = \frac{N}{N_0} n = \left(\frac{N}{nV}\right) n = \frac{N}{V} ,\tag{9}$$

or

$$pV = NT .\tag{10}$$

This is the equation of state of the lattice gas if there are no interactions among the atoms or sites, and if the fraction of occupied sites is $\ll 1$. In this problem the pressure arises entirely from the increase of entropy as the volume is increased. There is no kinetic energy. The entropy tends to increase on increasing the volume, and this tendency is restrained by the pressure.

Appendix C
Numerical Calculation of the
Chemical Potential of a Fermi Gas

The determination of the chemical potential as a function of temperature is important. It involves numerical integration which is lightened by the use of published tables[1] related to the FD distribution.

We saw in Chapter 14 that the number of electrons in a Fermi gas at temperature T is given by

$$N \equiv \langle N \rangle = \int_0^\infty d\epsilon \; \mathcal{D}(\epsilon) f(\epsilon, T) = \int_0^\infty d\epsilon \; \frac{\mathcal{D}(\epsilon)}{e^{(\epsilon - \mu)/T} + 1} . \tag{1}$$

But if the number of electrons in a specimen is constant, we must vary the value of μ with temperature in order to keep N fixed. This is the point of the present problem. The difficulty is that the problem has no simple explicit solution, but we must proceed numerically.

If we write

$$\mathcal{D}(\epsilon) = A\epsilon^{\frac{1}{2}} , \tag{2}$$

where the constant A is given in (6) below, then

$$N = A \int_0^\infty d\epsilon \; \frac{\epsilon^{\frac{1}{2}}}{e^{(\epsilon - \mu)/T} + 1} = AT^{\frac{3}{2}} \int_0^\infty dx \; \frac{x^{\frac{1}{2}}}{e^{x - \eta} + 1} , \tag{3}$$

where to write the integral in dimensionless form we have set

$$x \equiv \epsilon/T ; \qquad \eta = \mu/T .$$

---

[1] J. McDougall and E. C. Stoner, Philosophical Transactions of the Royal Society of London **237**, 67–104 (1938); E. C. Stoner, Philosophical Magazine **28**, 257–286 (1939). These tables also give quantities involved in the calculation of the energy and of the magnetic susceptibility. For further references consult A. Fletcher et al., *An index to mathematical tables*, 2nd ed., Addison-Wesley, 1962.

We write

$$\mathcal{I}(\eta) \equiv \int_0^\infty dx \, \frac{x^{\frac{1}{2}}}{e^{x-\eta}+1} \; ; \tag{4}$$

then by (3) the chemical potential $\mu \equiv T\eta$ must be determined so that

$$\mathcal{I}(\eta) \equiv \frac{N}{AT^{\frac{3}{2}}} \; . \tag{5}$$

In Problem (14.2) we found $\mathcal{D}(\epsilon_F) = 3N/2\epsilon_F$. On comparison with (2) evaluated at $\epsilon = \epsilon_F$ we find

$$A = \frac{3N}{2\epsilon_F^{\frac{3}{2}}} \; . \tag{6}$$

It follows that

$$\mathcal{I}(\eta) = \frac{2}{3}\left(\frac{\epsilon_F}{T}\right)^{\frac{3}{2}} \; . \tag{7}$$

We proceed to determine η. In Fig. 1 we have plotted $\mathcal{I}$ as a function of η, using Table 1. We calculate ϵ_F from the electron concentration:

$$\epsilon_F = \frac{\hbar^2}{2m}(3\pi^2 N/V)^{\frac{2}{3}} \; . \tag{8}$$

Then $T(\eta, \epsilon_F)$ is plotted versus η, as in Fig. 1 for the special value $\frac{2}{3}\epsilon_F^{\frac{3}{2}} = 1$. This special value, chosen for convenience, may be attained by suitable choice of the concentration. We multiply η by T to obtain the chemical potential μ, which is plotted against T in Fig. 14.9.

The energy, entropy, heat capacity, and free energy of a Fermi gas in three dimensions are plotted versus temperature in Fig. 2. Values of several thermodynamic functions are tabulated in Table 2.

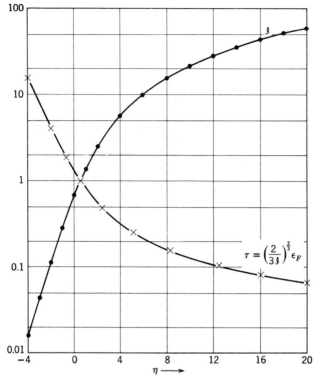

Figure 1 Plot of the Fermi-Dirac integral $\mathscr{I}$ versus $\eta\ (=\mu/T)$, from the tables by McDougall and Stoner. Also given is a plot of $\mathscr{I}^{-\frac{2}{3}}$, which is proportional to the temperature T.

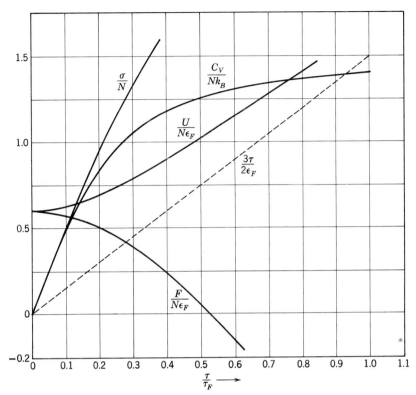

Figure 2 Thermodynamic functions for a Fermi gas. The plot gives the normalized energy, entropy, heat capacity, and free energy as functions of the normalized temperature. Also shown is the classical energy $\frac{3}{2}\tau$ normalized also to ϵ_F. (Courtesy of R. Cahn.)

Table 1 Fermi-Dirac Integral $\mathscr{I}$ as a Function of $\eta \equiv \mu/\tau$

η	$\mathscr{I}$	η	$\mathscr{I}$	η	$\mathscr{I}$	η	$\mathscr{I}$	η	$\mathscr{I}$
−4.0	0.016	0.0	0.678	4.0	5.770	8.0	15.380	12.0	27.951
−3.9	0.017	0.1	0.733	4.1	5.965	8.1	15.662	12.1	28.297
−3.8	0.019	0.2	0.792	4.2	6.163	8.2	15.945	12.2	28.645
−3.7	0.021	0.3	0.854	4.3	6.363	8.3	16.231	12.3	28.994
−3.6	0.023	0.4	0.920	4.4	6.566	8.4	16.518	12.4	29.344
−3.5	0.026	0.5	0.990	4.5	6.772	8.5	16.807	12.5	29.696
−3.4	0.029	0.6	1.063	4.6	6.980	8.6	17.097	12.6	30.050
−3.3	0.032	0.7	1.141	4.7	7.191	8.7	17.390	12.7	30.404
−3.2	0.035	0.8	1.222	4.8	7.404	8.8	17.684	12.8	30.760
−3.1	0.039	0.9	1.307	4.9	7.620	8.9	17.980	12.9	31.118
−3.0	0.043	1.0	1.396	5.0	7.837	9.0	18.277	13.0	31.477
−2.9	0.047	1.1	1.489	5.1	8.058	9.1	18.576	13.1	31.837
−2.8	0.052	1.2	1.586	5.2	8.281	9.2	18.877	13.2	32.199
−2.7	0.058	1.3	1.687	5.3	8.506	9.3	19.180	13.3	32.562
−2.6	0.064	1.4	1.792	5.4	8.733	9.4	19.484	13.4	32.927
−2.5	0.070	1.5	1.900	5.5	8.962	9.5	19.790	13.5	33.293
−2.4	0.077	1.6	2.013	5.6	9.194	9.6	20.097	13.6	33.660
−2.3	0.085	1.7	2.130	5.7	9.428	9.7	20.407	13.7	34.028
−2.2	0.094	1.8	2.250	5.8	9.665	9.8	20.717	13.8	34.398
−2.1	0.104	1.9	2.374	5.9	9.903	9.9	21.030	13.9	34.770
−2.0	0.114	2.0	2.502	6.0	10.144	10.0	21.344	14.0	35.142
−1.9	0.126	2.1	2.634	6.1	10.387	10.1	21.660	14.1	35.517
−1.8	0.138	2.2	2.769	6.2	10.631	10.2	21.977	14.2	35.892
−1.7	0.152	2.3	2.908	6.3	10.878	10.3	22.296	14.3	36.269
−1.6	0.167	2.4	3.050	6.4	11.127	10.4	22.616	14.4	36.647
−1.5	0.183	2.5	3.196	6.5	11.378	10.5	22.938	14.5	37.026
−1.4	0.201	2.6	3.345	6.6	11.632	10.6	23.262	14.6	37.407
−1.3	0.221	2.7	3.498	6.7	11.887	10.7	23.587	14.7	37.789
−1.2	0.242	2.8	3.654	6.8	12.144	10.8	23.913	14.8	38.172
−1.1	0.265	2.9	3.814	6.9	12.403	10.9	24.242	14.9	38.557
−1.0	0.290	3.0	3.976	7.0	12.664	11.0	24.571	15.0	38.943
−0.9	0.317	3.1	4.142	7.1	12.927	11.1	24.903	15.1	39.330
−0.8	0.346	3.2	4.311	7.2	13.192	11.2	25.235	15.2	39.718
−0.7	0.378	3.3	4.483	7.3	13.459	11.3	25.570	15.3	40.108
−0.6	0.412	3.4	4.658	7.4	13.728	11.4	25.905	15.4	40.499
−0.5	0.449	3.5	4.837	7.5	13.999	11.5	26.243	15.5	40.892
−0.4	0.489	3.6	5.018	7.6	14.271	11.6	26.581	15.6	41.285
−0.3	0.531	3.7	5.202	7.7	14.546	11.7	26.922	15.7	41.680
−0.2	0.577	3.8	5.388	7.8	14.822	11.8	27.263	15.8	42.076
−0.1	0.626	3.9	5.578	7.9	15.100	11.9	27.607	15.9	42.474
0.0	0.678	4.0	5.770	8.0	15.380	12.0	27.951	16.0	42.873

After J. McDougall and E. C. Stoner, Philosophical Transactions of the Royal Society of London **237**, 67–104 (1938).

Table 2   *Thermodynamic Functions of a Fermi Gas*

$\dfrac{\tau}{\tau_F}$	$\dfrac{\mu}{\epsilon_F}$	$\dfrac{U}{N\epsilon_F}$	$\dfrac{F}{N\epsilon_F}$	$\dfrac{\sigma}{N}$
0.00	1.00000	0.60000	0.60000	0.00000
0.05	0.99794	0.60615	0.59384	0.24612
0.10	0.99164	0.62428	0.57545	0.48831
0.15	0.98073	0.65336	0.54516	0.72132
0.20	0.96458	0.69152	0.50357	0.93975
0.25	0.94262	0.73671	0.45147	1.14095
0.30	0.91458	0.78724	0.38976	1.32493
0.35	0.88045	0.84181	0.31925	1.49303
0.40	0.84035	0.89949	0.24069	1.64699
0.45	0.79449	0.95960	0.15475	1.78854
0.50	0.74311	1.02164	0.06202	1.91925
0.55	0.68649	1.08525	−0.03701	2.04048
0.60	0.62487	1.15014	−0.14189	2.15338
0.65	0.55850	1.21609	−0.25223	2.25895
0.70	0.48761	1.28294	−0.36768	2.35802
0.75	0.41242	1.35054	−0.48793	2.45130
0.80	0.33314	1.41879	−0.61272	2.53939
0.85	0.24994	1.48760	−0.74180	2.62282
0.90	0.16300	1.55690	−0.87493	2.70204
0.95	0.07248	1.62663	−1.01194	2.77744
1.00	−0.02146	1.69674	−1.15262	2.84936
1.05	−0.11305	1.68303	−1.23507	2.91810
1.10	−0.19918	1.67084	−1.31307	2.98391
1.15	−0.28047	1.65994	−1.38710	3.04704
1.20	−0.35743	1.65015	−1.45754	3.10769
1.25	−0.43050	1.64132	−1.52472	3.16604
1.30	−0.50005	1.63332	−1.58893	3.22225
1.35	−0.56640	1.62605	−1.65044	3.27648
1.40	−0.62984	1.61941	−1.70945	3.32886
1.45	−0.69061	1.61334	−1.76617	3.37951
1.50	−0.74893	1.60776	−1.82077	3.42853
1.55	−0.80499	1.60262	−1.87340	3.47603
1.60	−0.85895	1.59788	−1.92421	3.52209
1.65	−0.91098	1.59350	−1.97331	3.56681
1.70	−0.96120	1.58943	−2.02082	3.61025
1.75	−1.00974	1.58565	−2.06684	3.65248
1.80	−1.05671	1.58212	−2.11146	3.69358
1.85	−1.10220	1.57883	−2.15476	3.73359
1.90	−1.14632	1.57576	−2.19682	3.77258
1.95	−1.18913	1.57288	−2.23772	3.81060
2.00	−1.23072	1.57018	−2.27750	3.84768

After E. C. Stoner, Philosophical Magazine **28**, 257–286 (1939).

Appendix D
Proof of the Virial Theorem

The **virial theorem** relates the average kinetic energy of particles bound by inverse-square-law forces to the average potential energy:

$$\langle \text{kinetic energy} \rangle = -\tfrac{1}{2} \langle \text{potential energy} \rangle , \tag{1}$$

where now the angle brackets denote the average for one system over very long times. The potential energy is defined to be zero at infinite separation of all particles.

We prove the virial theorem for inverse-square-law forces, such as gravitational forces. We consider N particles designated $1, 2, 3, \ldots, N$ of masses $M_1, M_2, \ldots, M_N$. The potential energy of the ith and jth particles is

$$U_{ij} = \frac{C_{ij}}{r_{ij}} , \tag{2}$$

where $C_{ij} \equiv -GM_i M_j = C_{ji}$ and $\mathbf{r}_{ij} = \mathbf{r}_i - \mathbf{r}_j$. The gravitational constant is G. The total potential energy is

$$U = \tfrac{1}{2} \sum_{i=1}^{N}{}' \sum_{j=1}^{N} \frac{C_{ij}}{r_{jj}} , \tag{3}$$

where the fraction $\tfrac{1}{2}$ enters because in the double sum each pair is counted twice. The prime on the Σ means that we exclude the terms $i = j$ from the sum.

We now write the N equations of motion, one for each particle:

$$M_i \frac{d\mathbf{v}_i}{dt} = \sum_{j=1}^{N}{}' \frac{C_{ij}}{r_{ij}^3} (\mathbf{r}_i - \mathbf{r}_j) . \tag{4}$$

We take the dot product of $\mathbf{r}_i$ with both sides and sum over all particles:

$$\sum_{i=1}^{N} M_i \left(\mathbf{r}_i \cdot \frac{d\mathbf{v}_i}{dt} \right) = \sum_{i=1}^{N}{}' \sum_{j=1}^{N} \frac{C_{ij}}{r_{ij}^3} (\mathbf{r}_i - \mathbf{r}_j) \cdot \mathbf{r}_i . \tag{5}$$

The left-hand side can be written as

$$\sum M_i\left(\mathbf{r}_i \cdot \frac{d\mathbf{v}_i}{dt}\right) = \sum\left[-M_i v_i^2 + \frac{d}{dt} M_i(\mathbf{r}_i \cdot \mathbf{v}_i)\right] \tag{6}$$

by virtue of the identity

$$\frac{d}{dt}(\mathbf{r} \cdot \mathbf{v}) = v^2 + \mathbf{r} \cdot \frac{d\mathbf{v}}{dt} \ . \tag{7}$$

On the right-hand side of (5) we have

$$\sum_i' \sum_j \frac{C_{ij}(\mathbf{r}_i - \mathbf{r}_j) \cdot \mathbf{r}_i}{r_{ij}^3} \ . \tag{8}$$

In this expression i and j are dummy indices which may be replaced by any other symbol without altering the value. Thus (8) is equal to

$$\sum_i' \sum_j \frac{C_{ij}(\mathbf{r}_j - \mathbf{r}_i) \cdot \mathbf{r}_j}{r_{ji}^3} \ . \tag{9}$$

But $r_{ij} = r_{ji}$ and $C_{ij} = C_{ji}$, so we can rewrite (9) as

$$-\sum_i' \sum_j \frac{C_{ij}}{r_{ij}^3}(\mathbf{r}_i - \mathbf{r}_j) \cdot \mathbf{r}_j \ . \tag{10}$$

Adding one-half of (8) to one-half of (10), and using

$$(\mathbf{r}_i - \mathbf{r}_j) \cdot (\mathbf{r}_i - \mathbf{r}_j) = r_{ij}^2 \ ,$$

we obtain for the right-hand side of (5)

$$\tfrac{1}{2}\sum_i' \sum_j \frac{C_{ij}}{r_{ij}} = U \ , \tag{11}$$

the potential energy.

The equality of (6) and (11) gives:

$$-\frac{1}{2}\frac{d}{dt}\left(\sum M_i \mathbf{r}_i \cdot \mathbf{v}_i\right) + (\text{total kinetic energy}) = -\tfrac{1}{2}(\text{total potential energy}) \ . \tag{12}$$

The long-time average of the first term on the left-hand side of (12) is zero if the particles remain indefinitely within some finite volume. The average involves many cycles of the motion, and $\mathbf{r} \cdot \mathbf{v}_1$ is positive as often as it is negative. We are left with the equality

$$\boxed{\ \langle\text{total kinetic energy}\rangle = -\tfrac{1}{2}\langle\text{total potential energy}\rangle \ , \ } \tag{13}$$

for a collection of particles bound by gravitational or electrostatic forces.

Appendix E
Statistical Mechanics in the Classical Limit

Statistical mechanics is simple to develop in terms of the quantum states of a system because the entropy of a closed system can be defined in a clear and decisive manner as the logarithm of the number of quantum states accessible to the system:

$$\sigma = \log g , \tag{1}$$

as in Chapter 4. When we try to formulate a classical, nonquantum version of statistical mechanics, we run into an immediate problem because a quantum state has no exact analogue in classical mechanics. In the absence of quantization we do not know what it is we are to count. This problem is very serious, but its solution is known.

The development of a single system of N atoms in the course of time is known when we know the values of the $6N$ coordinate and momentum variables p and q as functions of time. We can represent the evolution graphically as a single orbit in the $6N$ dimensional space of the p's and q's. This space is known as the **phase space** of the system. In Fig. 1 the notation $[p]$, $[q]$ indicates schematically the $3N$ momenta and the $3N$ coordinates.

The point that represents the system moves along the orbit. Physical properties of the system are calculated as time averages along the orbit. We may also construct an ensemble, as in Fig. 2, and calculate averages at a single time over the ensemble.

The closest classical analogue of a quantum state is a volume element

$$dp_1 \cdots dp_{3N} \, dq_1 \cdots dq_{3N} \equiv d\mathbf{p} \, d\mathbf{q} \tag{2}$$

in the phase space of the system. The volume in phase space associated with a given number of systems of the ensemble can be shown to be independent of time, for closed systems. (This result is a statement of the theorem of Liouville, a theorem proved in most books on statistical mechanics.) More precisely, let an ensemble be specified by giving the number of systems

$$P(\mathbf{p}, \mathbf{q}) \, d\mathbf{p} \, d\mathbf{q} \tag{3}$$

in the volume element $d\mathbf{p} \, d\mathbf{q}$ of phase space. In the course of time the repre-

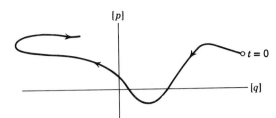

Figure 1   Orbit of a system in phase space.

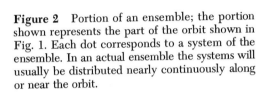

Figure 2   Portion of an ensemble; the portion shown represents the part of the orbit shown in Fig. 1. Each dot corresponds to a system of the ensemble. In an actual ensemble the systems will usually be distributed nearly continuously along or near the orbit.

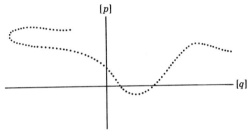

sentative points move in phase space; we say that they move along flow lines. The Liouville theorem is that the time rate of change of P along a flow line is zero; thus the volume associated with a given number of representative points (points that represent systems of an ensemble) is conserved in the motion. The conservation of volume in phase space is the closest thing in classical mechanics to the conservation of the number of accessible quantum states.

This argument suggests that in a classical transcription we should associate some volume of phase space with every quantum state. But what is the size of the volume element? We now look for a volume element V_p in phase space that will reproduce the result (18.58) for a single free particle in a volume V in three dimensions:

$$Z = \frac{V}{(2\pi\hbar)^3} (2\pi MT)^{\frac{3}{2}} . \tag{4}$$

This quantum result is correct as written for a single particle. We try to reproduce this with a classical partition function

$$Z = \frac{1}{V_P} \int_{-\infty}^{\infty}\int_{-\infty}^{\infty}\int_{-\infty}^{\infty} dp_x \, dp_y \, dp_z \int_V dq_x \, dq_y \, dq_z \, e^{-p^2/2MT} , \tag{5}$$

where now an integral over the volume element $dp \, dq$ replaces the quantum sum over states. In the exponent $p^2/2M$ is the energy ϵ of the system. The volume element V_P makes the partition function dimensionless, and our object is to find V_P.

We integrate over $d\mathbf{q}$ to obtain the volume V and rearrange the integral over $d\mathbf{p}$ to obtain

$$Z = \frac{V}{V_P} \cdot 4\pi \int_0^\infty p^2 e^{-p^2/2MT} \, dp = \frac{V}{V_P} \cdot 4\pi (2MT)^{\frac{3}{2}} \int_0^\infty s^2 e^{-s^2} \, ds$$

$$= \frac{V}{V_P} \cdot (2\pi MT)^{\frac{3}{2}} \ . \tag{6}$$

This is identical with the quantum result (4) if

$$V_P = (2\pi\hbar)^3 \ . \tag{7}$$

For a system of N particles we infer that

$$V_P = (2\pi\hbar)^{3N} \ . \tag{8}$$

This ensures that Z is dimensionless. (The integration is carried out over the $6N$ dimensional phase space). For N identical particles in the classical limit of $V \gg NV_Q$ we must also include in the partition function the factor $1/N!$ as derived in (18.59). Thus the classical partition function is

$$Z = \frac{1}{N!(2\pi\hbar)^{3N}} \iint e^{-\mathcal{H}(\mathbf{p},\,\mathbf{q})/T} \, d\mathbf{p} \, d\mathbf{q} \ . \tag{9}$$

This approximation will give correct results only in the classical limit.

Appendix F
Work and the Hamiltonian in an Electric Field

The motion of a system of charges q_i in a static external electric field is described by the hamiltonian

$$\mathcal{H} = \mathcal{H}_0 + \sum_i q_i \varphi(\mathbf{r}_i) , \tag{1}$$

where $\varphi(\mathbf{r})$ is the electrostatic potential function of the field:

$$\mathbf{E}(\mathbf{r}) = -\text{grad } \varphi(\mathbf{r}) . \tag{2}$$

We may expand $\varphi(\mathbf{r}_i)$ about an origin at $\mathbf{r}_0$:

$$\varphi(\mathbf{r}_i) = \varphi(\mathbf{r}_0) + (\mathbf{r}_i - \mathbf{r}_0) \cdot \nabla \varphi(\mathbf{r}_0) + \cdots , \tag{3}$$

so that

$$\Sigma q_i \varphi(\mathbf{r}_i) = (\Sigma q_i)[\varphi(\mathbf{r}_0) - \mathbf{r}_0 \cdot \nabla \varphi(\mathbf{r}_0)] + (\Sigma q_i \mathbf{r}_i) \cdot \nabla \varphi(\mathbf{r}_0) + \cdots . \tag{4}$$

For a neutral system the total charge $\Sigma q_i = 0$. By the definition of the total dipole moment we have

$$\boldsymbol{\mathcal{P}} = \Sigma q_i \mathbf{r}_i , \tag{5}$$

whence, using $\mathbf{E} = -\nabla \varphi$,

$$\mathcal{H}(\mathbf{E}) = \mathcal{H}_0 - \boldsymbol{\mathcal{P}} \cdot \mathbf{E} + \cdots . \tag{6}$$

The neglected terms are of higher order in the expansion (4) and are called quadrupole terms, octopole terms, and so on. Here $\mathcal{H}_0$ contains those parts of the hamiltonian that do not refer to the electric field of fixed charges external to the system.

The energy eigenvalue $\epsilon_l(E)$ of the state l of the system when in the electric field is given by the Schrödinger equation

$$\mathcal{H}(E)\psi_l(E) = \epsilon_l(E)\psi_l(E) . \tag{7}$$

We use this to obtain an expression for the energy: the diagonal matrix elements of (7) are

$$\epsilon_l(E) = (\psi_l(E), \mathcal{H}_0\psi_l(E)) - \mathbf{E} \cdot (\psi_l(E), \boldsymbol{P}\psi_l(E)) , \tag{8}$$

and, for $E = 0$,

$$\epsilon_l(0) = (\psi_l(0), \mathcal{H}_0\psi_l(0)) . \tag{9}$$

The energy change on application of the electric field is

$$\Delta U_A(l) = \epsilon_l(E) - \epsilon_l(0) = (\psi_l(E), \mathcal{H}_0\psi_l(E))$$
$$- (\psi_l(0), \mathcal{H}_0\psi_l(0)) - \mathbf{E} \cdot (\psi_l(E), \boldsymbol{P}\psi(E)) . \tag{10}$$

The first two terms on the right-hand side correspond to ΔU_B, the polarization energy of the frozen polarized system as observed in zero field, as discussed in Chapter 22. The third term is the energy of interaction of the dipole $\boldsymbol{P}$ with the field $\mathbf{E}$. Thus (10) may be written as

$$\epsilon_l(E) - \epsilon_l(0) = \Delta U_A(l) = \Delta U_B(l) - \overline{\boldsymbol{P}} \cdot \mathbf{E} , \tag{11}$$

where $\overline{\boldsymbol{P}}$ is the quantum average of the dipole moment.

We see the importance of the measurement process A, for in this process the work ΔU_A done on the system is equal to the change in the energy eigenvalue of the system on application of the electric field. On taking thermal averages we have

$$\Delta U_A = \langle \epsilon(E) - \epsilon(0) \rangle = \langle \mathcal{H}(E) - \mathcal{H}_0 \rangle . \tag{12}$$

The partition function for the system in the electric field is closely related to W_A, for the partition function involves the energy eigenvalues in the electric field:

$$Z(E) = \sum_l e^{-\epsilon_l(E)/T} . \tag{13}$$

The free energy of the system in the electric field is related to the partition function in the usual way:

$$F(E) = -T \log Z(E) . \tag{14}$$

We might write this as $F_A(E)$ because it is based on the definition of the energy by process A.

By definition W_A is the work done on a dielectric system in moving it from infinity to a point r_1 in the electric field of fixed charges. This work may be determined in principle by the rudimentary means sketched in Chapter 22, or from the effect of an electric field on the frequency of a photon emitted by an atom in an electric field $\mathbf{E}$.

Appendix G
Poisson Distribution

A famous result of probability theory is known as the Poisson distribution law. The result is exceedingly useful in the design and analysis of counting experiments in physics, biology, operations research, and engineering. The statistical methods we have developed lend themselves to an elegant derivation of the Poisson law, which is concerned with the occurrence of small numbers of objects in random sampling processes. For example, if on the average there is one bad penny in a thousand, what is the probability that no bad pennies will be found in a given sample of one hundred pennies? The problem was first considered and solved[1] in a curious and remarkable study of the role of luck in criminal and civil law trials in France in the early nineteenth century.

We may derive the Poisson distribution law with the aid of the modified lattice gas of Fig. 1. As a model system we consider a large number R of independent lattice sites in thermal and diffusive contact with a gas. The gas serves as a reservoir. Each lattice site may adsorb zero or one atom.

We want to find the probabilities

$$P(0), P(1), P(2), \ldots, P(n), \ldots ,$$

that a total of $0, 1, 2, \ldots, n, \ldots$, atoms are adsorbed on the R sites, if we are given the average number $\langle n \rangle$ of adsorbed atoms over an ensemble of similar systems. The **Poisson distribution** is the solution to the problem.

Consider a system composed of a single site, as in Fig. 2. It is convenient to set the binding energy of an atom to the site as zero. The identical form for the distribution is found if a binding energy is included in the calculation. The grand sum is

$$\mathfrak{Z}_1 = 1 + \lambda , \tag{1}$$

where the term λ is proportional to the probability the site is occupied, and

[1] S. D. Poisson, *Recherches sur la probabilité des jugements en matière criminelle et en matière civile*, Paris, 1837.

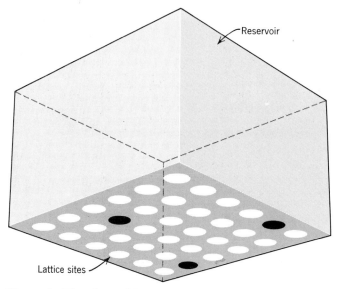

Figure 1 The plane of lattice sites is in thermal and diffusive contact with the gas in the container. The atoms in the gas are not pictured. The solid circles represent lattice sites, each with one atom adsorbed.

the term 1 is proportional to the probability the site is vacant. Thus the absolute probability that the site is occupied is

$$f = \frac{\lambda}{1 + \lambda} . \tag{2}$$

We note that if $\lambda \ll 1$, then $f \cong \lambda$. The actual value of λ is determined by the condition of the gas in the reservoir, because for diffusive contact between the lattice and the reservoir we must have

$$\lambda(\text{lattice}) = \lambda(\text{gas}) , \tag{3}$$

by the argument of Chapter 5. The evaluation of $\lambda(\text{gas})$ for an ideal gas was given in Chapter 11.

We now extend the treatment to R independent sites, as in Fig. 1. Then

$$\mathfrak{Z}_{\text{tot}} = \mathfrak{Z}_1 \mathfrak{Z}_2 \cdots \mathfrak{Z}_R = (1 + \lambda)^R . \tag{4}$$

By the argument used in Chapter 2 we know that the binomial expansion of $(\bigcirc + \bullet)^R$ or $(1 + \lambda)^R$ is a device that counts once and only once every state of the system of R sites. Each site has two alternative states, namely $\bigcirc$ for vacant or $\bullet$ for occupied, which corresponds in the grand sum to the term 1 for λ^0 and the term λ for λ^1.

In the low-occupancy limit of $f \ll 1$ we saw that $f \cong \lambda$, whence

$$\langle n \rangle = fR = \lambda R \tag{5}$$

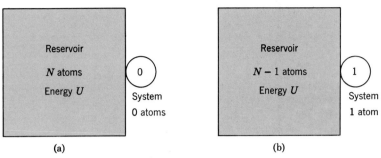

(a)   (b)

Figure 2 System of one site in contact with a reservoir. The binding energy of an atom to the site has been taken as zero. In (a) the site is vacant; in (b) the site is occupied by one atom.

is the average total number of adsorbed atoms. The Poisson distribution is concerned with this low-occupancy limit. We can now write (4) as

$$\mathcal{Z}_{tot} = \left(1 + \frac{\lambda R}{R}\right)^R = \left(1 + \frac{\langle n \rangle}{R}\right)^R . \tag{6}$$

Next we let the number of sites R increase without limit, while holding the average number of occupied sites $\langle n \rangle$ constant. (Remember that the Poisson distribution is concerned with infrequent events!) By the definition of the exponential function we have

$$\lim_{R \to \infty} \left(1 + \frac{\langle n \rangle}{R}\right)^R = e^{\langle n \rangle} , \tag{7}$$

so that

$$\mathcal{Z}_{tot} \cong e^{\langle n \rangle} = e^{\lambda R} = \sum_n \frac{(\lambda R)^n}{n!} . \tag{8}$$

The last step here is the expansion of the exponential function in a power series.

The term in λ^n in $\mathcal{Z}_{tot}$ is proportional to the probability $P(n)$ that n sites are occupied. With the grand sum as the normalization factor we have in the limit $R \to \infty$:

$$P(n) = \frac{\lambda^n R^n}{n!} \cdot \frac{1}{\mathcal{Z}_{tot}} = \frac{\lambda^n R^n e^{-\lambda R}}{n!} , \tag{9}$$

or, because $\lambda R = \langle n \rangle$,

$$P(n) = \frac{\langle n \rangle^n e^{-\langle n \rangle}}{n!} . \tag{10}$$

This is the **Poisson distribution law.**

Table 1 Values of the Poisson Distribution Function $P(n) = \dfrac{\langle n \rangle^n e^{-\langle n \rangle}}{n!}$

	$\langle n \rangle$									
	0.1	0.3	0.5	0.7	0.9	1	2	3	4	5
$P(0)$	0.9048	0.7408	0.6065	0.4966	0.4066	0.3679	0.1353	0.0498	0.0183	0.0067
$P(1)$	0.0905	0.2222	0.3033	0.3476	0.3659	0.3679	0.2707	0.1494	0.0733	0.0337
$P(2)$	0.0045	0.0333	0.0758	0.1217	0.1647	0.1839	0.2707	0.2240	0.1465	0.0842
$P(3)$	0.0002	0.0033	0.0126	0.0284	0.0494	0.0613	0.1805	0.2240	0.1954	0.1404
$P(4)$		0.0003	0.0016	0.0050	0.0111	0.0153	0.0902	0.1680	0.1954	0.1755
$P(5)$			0.0002	0.0007	0.0020	0.0031	0.0361	0.1008	0.1563	0.1755
$P(6)$				0.0001	0.0003	0.0005	0.0120	0.0504	0.1042	0.1462
$P(7)$						0.0001	0.0034	0.0216	0.0595	0.1044
$P(8)$							0.0009	0.0081	0.0298	0.0653
$P(9)$							0.0002	0.0027	0.0132	0.0363
$P(10)$								0.0008	0.0053	0.0181

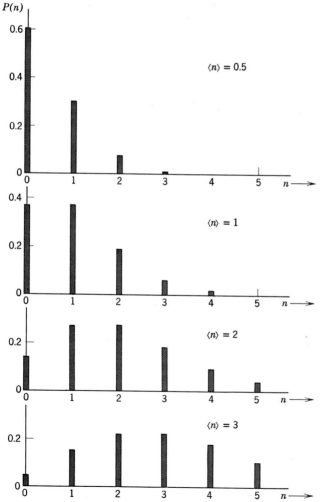

Figure 3 Poisson distribution, P versus n, for several values of $\langle n \rangle$.

Particular interest attaches to the probability $P(0)$ that none of the sites is occupied. From (11) we find, with $\langle n \rangle^0 = 1$ and $0! = 1$,

$$P(0) = e^{-\langle n \rangle} ; \qquad \log P(0) = -\langle n \rangle . \qquad (11)$$

Thus the probability of zero occupancy is simply related to the average number $\langle n \rangle$ of occupied sites. This suggests a simple experimental procedure for the determination of $\langle n \rangle$: just count the *systems* that have no adsorbed atoms.

Values of $P(n)$ for several values of $\langle n \rangle$ are given in Table 1. Plots are given in Fig. 3 for $\langle n \rangle = 0.5, 1, 2,$ and 3.

EXAMPLE. *Incorrect and correct counting of states.* (a) The grand sum for the R sites is *not*

$$\mathcal{Z}_{tot} = 1 + \lambda + \lambda^2 + \lambda^3 + \cdots + \lambda^R . \qquad (12)$$

Why not?

(b) The grand sum is

$$\mathcal{Z}_{tot} = (1 + \lambda)^R = 1 + R\lambda + \frac{R(R-1)}{2!}\lambda^2 + \cdots + \lambda^R = \sum_{n=0}^{R} g(R, n)\lambda^n , \qquad (13)$$

where

$$g(R, n) = \frac{R!}{(R - n)! \, n!}$$

is the binomial coefficient. Note that $g(R, n)$ is the number of independent *states* of the system for a given number of atoms n. The grand sum is a sum over all states and not over all energy levels.

Problem 1. *Poisson distribution in molecular biology.*

References. E. L. Ellis and M. Delbrück, "The growth of bacteriophage," Journal of General Physiology **22**, 365 (1939); G. Stent, *Molecular biology of bacterial viruses,* Freeman, 1963. A great textbook.

The classical simple system of molecular biology is the interaction of certain viruses (called bacteriophage) with *E. coli B* bacteria from sewage. A virus particle can only multiply inside a bacterium. The virus enters the bacterium as in Fig. 4 and takes over the biochemical machinery of the cell. The virus multiplies within the cell. After about 20 min at 37 C the cell wall of the bacterium breaks up. This destroys the bacterium and releases approximately 100 new virus particles, all duplicates of the single original virus that infected the bacterium.

The first experiment in laboratory courses in molecular biology is often

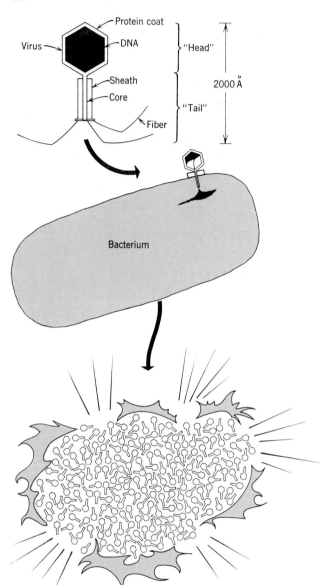

Figure 4 The infectious cycle of a virulent bacteriophage. The mature particle, shown in longitudinal section, attaches by its tail fibers to a bacterium. The sheath contracts, the core penetrates the cell surface, and the DNA of the particle passes into the bacterium. About twenty minutes later at 37 C the cell bursts and a few hundred new mature particles are released. (Franklin W. Stahl, The Mechanics of Inheritance, © 1964. Reprinted by permission of Prentice-Hall, Inc., Englewood Cliffs, New Jersey.)

the determination of the absolute concentration of virus particles in a solution. This may be done by a clever application of the Poisson distribution.

We describe with a minor modification the Ellis-Delbrück experiment. Suppose that we have 100 test tubes filled with *E. coli* bacteria in a nutrient solution. If left to incubate overnight at 37 C, all 100 test tubes should appear quite cloudy in the morning, because the bacteria multiply considerably in the nutrient solution and their dimensions ($\sim 1\ \mu$) are favorable for scattering visible light.

If a single virus particle is added to a test tube that contains *E. coli*, and then the tube is incubated overnight, the result will be a clear test tube. Why? The virus multiplies much more rapidly than the bacteria. The virus particles on incubation can break up and destroy every bacterium present in the tube. The viruses themselves and the bacterial fragments are too small to scatter visible light efficiently, and thus the resulting solution is clear. If two or more viruses are present initially in a tube, the result is indistinguishable from one virus particle—the tube is clear.

We imagine that 1 ml from a large vessel of virus solution is added to each of 100 tubes, and that after incubation 39 of these tubes are found to be cloudy. What is the average number of virus particles in 1 ml of the original virus solution?

EXAMPLE. *Elementary derivation of P(0)*. Let a total of N bacteria be distributed at random among L dishes. Each dish is viewed as a system of many sites to which a bacterium may attach. The L dishes represent an ensemble of L identical systems. The average number of bacteria per dish is

$$\langle n \rangle = \frac{N}{L} . \tag{14}$$

Each time a bacterium is distributed, the probability that a given dish will receive that bacterium is $1/L$. The probability the given dish will not receive the bacterium is

$$\left(1 - \frac{1}{L}\right) . \tag{15}$$

The probability in N tries that the given dish will receive no bacteria is

$$P(0) = \left(1 - \frac{1}{L}\right)^N \tag{16}$$

because the factor (15) enters on each try.

We may write (16) as

$$P(0) = \left(1 - \frac{\langle n \rangle}{N}\right)^N , \tag{17}$$

by use of $\langle n \rangle = N/L$. We know that in the limit of large N,

$$e^{-\langle n \rangle} = \lim_{N \to \infty} \left(1 - \frac{\langle n \rangle}{N}\right)^N , \tag{18}$$

by the definition of the exponential function. Thus for $N \gg 1$ and $L \gg 1$ we have

$$P(0) = e^{-\langle n \rangle} , \tag{19}$$

in agreement with (11).

Problem 2. *Random pulses.* A radioactive source emits alpha particles which are counted at an average rate of one per second. (a) What is the probability of counting exactly 10 alpha particles in 5 sec? (b) Of counting 2 in 1 sec? (c) Of counting none in 5 sec? (Notice that the answers to (a) and (b) are not identical.)

REFERENCE

T. C. Fry, *Probability and its engineering uses,* Van Nostrand, 1928. A classic elementary textbook.

Appendix H
Nyquist Theorem

The Nyquist theorem is concerned with the spontaneous thermal fluctuations of voltage across an electric circuit element. It is of great importance in experimental physics and in electronics. The theorem gives a quantitative expression for the thermal noise voltage generated by a resistor in thermal equilibrium. The theorem is therefore needed in any estimate of the limiting signal-to-noise ratio of an experimental apparatus. In the original form[1] the Nyquist theorem states that the mean square voltage across a resistor of resistance R in thermal equilibrium at temperature T is given by

$$\langle V^2 \rangle = 4Rk_BT\,\Delta f \, , \tag{1}$$

where Δf is the frequency[2] bandwidth within which the voltage fluctuations are measured; all frequency components outside the given range are ignored. We show below that the thermal noise power per unit frequency range delivered by a resistor to a matched load is k_BT; the factor 4 enters (1) where it does because in the circuit of Fig. 1 the power delivered to an arbitrary resistive load R' is

$$\langle I^2 \rangle R' = \frac{\langle V^2 \rangle R'}{(R + R')^2} \, , \tag{2}$$

which at match $(R' = R)$ is $\langle V^2 \rangle / 4R$.

Consider as in Fig. 2 a lossless transmission line of length l and characteristic impedance $Z_c = R$ terminated at each end by a resistance R. The line is therefore matched at each end, in the sense that all energy traveling down the line will be absorbed without reflection in the appropriate resistance. The entire circuit is maintained at temperature T.

A transmission line is essentially an electromagnetic cavity in one dimension. We follow the argument given in Chapter 15 for the distribution of

---

[1] H. Nyquist, Physical Review **32**, 110 (1928); a deeper discussion is given by C. Kittel, *Elementary statistical physics*, Wiley, 1958, Sections 27–30.

[2] In this appendix the word frequency refers to cycles per unit time, and not to radians per unit time.

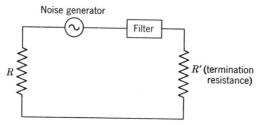

Figure 1 Equivalent circuit for a resistance R with a generator of thermal noise that delivers power to a load R'. The current

$$I = \frac{V}{R + R'} \; ,$$

so that the mean square power dissipated in the load is

$$\mathcal{P} = \langle I^2 \rangle R' = \frac{\langle V^2 \rangle R'}{(R' + R)^2} \; ,$$

which is a maximum with respect to R' when $R' = R$. In this condition the load is said to be matched to the power supply. At match, $\mathcal{P} = \langle V^2 \rangle / 4R$. The filter enables us to limit the frequency bandwidth under consideration; that is, the bandwidth to which the mean square voltage fluctuation applies.

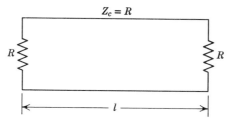

Figure 2 Transmission line of length l with matched terminations, as conceived for the derivation of the Nyquist theorem. The characteristic impedance Z_c of the transmission line has the value R. According to the fundamental theorem of transmission lines, the terminal resistors are matched to the line when their resistance has the same value R.

photons in thermal equilibrium, but now we treat a space of one dimension instead of three dimensions. The transmission line has two phonon or electromagnetic modes (one propagating in each direction) in the frequency range

$$\delta f = \frac{c'}{l} \; , \tag{3}$$

where c' is the propagation velocity on the line. There is only one polarization mode on the line. Each mode has energy

$$\frac{\hbar \omega}{e^{\hbar \omega / k_B T} - 1} \tag{4}$$

in equilibrium, according to the Planck distribution (15.7). We are usually concerned with circuits in the classical limit $\hbar \omega \ll k_B T$, so that the thermal en-

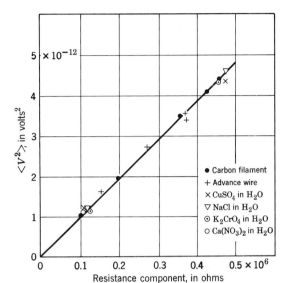

Figure 3   Voltage squared versus resistance for various kinds of conductors, including electrolytes. (After J. B. Johnson.)

ergy per mode is $k_B T$. It follows that the energy on the line in the frequency range Δf is

$$2k_B T \frac{\Delta f}{\delta f} = \frac{2k_B T l}{c'} \Delta f .\tag{5}$$

The rate at which energy comes off the line in *one* direction is

$$k_B T \, \Delta f .\tag{6}$$

The power coming off the line at one end is all absorbed in the terminal impedance R at that end; there are no reflections when the terminal impedance is matched to the line. The power input[3] to the load is

$$\mathcal{P} = \langle I^2 \rangle R = k_B T \, \Delta f ,\tag{7}$$

but $V = 2RI$, so that

$$\boxed{\langle V^2 \rangle = 4k_B T R \, \Delta f .}\tag{8}$$

This result is the Nyquist theorem on the thermal noise voltage across a resistor.

The dependence of $\langle V^2 \rangle$ on R and on T was discovered experimentally by J. B. Johnson.[4] He determined the Boltzmann constant k_B from the observed noise power and obtained a value within 8 percent of the correct value. His results exhibiting the dependence of $\langle V^2 \rangle$ on R at constant temperature and Δf are shown in Fig. 3.

[3] In thermal equilibrium the load must emit energy to the line at the same rate, or else its temperature would rise.

[4] J. B. Johnson, Physical Review **32**, 97 (1928).

Appendix I
Boltzmann Transport Equation

We introduce the classical theory of transport processes, using the Boltzmann transport equation. The transport equation method is very useful in dealing with flow processes, and in many circumstances the method is easy to use.

We work in the six-dimensional space of cartesian coordinates $\mathbf{r}$ and velocity $\mathbf{v}$. The distribution function $f(\mathbf{r}, \mathbf{v})$ is defined by the relation

$$f(\mathbf{r}, \mathbf{v}) \, d\mathbf{r} \, d\mathbf{v} = \text{number of particles in } d\mathbf{r} \, d\mathbf{v} \; . \tag{1}$$

At a point $\mathbf{r}, \mathbf{v}$ the time rate of change $\partial f/\partial t$ may be caused by the drift of particles in and out of the volume element and also by collisions among the particles:

$$\frac{\partial f}{\partial t} = \left(\frac{\partial f}{\partial t}\right)_{\text{drift}} + \left(\frac{\partial f}{\partial t}\right)_{\text{collisions}} \; . \tag{2}$$

We assume that the number of particles is conserved; if not, terms representing the generation and recombination of particles must be added to the right-hand side of (2). Such additional terms are required, for example, in the theory of the transistor and in nuclear reactor theory.

The simplest way to derive the Boltzmann equation is by the following argument. Consider the effect of a time displacement dt on the distribution function $f(t, \mathbf{r}, \mathbf{v})$. If we follow along a flow line, the Liouville theorem of mechanics tells us that the distribution is conserved:

$$f(t + dt, \mathbf{r} + d\mathbf{r}, \mathbf{v} + d\mathbf{v}) = f(t, \mathbf{r}, \mathbf{v}) \; , \tag{3}$$

apart from the effect of collisions. We have

$$f(t + dt, \mathbf{r} + d\mathbf{r}, \mathbf{v} + d\mathbf{v}) - f(t, \mathbf{r}, \mathbf{v}) = dt \left(\frac{\partial f}{\partial t}\right)_{\text{collisions}} \; , \tag{4}$$

where the right-hand side gives the effect of collisions. Thus

$$dt \frac{\partial f}{\partial t} + d\mathbf{r} \cdot \text{grad}_{\mathbf{r}} f + d\mathbf{v} \cdot \text{grad}_{v} f = \left(\frac{\partial f}{\partial t}\right)_{\text{collisions}} \; , \tag{5}$$

or, letting $\boldsymbol{\alpha}$ denote the acceleration dv/dt,

$$\boxed{\frac{\partial f}{\partial t} + \mathbf{v} \cdot \mathrm{grad}_r\, f + \boldsymbol{\alpha} \cdot \mathrm{grad}_v\, f = dt \left(\frac{\partial f}{\partial t}\right)_{\text{collisions}}.} \tag{6}$$

This is the Boltzmann equation in a general form. The acceleration will be expressed in terms of the external forces that act on a particle.

We may give the derivation in somewhat different language. If the number of particles is to be conserved, we must have

$$\left(\frac{\partial f}{\partial t}\right)_{\text{drift}} + \mathrm{div}\, f\mathbf{u} = 0 , \tag{7}$$

where $\mathbf{u}$ is the velocity vector in the six-dimensional space:

$$\mathbf{u} \equiv (\alpha_x, \alpha_y, \alpha_z, v_x, v_y, v_z) .$$

Now by a vector identity

$$\mathrm{div}\, f\mathbf{u} = f\, \mathrm{div}\, \mathbf{u} + \mathbf{u} \cdot \mathrm{grad}\, f .$$

The Liouville theorem tells us that $\mathrm{div}\, \mathbf{u} = 0$. Thus

$$\left(\frac{\partial f}{\partial t}\right)_{\text{drift}} = -\boldsymbol{\alpha} \cdot \mathrm{grad}_v\, f - \mathbf{v} \cdot \mathrm{grad}_r\, f , \tag{8}$$

in agreement with (6).

The collision term $(\partial f/\partial t)_{\text{collisions}}$ may require special treatment, but in many problems it is possible to justify approximately the introduction of a relaxation time[1] $\mathcal{T}_c(\mathbf{r}, \mathbf{v})$ defined by the equation

$$\left(\frac{\partial f}{\partial t}\right)_{\text{collisions}} = -\frac{(f - f_0)}{\mathcal{T}_c} , \tag{9}$$

where f_0 is the distribution function in thermal equilibrium. Let us suppose that a nonequilibrium distribution of velocities is set up by external forces suddenly removed. The decay of the distribution toward equilibrium is then obtained from (9). We note that by definition $\partial f_0/\partial t = 0$, so that

$$\frac{\partial(f - f_0)}{\partial t} = -\frac{f - f_0}{\mathcal{T}_c} , \tag{10}$$

which has the solution

$$(f - f_0)_t = (f - f_0)_{t=0}\, e^{-t/\mathcal{T}_c} . \tag{11}$$

---

[1] Here $\mathcal{T}_c$ is a time and not the temperature.

Combining (2), (8), and (10), we have the Boltzmann transport equation in the relaxation time approximation:[2]

$$\frac{\partial f}{\partial t} + \boldsymbol{\alpha} \cdot \text{grad}_v\, f + \mathbf{v} \cdot \text{grad}_r\, f = -\frac{f - f_0}{T_c} \, . \tag{12}$$

In the steady state, $\partial f / \partial t = 0$.

Electrical conductivity in an electron gas

We consider a specimen with an electric field E in the x direction and a temperature gradient dT/dx. Our program is to solve the Boltzmann equation approximately for the distribution function and then to find the flux of electric charge and of energy. We restrict ourselves to the steady state (dc conditions), so that $\partial f / \partial t = 0$. Then the transport equation (12) becomes, for particles of charge q and mass m in an electric field,

$$\frac{qE}{m}\frac{\partial f}{\partial u} + u\frac{\partial f}{\partial x} = -\frac{f - f_0}{T_c} \, , \tag{13}$$

because the acceleration is

$$\alpha = \frac{qE}{m} \, . \tag{14}$$

Here u is the x component of the velocity. Rewriting (13) we have

$$f = f_0 - T_c\left(\frac{qE}{m}\frac{\partial f}{\partial u} + u\frac{\partial f}{\partial x}\right), \tag{15}$$

where f_0 is the distribution function in thermal equilibrium. The subscript c on the relaxation time T_c distinguishes it from the temperature T. We now assume weak fields and small temperature gradients, so that the change in the distribution function will be small and terms in f involving squares and cross products of $f - f_0$ may be neglected. That is, we assume $(f - f_0) \ll 1$. To this approximation

$$f = f_0 - T_c\left(\frac{qE}{m}\frac{\partial f_0}{\partial u} + u\frac{\partial f_0}{\partial x}\right). \tag{16}$$

Higher-order effects may be found by an iterative procedure, using in each order the solution to the next lower order when evaluating the parentheses on the right-hand side of (15).

---

[2] If the introduction of a relaxation time is not justified, we have to treat the collision term in detail, introducing the transition probabilities for processes that take particles out of $d\mathbf{r}\, d\mathbf{v}$ and for processes that bring particles into this volume element. We are led in general to an integrodifferential equation.

Now f_0 is a function of the particle energy ϵ, the temperature T, and the chemical potential μ; the energy is a function of the velocity. Thus

$$\frac{\partial f_0}{dx} = \frac{\partial f_0}{\partial \mu}\frac{d\mu}{dx} + \frac{\partial f_0}{\partial T}\frac{dT}{dx} , \tag{17}$$

and

$$\frac{\partial f}{\partial u} = \frac{\partial f_0}{\partial \epsilon}\frac{d\epsilon}{du} = mu\frac{\partial f_0}{\partial \epsilon} . \tag{18}$$

The electrical conductivity is usually defined under the conditions $dT/dx = 0$ and $dn/dx = 0$, where n is the carrier concentration. Then $\partial f_0/\partial x = 0$, and (16) reduces to

$$f = f_0 - T_c qEu\frac{\partial f_0}{\partial \epsilon} . \tag{19}$$

The electric current density is given by, for particles of charge q,

$$j_q = \int quf\, dv = -T_c q^2 E \int u^2\left(\frac{\partial f_0}{\partial \epsilon}\right) dv , \tag{20}$$

as $\int uf_0\, dv = 0$ because f_0 is an even function of the velocity component u. In taking T_c out of the integral we are assuming that the relaxation time is independent of the velocity, but the theory is easily freed from this restriction. For the Maxwellian distribution

$$f_0 = n\left(\frac{m}{2\pi T}\right)^{\frac{3}{2}} e^{-mv^2/2T} , \tag{21}$$

where v is the magnitude of the velocity: $v^2 = v_x{}^2 + v_y{}^2 + v_z{}^2$. Here T is the temperature. The distribution function in the form (21) refers to the definition (1) of f_0; this definition is different from those used for f in Chapter 11 or in Chapter 13. We note that for the Maxwellian distribution

$$\frac{\partial f_0}{\partial \epsilon} = -\frac{1}{T} f_0 , \tag{22}$$

so that, from (20),

$$j_q = \frac{T_c e^2 E}{T} \int u^2 f_0\, dv . \tag{23}$$

But the kinetic energy of the x-motion is

$$\tfrac{1}{2}m \int u^2 f_0\, dv = \tfrac{1}{2}nT , \tag{24}$$

and thus

$$j_q = \frac{nq^2 T_c}{m} E . \tag{25}$$

The electrical conductivity is $\sigma = j_q/E$, or

$$\sigma = \frac{ne^2 \mathcal{T}_c}{m} \; .$$

(26)

Author Index

Subject Index